Pearson Education

AP* Test Prep Series

AP* CHEMISTRY

For:
Chemistry: The Central Science, Tenth Edition

By Theodore L. Brown, H. Eugene LeMay, Jr., and Bruce E. Bursten

Edward L. Waterman

Rocky Mountain High School

* AP and Advanced Placement are registered trademarks of The College Board, which was not involved in the production of, and does not endorse, this product.

Upper Saddle River, NJ 07458

Assistant Editor: Jennifer Hart
Executive Managing Editor: Kathleen Schiaparelli
Executive Editor: Nicole Folchetti
Assistant Managing Editor: Karen Bosch Petrov
Production Editor: Robert Merenoff
Supplement Cover Manager: Paul Gourhan
Supplement Cover Designer: Victoria Colotta
Manufacturing Buyer: Ilene Kahn
Manufacturing Buyer: Alexis Heydt-Long

© 2007 Pearson Education, Inc.
Pearson Prentice Hall
Pearson Education, Inc.
Upper Saddle River, NJ 07458

All rights reserved. No part of this book may be reproduced in any form or by any means, without permission in writing from the publisher.

Pearson Prentice Hall™ is a trademark of Pearson Education, Inc.

The author and publisher of this book have used their best efforts in preparing this book. These efforts include the development, research, and testing of the theories and programs to determine their effectiveness. The author and publisher make no warranty of any kind, expressed or implied, with regard to these programs or the documentation contained in this book. The author and publisher shall not be liable in any event for incidental or consequential damages in connection with, or arising out of, the furnishing, performance, or use of these programs.

This work is protected by United States copyright laws and is provided solely for teaching courses and assessing student learning. Dissemination or sale of any part of this work (including on the World Wide Web) will destroy the integrity of the work and is not permitted. The work and materials from it should never be made available except by instructors using the accompanying text in their classes. All recipients of this work are expected to abide by these restrictions and to honor the intended pedagogical purposes and the needs of other instructors who rely on these materials.

Printed in the United States of America

10 9 8 7 6 5 4 3 2

ISBN 0-13-236721-1

Pearson Education Ltd., *London*
Pearson Education Australia Pty. Ltd., *Sydney*
Pearson Education Singapore, Pte. Ltd.
Pearson Education North Asia Ltd., *Hong Kong*
Pearson Education Canada, Inc., *Toronto*
Pearson Educación de Mexico, S.A. de C.V.
Pearson Education—Japan, *Tokyo*
Pearson Education Malaysia, Pte. Ltd.

About the Author

Ed Waterman has taught chemistry and Advanced Placement Chemistry at Rocky Mountain High School in Fort Collins, Colorado since 1976. Mr. Waterman conducts workshops for teachers and administrators on the topics of small-scale chemistry, block scheduling, performance-based assessment, technology in science education, inquiry-based science education and Advanced Placement Chemistry. Mr. Waterman has co-authored three other books including *Prentice Hall Chemistry*, a popular text for first-year high school chemistry and *Small-Scale Chemistry Laboratory*, also published by Prentice Hall. He has also published numerous papers in peer reviewed journals including the *Journal of the American Chemical Society*, the *Journal of Organic Chemistry*, the *Journal of Chemical Education* and *The Science Teacher*. Mr. Waterman holds a Bachelor of Science Degree in Chemistry from Montana State University and a Master of Science Degree in Chemistry from Colorado State University.

This book is dedicated to all the hard-working teachers of Advanced Placement Chemistry across the United States and Canada. Your important contributions to chemical education, your dedication to your students and the significant and beneficial effects you continue to build into the fabric of our society can never be measured, only acknowledged. The students who choose to take up the challenge of the Advanced Placement Chemistry curriculum are indeed fortunate to have you as a teacher, a coach and a mentor. I hope that you will continue to find only the greatest satisfaction in helping young people reach for their dreams. Please accept my heartfelt thanks for your efforts and my best wishes for your continued success.

Ed Waterman
Rocky Mountain High School
Fort Collins, Colorado

Acknowledgement

We would like to thank James Mayhugh for accuracy reviewing the manuscript. We are grateful for his attention to detail, as well as his constructive comments and suggestions.

Table of Contents

Introduction

Advanced Placement Chemistry is more than just a course in first-year college general chemistry. Whether or not your AP exam score qualifies you for college credit, there are many advantages of taking Advanced Placement Chemistry. It is an opportunity, while you're still in high school, to prepare for college by challenging yourself with rigorous college-level work. Your classmates will often be some of the best and brightest students at your school and the peer group you study with will enhance your own abilities as a student. It is likely that your teacher will be among the best at your school and he or she will have invaluable knowledge and insight. Besides acquiring advanced knowledge of chemistry, the science central to all other scientific disciplines, you will also develop your skills in analytical thinking, abstract reasoning, problem solving and effective communication. You will enhance your study skills, both as an individual and within a group, and you will increase your own ability to learn how to learn. A second year of chemistry in high school will give you a decided competitive advantage over your future college classmates who have not taken Advanced Placement classes in high school. Advanced Placement Chemistry can serve as a measure of "survival insurance" for that upcoming pivotal year in life: the first year of college.

Advanced Placement Chemistry and College Credit

By taking the Advanced Placement test, your could qualify for college chemistry credit. Many public colleges and universities grant credit for scores of 3 or higher. More competitive public institutions and many private colleges have higher standards. Generally, the higher your score the more credit you receive. Because all colleges set their own standards, be sure to check the web site of any college you are considering attending or call or write the office of admissions for details about how that institution grants credit for Advanced Placement scores. To expedite the process, go to the following College Board link, type in the name of the college or university you are interested in and the site will take you directly to that institution's credit policy for Advanced Placement. *http://apps.collegeboard.com/apcreditpolicy/index.jsp*

How to Use this Book

This book is designed to help you score well on the Advanced Placement Examination in Chemistry. Each numbered topic is a chapter summary and correlates directly with the chapter of the same number in *Chemistry the Central Science* published by Prentice Hall. Because many other college chemistry texts are suitable for Advanced Placement Chemistry, you can use this book even if you do not have access to *Chemistry the Central Science.*

During the first half of your Advanced Placement course, focus on the course work your teacher emphasizes and what's in your text. Especially focus on solving the challenging quantitative problems and writing short, concise, directed answers to qualitative questions based on chemical principles.

Halfway through the course, in about December or January, as you continue with your class work, begin reviewing about two topics a week using this book. Read each topic summary and answer each Your Turn question as you encounter it. The Your Turn questions, while not always at the AP level, are each designed to focus your attention on one specific point and provide you with practice in writing clear, concise, directed responses, much as the AP exam requires you to do. At the end of each topic summary, you will find multiple choice and free response questions, as well as suggested additional end of chapter problems and questions from *Chemistry the Central Science*. Work all of these questions and check your answers with those at the end of each topic. If you don't know how to do a problem, or if you get a question wrong, go back and review the topic summary and/or the corresponding chapter in *Chemistry the Central Science.* Be sure to read the explanation for each question, even if you answered it correctly the first time.

Topic 12 is a diagnostic test, designed to simulate an Advanced Placement test, but it includes only questions from the first eleven topics. Work this test with the time limits suggested and check your answers. This will give you a measure of your progress in mastering the first eleven topics. Again, go back and review the material you need to. Be sure to read the explanations for each question. This is a good time to start becoming familiar with the format and scoring of the Advanced Placement Chemistry Exam detailed later in this section.

In February, ask your teacher about the procedure for ordering and paying for the Advanced Placement exam. This needs to be done well in advance of the exam and different high schools have different procedures.

In February or March continue working through your review, two topics per week. Keep in mind that Topics 14 through 20, excluding Topic 18, constitute a major portion of the Advanced Placement Examination. Focus especially on mastering these important topics, both in class and in your topic review.

Finally, about a month before the exam, work through Practice Test 1 within the suggested time limits, check your answers and calculate your score. Review the material you need to and then work Practice Test 2. These two practice tests are designed to simulate the Advanced Placement exam, each emphasizing different topics. Be sure to download and work some of the past Advanced Placement exams posted on the College Board website: *http://www.collegeboard.com/student/testing/ap/chemistry/samp.html?chem.*

A week before the exam, read through the equations list that you will use on exam day. Don't memorize what's on the equation sheet because you will have it to refer to during the free response portion of the exam. The point is to know what's on the equation sheet and where to find specific information as you work the exam. Look at each item and ask yourself, "What would I use this for?"

Also skim through the answers you wrote for the Your Turn questions and the multiple choice and free response questions at the end of each topic. Gather your materials together and prepare to go into the Advanced Placement exam, recognizing that you will not know everything, but confident that you will score high because you have worked hard and have prepared well.

What to Do on Exam Day

- Be sure to get a couple of good night's sleep before the exam.
- Eat a healthy breakfast the day of the exam.
- Bring a calculator with fresh batteries and a spare calculator, just in case. Know how to use the spare calculator!
- Bring at least six sharpened number two pencils with good erasers.
- Bring a water bottle and a nutritious energy snack to consume during the short break.
- Bring a photo ID and an admission ticket, if required.
- Bring a watch to keep track of the time. Be sure to turn off the alarm, if it has one.
- Arrive at the examination site at least 20 minutes early. When the door closes and you're not there, you don't get another chance.

- Be prepared for poor working conditions – some rooms provide only arm-chair desks with little room to work. Sometimes right-handed students will be assigned left-handed desks and vise-versa!
- Dress in layers. Some rooms are overly air conditioned, others are not air conditioned at all.
- Leave backpacks and personal belongings at home. All you need is a photo ID, two reliable calculators, a sharp pencil, and a mind to match.

Calculator Use

You can use a calculator only for Questions 1, 2 and 3 of Section II, the free-response section. Most types of scientific, programmable, and graphing calculators are permitted if they do not have typewriter-style (QWERTY) keyboards. You need not clear your calculator memory. You may not share a calculator with another student during the exam. You may not use a calculator on Section I, the multiple-choice section of the exam, or Part B of Section II.

The Content and Nature of the AP Exam in Chemistry

The Advanced Placement Examination in Chemistry is a comprehensive evaluation of student's knowledge of all areas of general chemistry at the first year college level. It is a three-hour exam comprised of two 90-minute sessions. The multiple-choice section consists of 75 multiple choice questions worth 50% of the total score. There are also six free-response questions that count for the other 50% of the total grade. For more information and published examples of recent exams please refer to the College Board's Advanced Placement website at *http://apcentral.collegeboard.com/*. Also, two complete practice exams, similar to the Advanced Placement Examination in Chemistry are included at the end of this book. Complete answers and explanations are given. The table shows the format and scoring of the AP exam.

The Multiple Choice Section (50%)

Because of the breadth and depth of the topics on the multiple choice section, this part of the exam encompasses the entire range of the Advanced Placement Syllabus. Hence, it is necessary to master the same content as that for the free response section. Calculators are not permitted on this section of the exam. The multiple choice questions are scored on the basis of the number right minus one quarter the number wrong so there is a penalty for guessing.

Format and Scoring for the Advanced Placement Exam in Chemistry

Section	**Types of Questions**	**Topics Usually Include:**	**Each Question is Worth:**	**Time Allowed:**
Section I Parts A and B 50%	75 multiple choice questions	All topics	0.667% for each right answer minus 0.167% for each wrong answer	90 minutes no calculators
Section II 50%	6 free response questions			
Part A 30%	Question 1 Quantitative Equilibrium problem in several parts	Ka/Kb-buffer, Ksp of Kp/Kc or a combination	10% partial credit for partial answers	55 minutes calculators allowed
	Question 2 Quantitative problem in several parts	Kinetics Gas Laws Electrochemistry Thermodynamics Stoichiometry	10% partial credit for partial answers	
	Question 3	Interpret data and analyze results of a labratory experiment	10% partial credit for partial answers	
Part B 20%	Question 4 Predicting products and writing chemical equations	Given the reactants, write the balance three chemical equations and answer a question about each.	5% partial credit for partial answers	40 minutes no calculators
	Question 5 Essay/short answer in several parts	Anatomic structure Periodicity Bonding Intermolecular forces	7.5% partial credit for partial answers	
	Question 6 Essay/short answer in several parts	Liquids and solids Solutions Laboratory Plus qualitative topics included in Questions 1,2, and 3.	7.5% partial credit for partial answers	

The Free Response Section (50%)

- **Part A** has 3 questions, each worth 10% of the total score. Fifty five minutes are allowed.

- **Part B** has 3 questions worth 20% of the total score. Forty minutes are allowed.

Students are required to answer all 6 of the free response questions. For the first three questions a calculator may be used. Calculators are not allowed when answering the final three free-response questions. Ample partial credit is given on the free response questions so students are encouraged to clearly and concisely demonstrate a written logical thought process and show important details of their work. A list of equations for each year's exam is available at least a year in advance, so students can refer to it for the duration of the course.

Section II, Part A consists of three questions for which calculators are permitted.

Question 1 is a quantitative chemical equilibrium problem. The exact topic varies from year to year but it is usually an acid-base equilibrium problem, which requires knowledge of weak acids and bases and buffer systems; a solubility equilibrium problem involving the solubility product constant, Ksp; a gaseous equilibria problem using Kc and/or Kp expressions or a combination problem using two or more equilibrium concepts.

Question 2 is another quantitative problem that is integrative in nature cutting across two or more different topics. Topics include stoichiometry, gas laws, solutions, kinetics, thermodynamics, redox and electrochemistry.

Stoichiometry includes limiting reactant stoichiometry, solution stoichiometry, acid-base and redox titrations, gases and partial pressures, colligative properties and empirical and molecular formulas.

Kinetics questions often ask students to derive a rate law from tabular initial rate data, use the rate law to calculate a rate constant and initial concentrations of reactants, match the rate law to a suitable mechanism and calculate reactant concentrations at various intervals. Rates are often reported as functions of changing molar concentrations, partial pressures or absorption via visible spectroscopy. Students must apply knowledge of the graphical derivation of rate constants to first order and second order reactions.

Thermodynamics includes calculations of enthalpy, free energy and entropy changes in chemical systems, and calculations of equilibrium constants from thermodynamic data.

Electrochemistry questions ask students to use standard reduction potentials to calculate voltaic cell EMF, free energy and equilibrium constants. Students must know how to apply the Nerst equation to calculate aqueous concentrations of chemical species within a voltaic cell, and solve quantitative problems of the chemistry of electrolytic cells using Faraday's constant, cell EMF, and the appropriate half reactions from standard reduction potential tables. Additionally, students use calculations to relate free energy, equilibrium constants, and cell EMF.

Question 3 is a laboratory-based question asking students to analyze and interpret data from a laboratory situation. Sometimes there are ample quantitative calculations. Other times only questions of a qualitative nature are asked. In a series of sub-questions, students are required to demonstrate knowledge of basic laboratory concepts, explain or do calculations, and analyze sample data. Although it is not expected that students perform every lab in the Advanced Placement Syllabus, students who do well on this question have a broad experience in doing experiments, recording and analyzing data, and thinking and writing critically about what their data means.

Section II, Part B consists of three questions to be answered in a 40 minute time limit and no calculators are permitted.

Question 4, worth 5% of the total score, asks students, given the reactants in words, to write and balance three net ionic equations for laboratory chemical reactions. There is one follow-up question about each chemical reaction. Reaction types represented on the exam always include acid-base, precipitation, oxidation-reduction, formation of complex ions, and combustion of simple organic molecules

Questions 5 and 6, worth 7.5% each, are essay/short answer questions. The questions draw from a variety of non-quantitative topics learned throughout the course. Topics often include periodicity, Lewis structures, bond energies, polarity of bonds and molecules, molecular geometry, hybrid orbitals, intermolecular forces, phase changes and equilibrium, especially Le Chatelier's Principal. When quantitative concepts are not asked in Questions 1 and 2, qualitative versions of the following topics may be addressed in Questions 5 and 6: behavior of gases, colligatiive properties, kinetics, acid behavior related to chemical structure, acid-base titrations, thermodynamics and electrochemistry.

Tips for Writing the Multiple Choice Section

- Make a note of the start time at the top of the question page. Add 90 minutes to define the stop time. Remember you have 90 minutes to answer 75 questions. Pace yourself accordingly.
- Read each question carefully. Don't assume you know what it asks without a careful reading.

- Pay close attention to words like EXCEPT and DOES NOT and look for the response that does not belong.

- When simple math is involved (remember, no calculators!), pay attention to factors of 2 and 3.

- Eliminate answers you know are wrong. Generally, if you can eliminate two of the five answers, it's to your advantage to guess. (Remember, scoring is the number correct minus one fourth the number wrong.)

- Never assume a question has two valid answers unless there is an "All of these" response. Closely analyze the two responses you think might be correct. How do they differ?

- Keep track of answers by marking the correct answer on both the test questions and the answer sheet to expedite a self-check later.

- Periodically check that you are marking the answer sheet correctly and check your watch to monitor your progress.

- Give yourself time to answer all the questions you know how to answer. The last question is worth the same number of points as the first question. Don't spend too much time on the harder or more time-consuming questions at the expense of not having enough time to answer the ones you can.

- As you read through and answer the questions consider using the **"answer/plus/minus"** method. It works like this:

  Answer:

  Answer any question you immediately know the answer to.

  Plus:

  Mark a "+"(on the exam, not on the answer sheet!) by any question you know how to answer but feel will take too much time to work. Be sure to skip the corresponding number on the answer sheet. You can come back to the questions marked "+" when you have finished reading all the questions and have answered as many as you know.

  Minus:

  Mark a "-" by any question your really don't understand. After you've finished the other questions, including the ones you've marked with a "+", if there is time, often a second reading of the "-" questions will give you fresh insights.

- Don't be afraid to leave a few questions un-answered. (Remember, scoring is number right minus one quarter the number wrong.) Try to answer at least 60 questions and as many as you can. Un-answered questions don't score any points! The more difficult questions tend to be toward the end of the exam so don't panic if, after your best effort, you don't finish.

- Finally, if there is still time, skim through all the questions, comparing the answers you've marked on the exam with those you have marked on the answer sheet. Correct any obvious errors, both in accurately marking the answer sheet and in choosing the best answer. Be sure to make any changes to the answer sheet by erasing cleanly and leaving no stray marks.

- When your watch shows that there are five minutes left, make a decision about how you can finish the question(s) that will score the most points. Be sure to stop when time is called.

Tips for Writing the Free Response Section

- Make a note of the start time at the top of the question page. Add 55 minutes to define the stop time for Part A. Remember you have 55 minutes to answer 3 questions. Pace yourself accordingly. Repeat for Part B. You have 40 minutes to answer three questions for Part B.

- Be sure to read each question carefully.

- Read all the questions once before you decide which question to answer first.

- You need not answer the questions in order but be sure to clearly label your answers in the answer booklet with the number and letter of each part.

- Fight off fear and apprehension. Of course you're not going to know everything! Remember, scoring well is a numbers game. You can make up for not knowing how to do one question by scoring well on another.

- Re-read the question that seems easiest to you. Determine what's asked and answer the question directly and specifically.

- With numerical problems, show your work clearly and

logically. You need not show any arithmetic, but the reader is looking for a logical progression of your ideas. Circle any numerical final answer.

- In written responses, be clear and concise.
- Write in complete sentences. Don't assume the reader knows what you are trying to say. Avoid one-word answers.
- Don't be afraid to state the obvious. What's obvious to you might be exactly what the reader is looking for.
- Avoid pronouns, especially the word, "it", even if your writing seems redundant.
- Underline and define key terms to make them stand out, if appropriate, but don't overdo it.
- Write what you know and stop. Two sentences often fully express an answer to any one part of a question. These are not essay questions. They are more like short answer questions requiring clear and concise answers with justifications. The reader doesn't want to see any more than what is specifically asked for. Avoid the word, "or". The reader is instructed to count an answer wrong if both a right answer and a wrong answer appear on the answer sheet.
- Don't hesitate to use a picture, a diagram, or an equation to illustrate your answer, if appropriate.
- Review your writing. Does it make sense to you? Will it make sense to the reader? Does it specifically answer the question that is asked? Have you used chemical terminology correctly?
- Keep in mind that partial credit is often given so make sure you put something down for every part.
- If you are not clear about an answer, re-write the question in your own words. Often this practice will jump-start your thinking and allow you to arrive at the answer. What you do write, even if it only partially answers the question, might score partial credit.
- When your watch shows that there are five minutes left, make a decision about how you can finish the question(s) that will score the most points. Be sure to stop when time is called.

TOPIC 1

Introduction: MATTER AND MEASUREMENT

Most of this material was probably learned in first-year chemistry. Quickly review the differences between elements, compounds and mixtures. The separation of mixtures based on their physical properties is important and so are SI units for measurement. Significant figures and dimensional analysis are also introduced here. Pay particular attention to these sections:

1.2 **Classification of Matter**

1.3 **Properties of Matter**

1.4 **Units of Measurement**

1.5 **Uncertainty in Measurement**

1.6 **Dimensional Analysis**

Classification of Matter

Section 1.2

Matter is classified as pure substances or mixtures.

Pure substances are either elements or compounds.

An **element** is a substance all of whose atoms contain the same number of protons.

A **compound** is a relatively stable combination of two or more chemically bonded elements in a specific ratio.

Mixtures consist of two or more substances.

Homogeneous mixtures are uniform throughout. Air, sea water, and a nickel coin (a mixture of copper and nickel metals called an alloy) are examples.

Heterogeneous mixtures vary in texture and appearance throughout the sample. Rocks, wood, polluted air, and muddy water are examples.

Section 1.3 Properties of Matter

Physical properties can be measured without changing the identity or composition of the substance. Physical properties include color, density, melting point, and hardness.

Chemical properties describe the way a substance changes (reacts) to form other substances. The flammability of gasoline is a chemical property because the gasoline reacts with oxygen to form carbon dioxide and water.

Intensive properties do not depend on the amount of substance in a sample. Temperature, density, and boiling point are intensive properties.

Extensive properties depend on the quantity of the sample. Energy content, mass, and volume are examples of extensive properties.

A **physical change** changes the appearance of a substance but does not change its composition. Changes of physical state, from solid to liquid or from liquid to gas are examples.

A **chemical change** (also called a chemical reaction) transforms a substance into a different substance or substances. When the chief component of natural gas, methane burns in air, the methane and the oxygen from the air are transformed into carbon dioxide and water.

Differences in properties are used to separate the components of mixtures.

Filtration separates a solid from a liquid.

Distillation separates substances based on their differences in boiling points.

Chromatography is a technique that separates substances based on their differences in intermolecular forces and their abilities to dissolve in various solvents.

Section 1.4 Units of Measurement

Chemists often use preferred units called **SI units** after the French *Système International d'Unités.* Table 1.1 lists the base SI units and their symbols.

Table 1.1 SI base units

Physical Quantity	Name of Unit	Abbreviation
Mass	Kilogram	kg
Length	Meter	m
Time	Second	s[a]
Temperature	Kelvin	K
Amount of substance	Mole	mol
Electric current	Ampere	A
Luminous intensity	Candela	cd

[a]The abbreviation sec is frequently used.

Table 1.2 lists metric prefixes which indicate decimal fractions or multiples of various units.

Table 1.2 Selected preferences used in the metric system

Prefix	Abbreviation	Meaning	Example
Giga	G	10^9	1 gigameter (Gm) = 1×10^9 m
Mega	M	10^6	1 megameter (Mm) = 1×10^6 m
Kilo	k	10^3	1 kilometer (km) = 1×10^3 m
Deci	d	10^{-1}	1 decimeter (dm) = 0.1 m
Centi	c	10^{-2}	1 centimeter (cm) = 0.01 m
Milli	m	10^{-3}	1 millimeter (mm) = 0.001 m
Micro	μ[a]	10^{-6}	1 micrometer (μm) = 1×10^{-6} m
Nano	n	10^{-9}	1 nanometer (nm) = 1×10^{-9} m
Pico	p	10^{-12}	1 picometer (pm) = 1×10^{-12} m
Femto	f	10^{-15}	1 femtometer (fm) = 1×10^{-15} m

[a]This is the Greek letter mu (pronounced "mew").

Temperature is commonly measured using either the Celsius scale or the Kelvin scale.

$$K = {}^{o}C + 273$$

Derived units are units derived from SI base units.

Volume, the space occupied by a substance, is commonly measured in cubic meters, m^3 or cubic centimeters, cm^3. A non-SI unit commonly used by chemists is the liter, L. One liter is the volume of a cube measuring exactly 10 centimeters on a side.

$$1\ L = 1000\ cm^3 = 1000\ mL. \quad 1\ cm^3 = 1\ mL$$

Density, the amount of matter packed into a given space, is often measured in g/cm^3 for liquids and solids and g/L for gases.

$$\text{Density} = \text{mass}/\text{volume}$$

Section 1.5

Uncertainty in Measurement

Exact numbers are known exactly and are usually defined or counted. One liter equals one thousand cubic centimeters describes a defined number. There are 32 students in this class describes a counted number.

Inexact numbers have some degree of error or uncertainty associated with them. All measured numbers are inexact.

Measured numbers are generally reported in such a way that only the last digit is uncertain. **Significant figures** are all digits of a measured number including the uncertain one.

Zeros in measured numbers are either significant or are merely there to locate the decimal place. The following guidelines describe when zeros are significant:

1. Zeros between nonzero digits are always significant.
2. Zeros at the beginning of a number are never significant.
3. Zeros at the end of a number are significant only when the number contains a decimal point.

In calculations involving measured quantities, the least certain measurement limits the certainty of the calculated quantity and determines the number of significant figures in the final answer.

For **multiplication and division**, the number of significant figures in the answer is determined by the measurement with the fewest number of significant figures.

For **addition and subtraction**, the result has the same number of decimal places as the measurement with the fewest number of decimal places.

Common misconception: The guidelines for determining the number of significant figures in a result obtained by carrying measured quantities through calculations sometimes do not give the correct number of significant figures. This is principally why the AP test allows full credit for answers reported to plus or minus one significant figure. In this book, all numerical answers are rounded to three significant figures.

Dimensional analysis is a way of converting a written question into an algebraic equation, followed by manipulating factors until the unit of the known quantity is converted into the unit of the unknown quantity.

Example:

"How many microseconds are there in one year?"

Solution:

The algebraic equivalent to the given sentence is:

x μs = 1 yr.

Now multiply the right side of the equation by what is known about a year in such a way that the unit of years cancels giving another unit. Continue to do this until the result has the unit of microseconds, μs.

x μs = 1yr (365 days/yr) (24 hr/day) (60 min/hr) (60s/min) (106 μs/s) = 3.15 x 1013 μs

Additional Practice in Chemistry the Central Science

For more practice answering questions in preparation for the Advanced Placement examination try these problems in Chapter 1 of Chemistry the Central Science:

Additional Exercises: 1.75, 1.76, 1.77, 1.78, 1.79.

TOPIC

2

ATOMS, MOLECULES AND IONS

Chemistry is the science that selects atoms, ions and molecules as the basic units of nature and uses them to explain the properties of materials and the changes that materials undergo. Chapter 2 examines a hidden structure of chemistry and answers a fundamental question: What is the structure of matter in terms of unseen atoms, ions, and molecules?

Because most of these topics were learned in the first-year course, little time should be spent reviewing and relearning the fundamental and important ideas of chemical formulas and nomenclature. Remember that it is imperative to know how to name and write formulas of common ions and ionic and covalent compounds in order to answer the questions on all sections of the AP exam, especially the reactions section.

These are the most important sections to focus on in Chemistry the Central Science:

2.6 Molecules and Molecular Compounds
2.7 Ions and Ionic Compounds
2.8 Naming Inorganic Compounds
2.9 Some Simple Organic Compounds

Important Contributions by Notable Scientists

Sections 2.1 and 2.2

John Dalton proposed that matter is composed of tiny particles called atoms. He postulated that all atoms of one element are the same but are different from atoms of another element. Atoms combine to form compounds.

J.J. Thomson showed that cathode rays are streams of negative particles. He is credited with discovering the electron and measured its charge to mass ratio. He postulated that all atoms contain electrons.

Robert Millikan measured the charge of an electron and calculated its mass.

Henri Becquerel discovered radioactivity, spontaneously emitted high-energy radiation.

Ernest Rutherford characterized radioactivity as three types of radiation: alpha and beta particles and gamma rays. He discovered the atomic nucleus and the proton and showed that atoms were mostly empty space.

James Chadwick discovered the neutron.

Section 2.3 The Modern View of Atomic Structure

In studying chemistry, it is often convenient to think of atoms as fundamental, indivisible units of matter. However, atoms are composed of three basic subatomic particles: protons, electrons and neutrons.

Atoms consist of a tiny dense positively charged nucleus surrounded by a cloud of negative electrons.

The **nucleus** contains positively charged protons and neutral neutrons.

Atoms are electrically neutral because each contains the same number of protons as electrons.

Atoms can gain electrons to form negatively charged ions called anions or they can lose electrons to become positively charged ions called cations.

The **atomic number** is the number of protons in the nucleus.

An **element** is a substance all of whose atoms contain the same number of protons. Each element is defined by its atomic number.

The **mass number** is the number of protons and neutrons in the nucleus of an atom.

Isotopes are atoms of identical atomic numbers but different mass numbers.

Symbols are often used to denote various elements and to distinguish isotopes. For example, the isotope of carbon containing 6 protons and 6 neutrons is designated like this:

$$^{12}_{6}\mathrm{C}$$

12 = **mass number = number of protons plus neutrons**

6 = **atomic number = number of protons**

Because carbon is the element that always contains six protons, often the atomic number designation is omitted and the following symbols all designate the same isotope of carbon:

$$^{12}_{6}\mathrm{C} = {}^{12}\mathrm{C} = \text{carbon-12} = \text{C-12}.$$

Carbon has several isotopes and each is distinguished by its mass number. Their respective number of neutrons is calculated by subtracting the atomic number from the mass number. Here are symbols for various isotopes of carbon with the number of neutrons each isotope possesses.

$^{11}_{6}C$	$^{12}_{6}C$	$^{13}_{6}C$	$^{14}_{6}C$
5 neutrons	6 neutrons	7 neutrons	8 neutrons

Your Turn 2.1

How many protons, electrons, and neutrons are in an oxygen-18 atom? Write your answer in the space provided.

The **atomic mass unit**, amu, is a convenient way to express the relative masses of tiny atoms. One amu equals 1.66054×10^{-24} g. However, it is more useful to compare the masses of atoms to the mass of one carbon-12 isotope. One carbon-12 atom has a defined mass of exactly 12 amu.

Atomic mass is the weighted average mass of all the isotopes of an element based on the abundance of each isotope found on earth. Atomic masses are expressed in amu. All atomic masses reported on the periodic table are based on the carbon-12 standard. For example, the atomic mass of magnesium is 24.3040 amu. This means that, the average mass of all the magnesium isotopes is a little more than twice the mass of a carbon-12 atom.

The Periodic Table

Section 2.5

The **periodic table** is an arrangement of elements in order of increasing atomic number with elements having similar properties placed in vertical columns. The vertical columns are called groups or families and the horizontal rows are called periods. Figure 2.1 shows the periodic table with the symbol, atomic number, and atomic mass of each element. It also shows two commonly used numbering systems for the groups. Elements are classified as metals, nonmetals and metalloids.

Main groups | Transition metals | Main groups

	1A[a] 1	2A 2	3B 3	4B 4	5B 5	6B 6	7B 7	8B 8	8B 9	8B 10	1B 11	2B 12	3A 13	4A 14	5A 15	6A 16	7A 17	8A 18
1	1 **H** 1.00794																	2 **He** 4.002602
2	3 **Li** 6.941	4 **Be** 9.012182											5 **B** 10.811	6 **C** 12.0107	7 **N** 14.0067	8 **O** 15.9994	9 **F** 18.998403	10 **Ne** 20.1797
3	11 **Na** 22.989770	12 **Mg** 24.3050											13 **Al** 26.981538	14 **Si** 28.0855	15 **P** 30.973761	16 **S** 32.065	17 **Cl** 35.453	18 **Ar** 39.948
4	19 **K** 39.0983	20 **Ca** 40.078	21 **Sc** 44.955910	22 **Ti** 47.867	23 **V** 50.9415	24 **Cr** 51.9961	25 **Mn** 54.938049	26 **Fe** 55.845	27 **Co** 58.933200	28 **Ni** 58.6934	29 **Cu** 63.546	30 **Zn** 65.39	31 **Ga** 69.723	32 **Ge** 72.64	33 **As** 74.92160	34 **Se** 78.96	35 **Br** 79.904	36 **Kr** 83.80
5	37 **Rb** 85.4678	38 **Sr** 87.62	39 **Y** 88.90585	40 **Zr** 91.224	41 **Nb** 92.90638	42 **Mo** 95.94	43 **Tc** [98]	44 **Ru** 101.07	45 **Rh** 102.90550	46 **Pd** 106.42	47 **Ag** 107.8682	48 **Cd** 112.411	49 **In** 114.818	50 **Sn** 118.710	51 **Sb** 121.760	52 **Te** 127.60	53 **I** 126.90447	54 **Xe** 131.293
6	55 **Cs** 132.90545	56 **Ba** 137.327	71 **Lu** 174.967	72 **Hf** 178.49	73 **Ta** 180.9479	74 **W** 183.84	75 **Re** 186.207	76 **Os** 190.23	77 **Ir** 192.217	78 **Pt** 195.078	79 **Au** 196.96655	80 **Hg** 200.59	81 **Tl** 204.3833	82 **Pb** 207.2	83 **Bi** 208.98038	84 **Po** [208.98]	85 **At** [209.99]	86 **Rn** [222.02]
7	87 **Fr** [223.02]	88 **Ra** [226.03]	103 **Lr** [262.11]	104 **Rf** [261.11]	105 **Db** [262.11]	106 **Sg** [266.12]	107 **Bh** [264.12]	108 **Hs** [269.13]	109 **Mt** [268.14]	110 [271.15]	111 [272.15]	112 [277]	113 [284]	114 [289]	115 [288]	116 [292]		

*Lanthanide series	57 ***La** 138.9055	58 **Ce** 140.116	59 **Pr** 140.90765	60 **Nd** 144.24	61 **Pm** [145]	62 **Sm** 150.36	63 **Eu** 151.964	64 **Gd** 157.25	65 **Tb** 158.92534	66 **Dy** 162.50	67 **Ho** 164.93032	68 **Er** 167.259	69 **Tm** 168.93421	70 **Yb** 173.04
†Actinide series	89 **†Ac** [227.03]	90 **Th** 232.0381	91 **Pa** 231.03588	92 **U** 238.02891	93 **Np** [237.05]	94 **Pu** [244.06]	95 **Am** [243.06]	96 **Cm** [247.07]	97 **Bk** [247.07]	98 **Cf** [251.08]	99 **Es** [252.08]	100 **Fm** [257.10]	101 **Md** [258.10]	102 **No** [259.10]

[a]The labels on top (1A, 2A, etc.) are common American usage. The labels below these (1, 2, etc.) are those recommended by the International Union of Pure and Applied Chemistry.
The names and symbols for elements 110 and above have not yet been decided.
Atomic weights in brackets are the masses of the longest-lived or most important isotope of radioactive elements.
Further information is available at http://www.webelements.com
The production of element 116 was reported in May 1999 by scientists at Lawrence Berkeley National Laboratory.

Figure 2.1. Periodic Table

Table 2.1 shows the special names given to four of the groups.

Table 2.1. Names given to four groups of elements on the periodic table.

Group Number	Name of Group
1 or 1A	alkali metals
2 or 2A	alkaline earth metals
17 or 7A	halogens
18 or 8A	noble gases

Section 2.6 Molecules and Molecular Compounds

Although the atom is the smallest representative particle of an element, most matter is composed of molecules or ions, which are combinations of atoms.

A **molecule** is an assembly of two or more atoms tightly bonded together. For example, a molecule that is made up of two atoms is called a diatomic molecule. Seven elements normally occur as diatomic molecules. They are hydrogen, oxygen, nitrogen, fluorine, chlorine, bromine, and iodine. The formulas for these diatomic molecules are written like this:

H_2, O_2, N_2, F_2, Cl_2, Br_2, I_2.

The subscript displayed in each formula indicates that two atoms are present in each molecule.

Common misconception: Chemists often use the name oxygen to refer to both O and O_2, even though the latter's official name is dioxygen to distinguish it from monatomic oxygen. For chemists, the correct species can easily be inferred by the context of the sentence. For example, the oxygen we breathe is O_2, and the oxygen in the water molecule is O. Most texts use the monatomic names for the diatomic elements. Pay close attention to the context in which these names are used to pick up the exact meaning.

Your Turn 2.2

Tell which chemical form of chlorine is implied in these two sentences: a) Chlorine is a toxic gas. b) Common table salt contains the element chlorine. Write your answer in the space provided.

Compounds are substances consisting of two or more different elements. Generally, there are two types of compounds, molecular compounds and ionic compounds.

Molecular compounds are composed of molecules and usually contain only nonmetals. A molecular formula indicates the actual number and type of atoms in the molecule and is the most often used formula for molecular compounds.

Ions and ionic compounds

Section 2.7

Ions are charged particles comprised of single atoms (called monatomic ions) or aggregates of atoms (called polyatomic ions). A cation is a positive ion, and an anion is a negative ion.

Ionic compounds are composed of ions and usually contain both metals and nonmetals.

An **empirical formula** gives only the relative number of atoms or each type in the compound. Empirical formulas are almost always used for ionic compounds and sometimes used for molecular compounds.

Metal atoms can lose electrons to become monatomic cations.

Nonmetal atoms gain electrons to become monatomic anions.

The periodic table is useful in remembering the charges of mon-atomic cations and anions. Figure 2.2 shows the common charges of ions derived from elements on the left and right sides of the periodic table. Notice that the cations of the A groups carry a positive charge equal to the group number, and the anions carry a negative charge equal to the group number minus 8. Transition metals tend to form cations of more than one charge.

1A	2A	3B	4B	5B	6B	7B	8B	8B	8B	9B	10B	3A	4A	5A	6A	7A	8A
H^+																H^-	
Li^+														N^{3-}	O^{2-}	F^-	
Na^+	Mg^{2+}											Al^{3+}			S^{2-}	Cl^-	
K^+	Ca^{2+}	Sc^{3+}	Ti^{2+} Ti^{4+}	V^{2+} V^{4+}	Cr^{2+} Cr^{3+}	Mn^{2+} Mn^{4+}	Fe^{2+} Fe^{3+}	Co^{2+} Co^{3+}	Ni^{2+} Ni^{3+}	Cu^{2+} Cu^+	Zn^{2+}				Se^{2-}	Br^-	
Rb^+	Sr^{2+}									Ag^+	Cd^{2+}		Sn^{2+} Sn^{4+}		Te^{2-}	I^-	
Cs^+	Ba^{2+}									Au^+ Au^{3+}	Hg^{2+} Hg_2^{2+}		Pb^{2+} Pb^{4+}				

Figure 2.2. Charges of some monatomic cations and anions are consistent within groups on the periodic table. Notice that some transition metals form ions having more than one charge.

Section 2.8 Naming Inorganic Compounds

Names of monatomic cations

Monatomic cations have the same name as the metal. If the metal forms cations of different charges, a Roman numeral in the name indicates the charge.

K^+ = potassium ion. **Ca^{2+}** = calcium ion.

Co^{2+} = cobalt(II) ion. **Co^{3+}** = cobalt(III) ion.

Notice that most transition metals form cations with more than one charge with four common exceptions:

Sc^{3+} = scandium ion. Ag^{+} = silver ion.

Zn^{2+} = zinc ion. Cd^{2+} = cadmium ion.

Most non-transition metals form cations having only a single charge with two common exceptions:

Sn^{2+} = tin(II) ion. Sn^{4+} = tin(IV) ion.

Pb^{2+} = lead(II) ion. Pb^{4+} = lead(IV) ion.

Names of monatomic anions

Monatomic anions replace the end of the element name with –ide.

Table 2.2.

Group 1A	Group 5A	Group 6A	Group 7A
H^- hydride	N^{3-} nitride	O^{2-} oxide S^{2-} sulfide Se^{2-} selenide Te^{2-} telluride	F^- fluoride Cl^- chloride Br^- bromide I^- iodide

Common misconception: The names and formulas of monatomic anions need not be memorized. Simply locate the atom on the periodic table, assign the charge based on the group number and change the ending to -ide.

A few common diatomic anions also end in –ide:

OH^- hydroxide CN^- cyanide O_2^{2-} peroxide

Names of polyatomic cations:

Polyatomic (more than one atom) cations formed from nonmetals end in –ium.

NH_4^+ = ammonium $CH_3NH_3^+$ = methylammonium

$(CH_3)_2NH_2^+$ = dimethylammonium

Polyatomic oxyanions (anions containing oxygen)

Some polyatomic anions end in -ate. These are called oxyanions because they contain oxygen. Because there is no logical way to predict their formulas or charges these must be memorized.

Table 2.3 lists the names and formulas of some common oxyanions (polyatomic anions containing oxygen).

Some polyatomic oxyanions have a hydrogen ion attached. The word hydrogen or dihydrogen is used to indicate anions derived by adding H^+ to an oxyanion. An added hydrogen ion changes the charge by 1+.

Table 2.3 Names and formulas of some common oxyanions listed by charge.

Charge	**3-**	**2-**	**1-**
Name and Formula	phosphate PO_4^{3-}	hydrogen phosphate HPO_4^{2-}	dihydrogen phosphate $H_2PO_4^-$
		sulfate SO_4^{2-}	hydrogen sulfate HSO_4^-
		carbonate CO_3^{2-}	hydrogen carbonate HCO_3^-
			nitrate NO_3^-
			acetate CH_3COO^- (also ethanoate)

Oxyanions are polyatomic anions that usually end in –ate or –ite. The ending –ate denotes the most common oxyanion of an element. The ending –ite refers to an oxyanion having the same charge but one fewer oxygen.

The halogens form a series of four oxyanions and use prefixes to distinguish them:

ClO^- hypochlorite Hypo- denotes one fewer oxygen.

ClO_2^- chlorite

ClO_3^- chlorate

ClO_4^- perchlorate Per- denotes one more oxygen.

Names of ionic compounds:

Names of ionic compounds consist of the cation name followed by the anion name:

$MgBr_2$ = magnesium bromide

$Ca_3(PO_4)_2$ = calcium phosphate

Fe_2O_3 = iron(III) oxide

FeO = iron(II) oxide

(Remember that the Roman numerals III and II refer the 3+ and 2+ charges of the iron cations, respectively. Notice that the charge on the iron atom in each case is inferred by the ratio in which iron combines with the 2- charged oxide ion.)

Your Turn 2.3.

Name the following ionic compounds: KI, $MgSO_4$, FeS, Al_2O_3, $Pb_3(PO_4)_2$
Write your answers in the space provided.

Names and Formulas of Acids

The names and formulas of some common acids, an important class of hydrogen-containing compounds, are listed in Table 2.4. In all cases, the formulas of acids are comprised of one or more hydrogen ions added to a common monatomic anion or oxyanion. Acids of monatomic ions are called **binary acids** and acids of oxyanions are called **oxyacids.**

Table 2.4. Names and Formulas of Some Common Acids.

Binary acids		Oxyacids	
HF	hydrofluoric acid	HNO_3	nitric acid
HCl	hydrochloric acid	H_2SO_4	sulfuric acid
HBr	hydrobromic acid	H_3PO_4	phosphoric acid
HI	hydroiodic acid	CH_3COOH	acetic acid

Table 2.4. continued

H_2S	hydrosulfuric acid	H_2CO_3	carbonic acid
H_2Se	hydroselenic acid	HNO_2	nitrous acid
		H_2SO_3	sulfurous acid
		$HClO_4$	perchloric acid
		$HClO_3$	chloric acid
		$HClO_2$	chlorous acid
		$HClO$	hypochlorous acid

To name binary acids replace the –ide ending of the anion with –ic acid and add the prefix hydro-.

_____ide becomes hydro_______ic acid.
Bromide, Br-, becomes hydrobromic acid, HBr.

To name oxyacids, replace the –ate ending of the oxyanion with –ic acid or the –ite ending of the oxyanion with –ous acid.

_____ate becomes _____ic acid.
Nitrate, NO_3^- becomes nitric acid, HNO_3.

_____ite becomes _____ous acid.
Hypochlorite, ClO^- becomes hypochlorous acid, HClO.

Some common exceptions:

Phosphate, PO_4^{3-} becomes phosPHORic acid H_3PO_4.
Sulfate, SO_4^{2-} becomes sulfURic acid, H_2SO_4.

Names of binary molecular compounds

A binary molecular compound contains two nonmetals. The rules for naming binary molecular compounds are:

1. Name the first element.

2. Name the second element giving it an –ide ending.

3. Use prefixes to denote how many of each element are in the formula. The prefixes are shown in Table 2.5.

Table 2.5 Prefixes used in naming binarycompounds formed between non metals

Prefix	*Meaning*
Mono-	*1*
Di-	*2*
Tri-	*3*
Tetra-	*4*
Penta-	*5*
Hexa-	*6*
Hepta-	*7*
Octa-	*8*
Nona-	*9*
Deca-	*10*

CO_2 = carbon dioxide

SO_3 = sulfur trioxide

N_2O_5 = dinitrogen pentoxide

P_4O_{10} = tertaphosphorus decoxide

Your Turn 2.4

Name the following compounds: SCl_2, CF_4, BrI_3, PBr_5, SF_6. Write your answers in the space provided.

Names of simple organic molecules

Organic chemistry is the study of carbon compounds.

Hydrocarbons are compounds containing only carbon and hydrogen.

Alkanes are hydrocarbons containing only C-C single bonds.

Table 2.6 shows the names and formulas of some common alkanes. Notice in each name a prefix indicates the number of carbon atoms in the formula and it is followed by the suffix –ane to indicate that the formula is an alkane

Table 2.6. Names and formulas of some simple alkanes. The prefix of the name tells how many carbon atoms are in the formula.

Number of carbon atoms	Prefix	Name	Formula
1	meth-	methane	CH_4
2	eth-	ethane	CH_3CH_3
3	prop-	propane	$CH_3CH_2CH_3$
4	but-	butane	$CH_3CH_2CH_2CH_3$
5	pent-	pentane	$CH_3CH_2CH_2CH_2CH_3$
6	hex	hexane	$CH_3CH_2CH_2CH_2CH_2CH_3$
7	hept-	heptane	$CH_3CH_2CH_2CH_2CH_2CH_2CH_3$
8	oct-	octane	$CH_3CH_2CH_2CH_2CH_2CH_2CH_2CH_3$
9	non-	nonane	$CH_3CH_2CH_2CH_2CH_2CH_2CH_2CH_2CH_3$
10	dec-	decane	$CH_3CH_2CH_2CH_2CH_2CH_2CH_2CH_2CH_2CH_3$

The same prefixes are used to name organic compounds having **functional groups,** groups of atoms that give rise to the structure and properties of an organic compound. For example, the functional group of a class of organic compounds classified as alcohols is –OH. That of a carboxylic acid is –COOH. Table 2.7 lists the names and formulas of simple alcohols and carboxylic acids. In Chapter 25 of Chemistry the Central Science, you will study organic compounds in more detail.

Table 2.7. Names and formulas of some simple alcohols and carboxylic acids. The prefix tells how many carbon atoms are in the formulas.

Alcohols		**Carboxylic acids**	
name	**formula**	**name**	**formula**
methanol	CH_3OH	methanoic acid	H_2COOH
ethanol	CH_3CH_2OH	ethanoic acid	CH_3COOH
1-propanol	$CH_3CH_2CH_2OH$	propanoic acid	CH_3CH_2COOH
2-porpanol	$CH_3CHOHCH_3$		
1-butanol	$CH_3CH_2CH_2CH_2OH$	butanoic acid	$CH_3CH_2CH_2COOH$
2-butanol	$CH_3CH_2CHOHCH_3$		

Multiple Choice Questions

1. *Fluorine gas is bubbled through an aqueous solution of calcium bromate. Besides water, this statement refers to which chemical formulas?*

A) F_2 *and* $CaBr_2$

B) F *and* $CaBr_2$

C) F_2 *and* $Ca(BrO_3)_2$

D) F_2 *and* $CaBrO_3$

E) F *and* $CaBrO_3$

2. *Magnesium nitride reacts with water to form ammonia and magnesium hydroxide. Besides water, this statement refers to which chemical formulas?*

A) Mg_3N_2, NH_3 *and* $Mg(OH)_2$

B) $Mg(NO_3)_2$, NH_4^+ *and* MgH_2

C) $Mg(NO_3)_2$, NH_3 *and* MgH_2

D) $Mg(NO_2)_2$, NH_4^+ *and* $Mg(OH)_2$

E) Mg_3N_2, NH_3 *and* $MgOH$

3. *The mineral spinel is an ionic compound containing only the elements magnesium, aluminum and oxygen. Its simplest formula is probably*

A) $MgAlO_3$

B) Mg_2AlO_4

C) $MgAl_2O_4$

D) $Mg_2Al_2O_3$

E) Mg_2AlO_3

4. *The mineral chromite,* $FeCr_2O_4$, *consists of a mixture of iron(II) oxide and chromium(III) oxide. What is the most likely ratio of iron(II) oxide to chromium(III) oxide in chromite?*

A) 1:1

B) 1:2

C) 2:3

D) 3:2

E) 2:1

5. *Which formula represents a peroxide?*

A) K_2O

B) K_2O_2

C) KO_2

D) CaO

E) Ca_2O_2

6. *What is the general formula for an alkaline earth metal hydride?*

A) MH

B) MOH

C) MH_2

D) $M(OH)_2$

E) M_2H

7. *Which of the following compounds contains four carbon atoms?*

A) propane

B) butanoic acid

C) ethyl methyl ether

D) 2-pentanol

E) heptanal

8. *Bromine has just two major isotopes giving it an atomic mass of 79.904 amu. Based on this information, which of the following statements can explain the atomic mass value?*

A) The isotope, Br-81, is more common than Br-79.

B) Br-79 and Br-81 exist in about equal proportions.

C) Br-78 is about twice as abundant as Br-81.

D) Br-82 is more abundant than Br-79.

E) The two major isotopes of Br have 45 and 46 neutrons.

9. *Which is a collection of only molecular compounds?*

A) NO, CS_2, PCl_3, HBr

B) $NaNO_3$, CCl_4, CuS

C) Ar, NH_3, SF_4, PCl_5

D) Cl_2, CCl_4, NO_2, SF_6

E) H_2O, CaO, CO, CO_2

10. *Which is true of the* $^{243}Am^{3+}$ *ion?*

	protons	*electrons*	*neutrons*
A)	148	148	243
B)	95	98	243
C)	95	95	148
D)	95	92	148
E)	92	95	148

Free Response Questions

1. *The two stable isotopes of chlorine have masses of 34.969 amu and 36.966 amu.*

   a. *What are the mass numbers of the two isotopes of chlorine?*

   b. *Calculate the % abundance of the lighter isotope.*

   c. *How many types of molecules with different masses exist in a sample of chlorine gas if the sample exists entirely as diatomic molecules? Explain your answer.*

   d. *Calculate the mass of the chlorine molecule having the largest molecular mass.*

   e. *What is the mass of the most abundant molecule? Calculate its % abundance.*

2. *Like chlorine, iodine is a halogen and forms similar compounds. Write the names and formulas of the four oxyanions and the four oxyacids of iodine.*

Additional Practice in Chemistry the Central Science

For more practice answering questions in preparation for the Advanced

Placement examination try these problems in Chapter 2 of Chemistry the Central Science:

Additional Exercises: 2.73, 2.84, 2.89,2.91, 2.92, 2.93, 2.94, 2.95, 2.96, 2.97.

Multiple Choice Answers and Explanations

1. *C. Fluorine is one of the seven diatomic elements and its formula is written F_2. Calcium, in Group 2A, forms a Ca^{2+} ion. Bromate is an oxyanion having the formula, BrO_3^-.*

2. *A. Magnesium, from Group 2A forms Mg^{2+} ions. Nitrogen, from Group 5A forms nitride ions, N^{3-}. Ammonia is the name given to NH_3. Hydroxide ion is OH^-. The correct formulas of magnesium nitride, ammonia, and magnesium hydroxide, respectively are, Mg_3N_2, NH_3 and $Mg(OH)_2$*

3. *C. Ions of magnesium, aluminum, and oxygen carry charges of 2+, 3+ and 2–, respectively Mg^{2+}, Al^{3+} and O^{2-} will combine in a way that the sum of the charges is zero. Of the choices given, only $MgAl_2O_4$ fits the criterion.*

4. *A. Iron(II) oxide is FeO and chromium(III) oxide is Cr_2O_3. Combining one formula unit of FeO with one formula unit of Cr_2O_3 yields $FeCr_2O_4$.*

5. *B. Peroxide is O_2^{2-}. Potassium is in Group 1A so it loses one electron to form a K^+ ion. K^+ and O_2^{2-} combine to form potassium peroxide, K_2O_2.*

6. *C. Hydride is H^-. Alkaline earth metals are metals of Group 2A, all of which lose 2 electrons to form ions with a 2+ charge. M^{2+} and H combine as MH_2.*

7. *B. The prefix but- signifies four carbon atoms. Note: Ethylmethyl ether has three carbon atoms because the prefix ethyl- means two carbon atoms and the prefix methyl- implies one carbon atom.*

8. *B. The atomic mass of an element is the weighted average of the mass of all the isotopes. If Br-79 and Br-81 are in about equal propor-*

tions, the atomic mass of bromine will be about 80 amu. If Br-81 is the most abundant, the mass would be greater than 80. If Br-78 is twice as abundant as Br-81, then the atomic mass would be about 79. (78+78+81)/3 = 79. If Br-82 is more abundant than Br-79, then the atomic mass would be greater than 80.5. (82+79)/2 = 80.5. Isotopes having 45 and 46 neutrons would have masses of 80 and 81, respectively. The average mass of those isotopes would fall between 80 and 81, depending on their relative abundances.

9. *A. Molecular compounds contain more than one type of atom and generally contain only nonmetals. Na and Ca are metals and Ar and Cl_2 are not compounds.*

10. *D. Americium has an atomic number of 95, meaning that it has 95 protons and 95 electrons in the neutral atom. However, the americium ion carries a 3+ charge meaning that there are three fewer electrons than protons giving it 92 electrons. The number of neutrons is the mass number minus the atomic number, 243-95 = 148 neutrons.*

Free Response Answers and Explanations

1. *a. 35 and 37. Mass number is the total number of protons and neutrons in the nucleus. The mass number of an isotope is very close to the mass of the isotope. The mass of the isotope is the total mass of the protons and neutrons. The mass is not a whole number because the masses of protons and neutrons are not integers. Also nuclear binding energies have mass equivalents.*

b. To solve this problem you need the atomic mass of chlorine listed on the periodic table: 35.4527 amu. Let y = % abundance of Cl-37. Let x = % abundance of Cl-35.

$x + y = 1$

$35.4527 = 36.966y + 34.969x$

$35.4527 = 36.966(1-x) + 34.969x$

$x = 0.7578 = 75.78\ \%$ *Cl-35*

c. Chlorine molecules have three different masses. Because there are two isotopes of chlorine atoms, there are three different ways they can combine to form diatomic molecules: Cl-35 with Cl-35, Cl-35 with Cl-37 and Cl-37 with Cl-37.

d. *The chlorine molecule having the largest molecular mass is ^{37}Cl-^{37}Cl. It's mass is 2 x 36.966 = 73.932 amu.*

e. *Because chlorine-35 is the more abundant isotope, the most abundant molecule is ^{35}Cl-^{35}Cl.*

It's mass is 2 x 34.969 = 69.938 amu.

2. *Chlorine and iodine are in the same family so they will be expected to form analogous oxyions and oxyacids.*

IO^-	*hypoiodite*	*HIO*	*hypoiodous acid*
IO_2^-	*iodite*	HIO_2	*iodous acid*
IO_3^-	*iodate*	HIO_3	*iodic acid*
IO_4^-	*periodate*	HIO_4	*periocid acid* (pronounced "per-iodic")

Your Turn Answers

2.1. *Oxygen-18 contains 8 protons, 8 electrons, and 10 neutrons.*

2.2. a) *Chlorine, a toxic gas, refers to the free element, dichlorine, Cl_2.*

b) *Common table salt containing the element chlorine refers to monatomic Cl.*

2.3. *KI = potassium iodide. $MgSO_4$ = magnesium sulfate. FeS = iron(II) sulfide. Al_2O_3 = aluminum oxide. $Pb_3(PO_4)_2$ = lead(II) phosphate.*

2.4. *SCl_2 = sulfur dichloride. CF_4 = carbon tetrafluoride. BrI_3 = bromine triiodide. PBr_5 = phosphorus pentabromide. SF_6 = sulfur hexafluoride.*

TOPIC 3

STOICHIOMETRY: Calculations with Chemical Formulas and Equations

This chapter introduces the concept of the mole, a quantitative model for chemical structure. It also considers how chemical structure, encoded in chemical formulas, changes through the process of chemical reactions, which are encoded in chemical equations. The chemical equation is the qualitative model for chemical change and stoichiometry is the quantitative model.

For success on the Advanced Placement Chemistry exam, students need to be able to think holistically about problem solving using the fundamental ideas of the mole and stoichiometry. Chemical problem solving is usually about the quantitative manipulation of moles and how they relate to chemical formulas and equations. Much of the content of this chapter may have been learned in the first year course, but mastery of complex problems involving limiting reactants, empirical and molecular formulas, and combustion and hydrate analyses is essential. Pay particular attention to these sections:

3.5 **Empirical Formulas**
3.6 **Quantitative Information from Balanced Equations**
3.7 **Limiting Reactants, Theoretical Yields**

Chemical Equations

Section 3.1

Stoichiometry is the area of study that examines the quantities of substances involved in chemical reactions.

A **chemical reaction** is a process by which one or more substances are converted to other substances.

Chemical equations use chemical formulas to symbolically represent chemical reactions. For example, the following chemical equation describes how butane, C_4H_{10}, burns in air:

$2C_4H_{10}(g) + 13O_2(g) \rightarrow 8CO_2(g) + 10H_2O(l)$

reactant $\rightarrow$ **products**

The equation is a "chemical sentence" written in the symbolic "words" of chemical formulas and symbols. The formulas on the left are reactants, and the formulas on the right are products. The equation reads:

"Two molecules of gaseous butane react with thirteen molecules of oxygen gas to produce eight molecules of carbon dioxide gas and ten molecules of liquid water.

A **balanced chemical equation** has an equal number of atoms of each element on each side of the arrow. The coefficients preceding each formula "balance" the equation. Notice that, because of the coefficients, on each side of the arrow there are eight carbon atoms, twenty hydrogen atoms, and twenty-six oxygen atoms.

The symbols (g), (l), (s) and (aq) are used to designate the physical state of each reactant and product: gas, liquid, solid and aqueous (dissolved in water).

Your Turn 3.1

Write a chemical sentence to illustrate how to "read" the following chemical equation:

$CO_2(g) + 2NH_3(g) + H_2O(l) \rightarrow 2NH_4^+(aq) + CO_3^{2-}(aq)$

Write your answer in the space provided.

To balance simple equations, start by balancing atoms other than hydrogen or oxygen. Balance hydrogen atoms next to last and balance oxygen atoms last.

Example:

Balance the following equation:

$H_2SO_4(aq) + NaHCO_3(s) \rightarrow CO_2(g) + H_2O(l) + Na_2SO_4(aq)$

Solution:

1. *S and C are already balanced so start with Na.*

$H_2SO_4(aq) + 2NaHCO_3(s) \rightarrow CO_2(g) + H_2O(l) + Na_2SO_4(aq)$

2. Now balance carbon.

$H_2SO_4(aq) + 2NaHCO_3(s) \rightarrow 2CO_2(g) + H_2O(l) + Na_2SO_4(aq)$

3. Next balance H.

$H_2SO_4(aq) + 2NaHCO_3(s) \rightarrow 2CO_2(g) + 2H_2O(l) + Na_2SO_4(aq)$

4. Check to see that oxygen is balanced and double check all other atoms.

(As in algebraic equations, the numeral 1 is usually not written.)

Simple Patterns of Reactivity

Section 3.2

Predicting the products of chemical reactions is an essential skill to acquire in the study of chemistry. Sometimes reactions fall into simple patterns and recognizing these patterns can be helpful in predicting which products will be produced from given reactants. Topic 4 addresses predicting products of chemical reactions in more depth. For now, here are a few patterns to learn to recognize.

1. A **combination reaction** is where two elements combine to form one compound. (Other combinations are possible and Chapter 4 describes better ways to predict what will happen.)

Example:

A metal reacts with a nonmetal to produce an ionic compound. Write an equation to describe what happens when:

Solid sodium is exposed to chlorine gas.

Solution:

$2Na(s) + Cl_2(g) \rightarrow 2NaCl(s)$

Your Turn 3.2

Write an equation to describe what happens when solid magnesium metal reacts at high temperature with nitrogen gas. Write your answer in the space provided.

2. A **decomposition reaction** is where one reactant changes to two or more products.

Example:

Upon heating, metal carbonates decompose to yield metal oxides and carbon dioxide. Write an equation to describe what happens when:

Solid magnesium carbonate is heated.

Solution:

$MgCO_3(s) \rightarrow MgO(s) + CO_2(g)$

Your Turn 3.3

Write an equation to describe what happens when liquid water is decomposed to its elements by an electrical current. Write your answer in the space provided.

3. A **combustion reaction** usually involves oxygen, often from air, reacting with hydrocarbons or other organic molecules containing carbon, hydrogen, and oxygen to produce carbon dioxide and water.

Example:

Write an equation to describe what happens when liquid ethanol burns in air.

Solution:

$CH_3CH_2OH(l) + 3O_2(g) \rightarrow 2CO_2(g) + 3H_2O(g)$

(Recall that encoded in the name "ethanol" is a two-carbon alcohol having the –OH functional group.)

Your Turn 3.4

Write an equation to describe what happens when liquid hexane is burned in air. Write your answer in the space provided.

Avogadro's Number, the Mole, and Molar Masses

Section 3.4

Avogadro's number is 6.02 X 10^{23}. It represents the number of atoms in exactly 12 grams of isotopically pure ^{12}C. Because all atomic masses are based on ^{12}C, the atomic mass of any element expressed in grams represents Avogadro's number (6.02 X 10^{23}) of atoms of that element.

A **mole** is the amount of matter that contains 6.02 X 10^{23} atoms, ions, molecules, or formula units.

The **molar mass** of a substance is the mass in grams of one mole of that substance. To calculate a molar mass of any substance, add the atomic masses of all the atoms in its formula. (For convenience atomic masses are often rounded to three significance figures.) Table 3.1 shows the molar masses of various substances.

Common misconception: "Molar mass" is a universal term that is often used to replace the terms atomic mass, molecular mass, and formula mass. Molar mass is used to express the mass of one mole of any substance, whether it is an atom (atomic mass), a molecule (molecular mass), an ion (formula mass) or an ionic compound (formula mass).

Table 3.1. The molar mass of any substances is the mass in grams of 6.02 X 10^{23} particles or one mole of that substance.

Formula	Number of particles	Representative particles	Molar mass	Alternate term
Ar	6.02 X 10^{23}	atoms	39.9 g/mol	Atomic mass
CO_2	6.02 X 10^{23}	molecules	44.0 g/mol	Molecular mass
NaBr	6.02 X 10^{23}	formula units	103 g/mol	Formula mass
CO_3^{2-}	6.02 X 10^{23}	ions	60.0 g/mol	Formula mass

Grams, moles and representative particles (atoms, molecules, ions or formula units) are converted one to another using the "Mole Road' described in Figure 3.1.

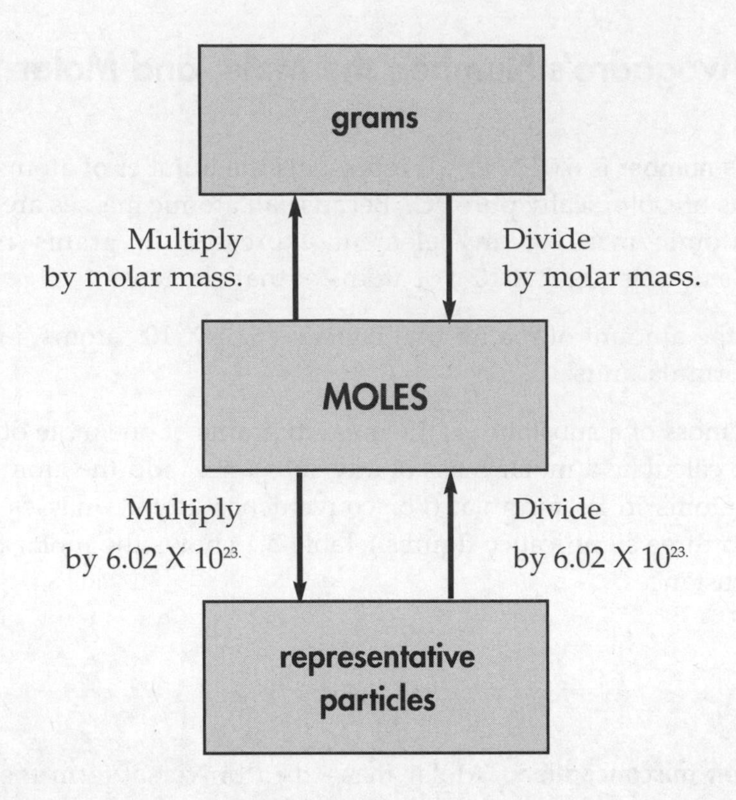

FIGURE 3.1. The "Mole Road". Divide to convert to moles. Multiply to convert from moles.

Calculating percentage composition of a compound

The **percentage composition** of a compound is the percentage by mass contributed by each element in the compound. To calculate the percentage composition of an element in any formula, divide the molar mass of the element multiplied by the number of times it appears in the formula, by the molar mass of the formula and multiply by 100.

% composition = 100 X (molar mass of element X subscript for element)/(molar mass of substance)

Example:

What is the % composition of Na_2CO_3?

Solution:

% Na = 100 X 2(23.0) g Na/[2(23.0)+12.0 +3(16.0) g] = 43.4% Na

% C = 100 X 12.0 g C/[2(23.0)+12.0 +3(16.0) g] = 11.3% C

% O = 100 X 3(16.0) g/[2(23.0)+12.0 +3(16.0) g] = 45.3% O

Calculating An Empirical Formula from Percentage Composition

Section 3.5

The **empirical formula** for a compound expresses the simplest ratio of atoms in the formula. The percentage composition of a compound can be determined experimentally by chemical analysis and the empirical formula can be calculated from the percentage composition.

Example:

What is the empirical formula of a compound containing 68.4% chromium and 31.6 % oxygen?

Solution:

In chemistry, percentage always means mass percentage, unless specified otherwise. The data means that for every 100 grams of compound, there are 68.4 g of Cr and 31.6 g of O.

1. Write the formula in terms of grams.

Cr *68.4* **O** *31.6*

2. Convert grams to moles by dividing by the molar mass of each element.

Cr *68.4/52.0* **O** *31.6/16.0 =* **Cr** *1.315* **O** *1.975*

3. Convert to small numbers by dividing each mole quantity by the smaller mole quantity.

Cr *1.315/1.315* **O** *1.975/1.315 =* **Cr** *1* **O** *1.5*

4. If necessary, multiply each mole quantity by a small whole number that converts all quantities to whole numbers.

Cr *1 x 2* **O** *1.5 x 2 =* Cr_2O_3

Molecular formulas from empirical formulas

A **molecular formula** tells exactly how many atoms are in one molecule of the compound. The subscripts in a molecular formula are always whole number multiples of the subscripts in the empirical formula. Molecular formulas can be determined from empirical formulas if the molar mass of the compound is known.

Example:

A compound containing only carbon, hydrogen and oxygen is 63.16% C and 8.77% H. It has a molar mass of 114 g/mol. What is its empirical formula and molecular formula?

Solution:

1. *Write the formula in terms of grams.*

C 63.16 **H** 8.77 **O** 28.07

(Calculate the grams of oxygen by subtracting the grams of carbon and grams of hydrogen from 100g. g O = 100 g - 63.16 g C - 8.77 g H = 28.07 g O.)

2. *Convert grams to moles by dividing by the molar mass of each element.*

C 63.16/12.0 **H** 8.77/1.00 **O** 28.07/16.0 = **C** 5.263 **H** 8.77 **O** 1.754

3. *Convert to small numbers by dividing each mole quantity by the smallest mole quantity.*

C 5.263/1.754 **H** 8.77/1.754 **O** 1.754/1.754 = $C_3H_5O_1$

empirical formula

4. *All quantities are whole numbers.*

5. *Divide the known molar mass by the mass of one mole of the empirical formula. The result produces the integer by which you multiply the empirical formula to obtain the molecular formula.*

(114 g/mol)/(57.0 g/mol) = 2

C 3x2 **H** 5x2 **O** 1x2 = $C_6H_{10}O_2$

molecular formula

Empirical formulas from combustion analysis

When a compound containing carbon, hydrogen, and oxygen is completely combusted, all the carbon is converted to carbon dioxide, and all the hydrogen becomes water. The empirical formula of the compound can be calculated from the measured masses of the products.

Example:

A 3.489 g sample of a compound containing C, H, and O yields 7.832 g of carbon dioxide, and 1.922 grams of water upon combustion. What is the simplest formula of the compound?

The unbalanced chemical equation is:

$CxHyOz + O_2 \rightarrow CO_2 + H_2O$

X *is the number of moles of carbon in the compound because all the carbon in the compound is converted to* CO_2*. X = the number of moles of* CO_2 *because there is one mole of carbon in one mole of* CO_2*.*

X = *mol C = 7.832 g* CO_2*/44.0 g/mol* = ***0.178 mol C***

***Y** is the number of moles of hydrogen in the compound because all the hydrogen becomes water. Y = twice the number of moles of water because there are two moles of hydrogen in one mole of water.*

Y = *mol H = (1.922 g* H_2O*/18.0 g/mol) x 2* = ***0.2136 mol H***

***Z** is the number of mol of O. To obtain the number of grams of O in the compound, subtract the number of grams of C (X mol x 12.0 g/mol) and the number of grams of H (Y mol x 1.00 g/mol) from the total grams of the compound. Convert the result to moles of O by dividing by 16.0 g/mol.*

Z = *mol O = (3.489 – 12.0 X – 1.00 Y)/16* = ***0.0712 mol O***

Convert to small numbers by dividing each mole quantity by the smallest mole quantity.

$C_{0.178}H_{0.2136}O_{0.0712} =$

$C_{0.178/0.0712}H_{0.2136/0.0712}O_{0.0712/0.0712} = C_{2.5}H_3O_1$

Finally, multiply each mole quantity by a small whole number that converts all quantities to whole numbers.

$C_{2.5x2}H_{3x2}O_{1x2} = C_5H_6O_2$

Formulas of hydrates

Ionic compounds often form crystal structures called **hydrates** by acquiring one or more water molecules per formula unit. For example, solid sodium thiosulfate decahydrate, $Na_2S_2O_3 \bullet 10H_2O$, has ten water molecules per formula unit of sodium thiosulfate. Heating a sample of hydrate causes it to lose water. The number of water molecules per formula unit can be calculated from the mass difference before and after heating.

Example:

When 21.91 g of a hydrate of Copper(II) sulfate is heated to drive off the water, 14.00 g of anhydrous copper(II) sulfate remain. What is the formula of the hydrate?

Solution:

This is a variation of an empirical formula problem. The solution lies in calculating the ratio of H_2O *moles to the moles of* $CuSO_4$*. The chemical equation is:*

$CuSO_4 \bullet X\,H_2O(s) \rightarrow CuSO_4(s) + X\,H_2O(g)$

21.91 g   *14.00 g*

Calculate the grams of water by subtracting the grams of copper(II) sulfate from the grams of the hydrate:

g H_2O *= 21.91g -14.00g = 7.91 g* H_2O

mol H_2O *= 7.91 g* H_2O*/18.0 g/mol = 0.439 mol* H_2O

mol $CuSO_4$ *= 14.00 g/159.5 g/mol = 0.08777 mol* $CuSO_4$.

mol H_2O*/ mol* $CuSO_4$ *= 0.439 mol/0.08777 mol = 5.00*

The formula has five moles of water per mole of copper(II) sulfate:

$CuSO_4 \bullet 5H_2O$

Section 3.6 Stoichiometry

Stoichiometry is the area of study that examines the quantities of substances involved in chemical reactions. The coefficients in a balanced chemical equation indicate both the relative number of molecules (or formula units) involved in the reaction, and the relative number of moles. For example, the equation for the combustion of octane, C_8H_{18}, a component of gasoline is:

$$2C_8H_{18}(l) + 25O_2(g) \longrightarrow 16CO_2(g) + 18H_2O(g)$$

The four coefficients that balance the equation are proportional to one another and can be used to relate mole quantities of reactants and/or products.

Example:

How many moles of octane will burn in the presence of 37.0 moles of oxygen gas?

Solution:

The balanced equation tells us that two moles of octane will burn in 25 moles of oxygen gas, so the answer to the question involves the ratio 2 mol C_8H_{18} *per 25 mol* O_2.

x mol C_8H_{18} *= 37.0 mol* O_2 *(2 mol* C_8H_{18}*/25 mol* O_2*) = 37.0 x 2/25 = 2.96 mol* C_8H_{18}.

Alternatively, we can use the mole road:

To convert mol O_2 *to mol* C_8H_{18}*, multiply by 2/25*

mol C_8H_{18} *= 37.0 mol (2/25) = 2.96 mol* C_8H_{18}

(Always multiply by the coefficient at the head of the arrow and divide by the coefficient at the tail of the arrow.)

x 2/25

mol C_8H_{18} ⇄ *mol* O_2

x 25/2

To convert mol C_8H_{18} to mol O_2, multiply by 25/2

Figure 3.2 shows an expanded mole road to include the relationship between the moles of any reactant or product in a balanced chemical equation. This stoichiometry mole road allows us to relate any quantity in a balanced equation to any other quantity in the same equation.

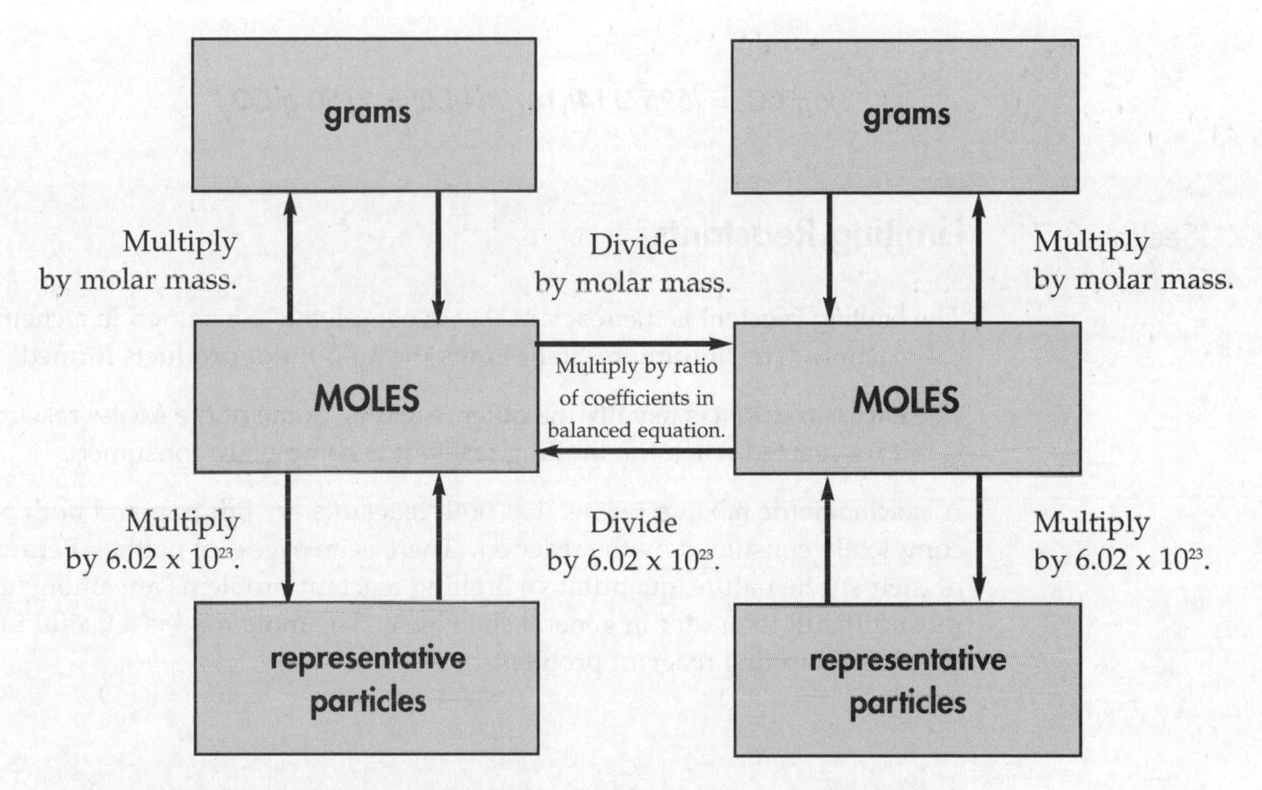

Figure 3.2. The stoichiometry "Mole Road". Divide to convert to moles. Multiply to convert from moles. Multiply by the ratio of coefficients in a balanced equation to convert moles of one substance to moles of another substance.

Example:

How many grams of carbon dioxide are obtained when 695 g (about one gallon) of octane are burned in oxygen?

Solution:

Follow the road map:

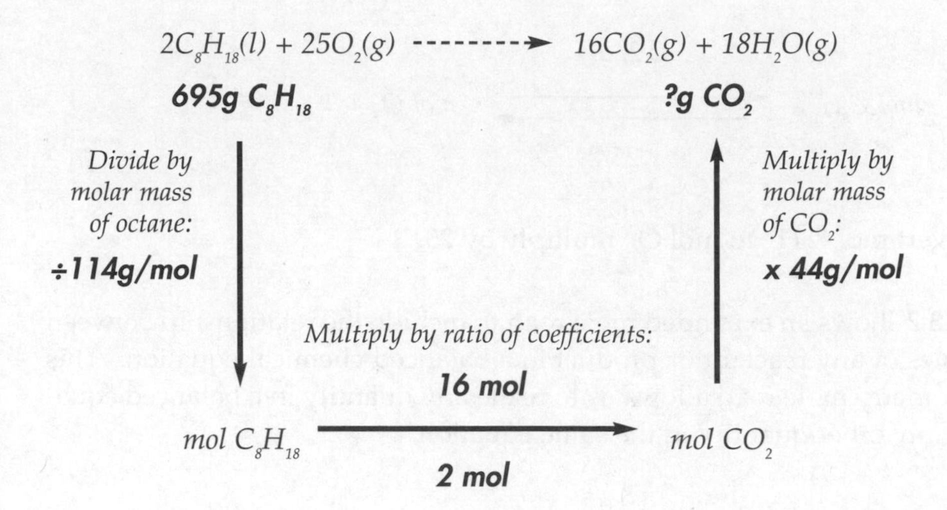

x g CO_2 = (695/114)(16/2)(44.0) = 2150 g CO_2

Section 3.7 Limiting Reactants

The **limiting reactant** is the reactant that is completely consumed in a chemical reaction. The limiting reactant limits the amount of products formed.

The **excess reactant** is usually the other reactant. Some of the excess reactant is left un-reacted when the limiting reactant is completely consumed.

A **stoichiometric mixture** means that both reactants are limiting and both are completely consumed by the reaction. There is an excess of neither. Because of their subtle nature, quantitative limiting reactant problems are among the most difficult to master in general chemistry. The mole road is a useful tool in solving limiting reactant problems.

Common misconception: Stoichiometry calculations can calculate only how much reactants react or how much products are formed. Stoichiometry cannot calculate how much excess reactant is left un-reacted. To calculate how much excess reactant remains after the reaction is complete, first calculate the how much is consumed and then subtract that amount from how much total reactant was initially present.

Example:

255 grams of octane and 1510 grams of oxygen gas are present at the beginning of a reaction that goes to completion and forms carbon dioxide and water according to the following equation.

$2C_8H_{18}(l) + 25O_2(g) \rightarrow 16CO_2(g) + 18H_2O(g)$

a. What is the limiting reactant?

b. How many grams of water are formed when the limiting reactant is completely consumed?

c. How many grams of excess reactant is consumed?

d. How many grams of excess reactant is left un-reacted?

Solution:

a. To find the limiting reactant, first compare the number of moles of reactants relative to the ratio in which they react. To do this, divide the number of moles of each reactant by its corresponding coefficient that balances the equation. The resulting lower number always identifies the limiting reactant. Then use the mole road to solve the problem based on the identified limiting reactant.

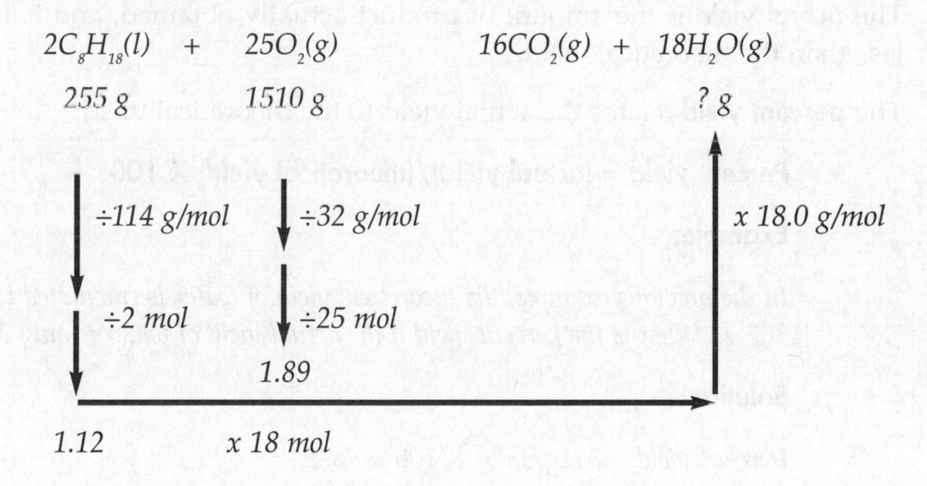

a. C_8H_{18} *limiting reactant because 1.12 < 1.89*

b. x mol H_2O *= (255g /114 g/mol)(18 mol/2 mol)(18.0 g/mol) = 362 g* H_2O

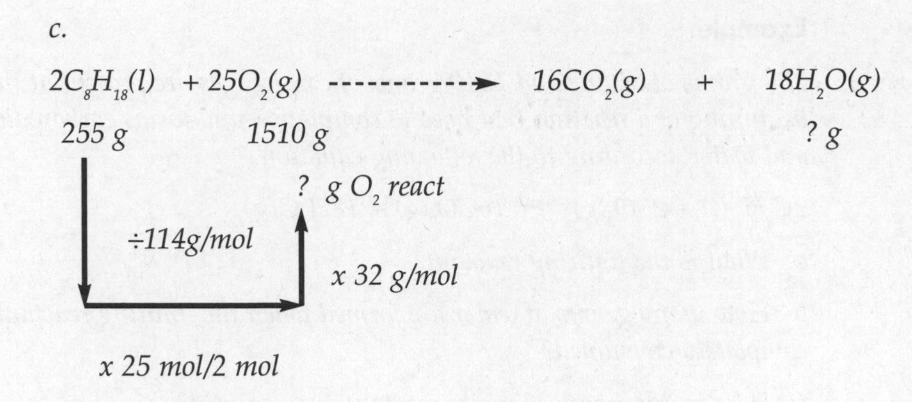

x *mol* O_2 *= (255 g)/114 g/mol)(25 mol/2 mol)(32.0 g/mol) = 895 g* O_2 *react*

d. g O_2 *left un-reacted = 1510 g – 895 g = 615 g* O_2 *un-reacted*

Theoretical, Actual, and Percent Yields

The **theoretical yield** of a reaction is the quantity of product that is calculated to form.

The **actual yield** is the amount of product actually obtained, and is usually less than the theoretical yield.

The **percent yield** relates the actual yield to the theoretical yield:

Percent yield = (actual yield)/(theoretical yield) X 100

Example:

In the previous example, the theoretical yield of water is calculated to be 362 g. What is the percent yield if the actual yield of water is only 312 g?

Solution:

Percent yield = 312g/362g X 100 = 86.2%

Multiple Choice Questions

1. *Ammonia forms when hydrogen gas reacts with nitrogen gas. If equal number of moles of nitrogen and hydrogen are combined, the maximum number of moles of ammonia that could be formed will be equal to*

A) *the number of moles of hydrogen.*

B) *the number of moles of nitrogen.*

C) *twice the number of moles of hydrogen*

D) *twice the number of moles of nitrogen*

E) *two thirds the number of moles of hydrogen*

2. *If $C_4H_{10}O$ undergoes complete combustion, what is the sum of the coefficients when the equation is completed and balanced using smallest whole numbers?*

A) 8

B) 16

C) 22

D) 25

E) 32

3. *What are the products when lithium carbonate is heated?*

A) $LiOH + CO_2$

B) $Li_2O + CO_2$

C) $LiO + CO_2$

D) $LiC + O_2$

E) $LiO + CO$

4. *Beginning with 48 moles of H_2, how many moles of $Cu(NH_3)_4Cl_2(aq)$ can be obtained if the synthesis of $Cu(NH_3)_4Cl_2(aq)$ is carried out through the following sequential reactions? Assume that a 50% yield of product(s) is(are) obtained in each reaction.*

1. $3H_2(g) + N_2(g) \longrightarrow 2NH_3(g)$

2. $4\ NH_3(g) + CuSO_4(aq) \longrightarrow Cu(NH_3)_4SO_4(aq)$

3. $Cu(NH_3)_4SO_4 + 2NaCl \longrightarrow Cu(NH_3)_4Cl_2(aq) + Na_2SO_4(aq)$

A) 1

B) 2

C) 4

D) 8

E) 12

5. *What mass of water can be obtained from 4.0 g of H_2 and 16 g of O_2?*
   $2H_2(g) + O_2(g) \dashrightarrow 2\,H_2O$

A) 9 g

B) 18 g

C) 36g

D) 54 g

E) 72 g

6. *The empirical formula of pyrogallol is C_2H_2O and its molar mass is 126. Its molecular formula is*

A) C_2H_2O

B) $C_4H_4O_2$

C) $C_2H_6O_3$

D) $C_6H_6O_3$

E) $C_2H_6O_6$

7. *What is the maximum amount of water that can be prepared from the reaction of 20.0 g of HBr with 20.0 g of $Ca(OH)_2$?*

   $2HBr + Ca(OH)_2 \dashrightarrow CaBr_2 + 2H_2O$

A) (20/81)(2/2)(18) g

B) (20/74)(2/2)(18) g

C) (20/81)(2/1)(18) g

D) (20/74)(2/1)(18) g

E) (20/74)(1/2)(18) g

8. *How many moles of of ozone, O_3, could be formed from 96.0 g of oxygen gas, O_2?*

A) 0.500 mol

B) 1.00 mol

C) 2.00 mol

D) 3.00 mol

E) 1/16 mol

9. *The percentage of oxygen in $C_8H_{12}O_2$ is:*

A) (16/140)(100)

B) (32/140)(100)

C) (16/124)(100)

D) (140/32)(100)

E) (32/124)(100)

10. *A compound contains 48% O, 40.0% Ca, and the remainder is C. What is its empirical formula?*

A) $O_3C_2Ca_2$

B) O_3CCa_2

C) O_3CCa

D) O_3CCa_2

E) O_2CCa

Free Response Question

1. *Combustion of 8.652 grams of a compound containing C, H, O, and N yields 11.088 g of CO_2 3.780 grams of H_2O and 3.864 grams of NO_2.*

a. *How many moles of C, H, and N are contained in the sample?*

b. *How many grams of oxygen are contained in the sample?*

c. *What is the simplest formula of the compound?*

d. *If the molar mass of the compound lies between 200 and 300, what is its molecular formula?*

e. *Write and balance a chemical equation for the combustion of the compound.*

Additional Practice in Chemistry the Central Science

For more practice answering questions in preparation for the Advanced Placement examination try these problems in Chapter 3 of Chemistry the Central Science:

Additional Exercises: 3.81, 3.84, 3.86, 3.88, 3.89, 3.90, 3.91, 3.96, 3.98, 3.99, 3.100.

Integrative Exercises: 3.103, 3.104, 3.105, 3.107.

Multiple Choice Answers and Explanations

1. *E. The equation is $3H_2(g) + N_2(g) \longrightarrow 2NH_3(g)$. If equal number of moles of reactants are mixed, the limiting reactant is H_2 because the reaction requires three times as much H_2 as it does N_2. The number of moles of ammonia depends on the limiting reactant, H_2, and is 2/3 the amount of H_2 present.*

2. *B. The products of a complete combustion of $C_4H_{10}O$ are carbon dioxide and water. The correctly completed and balanced equation is:*

   $C_4H_{10}O + 6O_2 \longrightarrow 4CO_2 + 5H_2O$

   The sum of the coefficients is 1+ 6 + 4 + 5 = 16.

3. *B. Upon heating, a metal carbonate decomposes to a metal oxide and carbon dioxide gas. $Li_2CO_3(s) \longrightarrow Li_2O(s) + CO_2(g)$*

4. *A. Reaction 1 will yield 48(2/3) X 0.50 = 16 mol of NH_3. Reaction 2 will yield 16(1/4) X 0.50 = 2 moles of $Cu(NH_3)_4SO_4(aq)$. Reation 3 will yield 2(1/1) x 0.50 = 1 mol of $Cu(NH_3)_4Cl_2(aq)$.*

5. *B. 4.0 g H_2 is 2 mol. 16 g O_2 is 0.5 mol. O_2 is the limiting reactant. 0.5 mol of O_2 will produce 1.0 mol H_20, which is 18 g.*

6. *D. The ratio of the molar mass to the mass of the empirical formula is 126/42 = 3. This means that the molecular formula has there three times as many atoms as the empirical formula.*

7. *A. The limiting reactant is HBr.*

   (20.0 g)/(81.0 g/mol)/2 < (20.0 g)/(74.0 g/mol)/1.

   The 20.0 grams of HBr will limit how much water will be formed.

   x g H_2O = 20.0 g HBr(1mol/81.0 g)(2 mol H_2O/2 mol HBr)(18.0 g/mol)

8. *C. The balanced equation is: $3O_2(g) \longrightarrow 2O_3(g)$*

   x mol O_3 = 96.0 g O_2 (1mol/32.0g)(2 mol O_3/3 mol O_2) = 2.00 mol

9. *B. % O = [(2 x atomic mass of O)/(molar mass of $C_8H_{12}O_2$)] x 100 = (2 x16)/(140) x 100 = (32/140)(100)*

10. *C. The percentage composition is the number of grams of each element contained in 100 grams of compound. The % of C is 100% -*

48.0% - 40.0% = 12.0% C. Convert grams of each atom to moles by dividing by the respective atomic mass.

$$O_{48.0/16.0}\,C_{12.0/12.0}\,Ca_{40.0/40.0} = O_3C_1Ca_1$$

This is calcium carbonate, $CaCO_3$.

Free Response Answer

a. *The moles of carbon are equal to the moles of carbon dioxide because there is one mole of C in every mole of CO_2. Divide the grams of CO_2 produced by the molar mass of CO_2 to obtain the moles of carbon. Divide the mass of water by the molar mass of water, and multiply by two, to obtain the moles of hydrogen. Repeat for the moles of N.*

mol C = 11.088 g CO_2/44.0 g/mol = 0.252 mol CO_2 = 0.252 mol C.

mol H = 3.780 g H_2O/18.0 g/mol = 0.210 mol H_2O x 2 = 0.420 mol H

mol N = 3.864 g NO_2/46.0 g/mol = 0.0840 mol NO_2 = 0.0840 mol N

b. *To obtain the number of grams of oxygen in the formula, subtract from the given sample mass, the number of grams of each of the other elements, C, H, and N. To obtain their masses, multiply the number of moles calculated in Part a by the atomic masses of the respective elements.*

g O = 8.652 g compound – g C – g H – g N

g O = 8.652 g – (0.252 mol C x 12.0 g/mol) – (0.420 mol H x 1.00 g/mol) – (0.0840 mol N x 14.0 g/mol) = 4.032 g O.

c. *To obtain the empirical formula convert the mass of O calculated in Part b. to moles by dividing by the atomic mass of oxygen. Then convert each mole quantity to small whole numbers by dividing each by the lowest of the mole quantities.*

Mol O = 4.032 g /16.0 g/mol = 0.252 mol O

$$C_{0.252}\,H_{0.420}\,N_{0.0840}\,O_{0.252}$$

$$C_{0.252/0.0840}\,H_{0.420/0.0840}\,N_{0.0840/0.0840}\,O_{0.252/0.0840}$$

$$C_3H_5N_1O_3$$

d. *To obtain the molecular formula, calculate the molar mass of the empirical formula determined in Part c. Multiply the result by small integers until an integer multiple of the molar mass is a number between 200 and 300. Multiply each subscript in the empirical formula by this integer. Check your work by calculating the molar mass of the resulting molecular formula to see that it lies between 200 and 300 g/mol.*

Molar mass of $C_3H_5N_1O_3$ = *3(12.0) + 5(1.00) + 1(14.0) + 3(16.0) = 103 g/mol*

2 x 103 g/mol = 206 g/mol.

$C_{3x2}H_{5x2}N_{1x2}O_{3x2} = C_6H_{10}N_2O_6$

e. *Balance in order, C, N, H and O:*

$2C_6H_{10}N_2O_6 + 15O_2(g) \rightarrow 12CO_2(g) + 10H_2O(g) + 4NO_2(g)$

Answers to Your Turn

3.1 *One molecule of gaseous carbon dioxide reacts with two molecules of ammonia gas and one molecule of liquid water to produce two aqueous ammonium ions and one aqueous carbonate ion.*

3.2. $3Mg(s) + N_2(g) \rightarrow Mg_3N_2(s)$

3.3. $2H_2O(l) \rightarrow 2H_2(g) + O_2(g)$

3.4. $2C_6H_{14}(l) + 19O_2(g) \rightarrow 12CO_2(g) + 14H_2O(l)$

Alternatively: $C_6H_{14}(l) + 19/2\ O_2(g) \rightarrow 6CO_2(g) + 7H_2O(l)$

TOPIC 4

AQUEOUS REACTIONS AND SOLUTION STOICHIOMETRY

This chapter lays the foundation to successfully predict the products of a reaction, given the reactants. The content of this chapter also helps you to understand and apply more complex concepts of chemical reactions including quantitative calculations. Fully mastering later topics including buffer solutions, solubility product constants, thermodynamics and electrochemistry depends on a thorough knowledge of chemical reactions. All six sections of this chapter are relevant to the AP exam.

General Properties of Aqueous Solutions

Section 4.1

A **solution** is a homogeneous mixture of two or more substances.

The **solvent** is the dissolving medium, usually the substance present in the greatest quantity in a solution.

Aqueous solutions are solutions in which water is the solvent.

Solutes are other substances in the solution.

An **electrolyte** is a substance whose aqueous solutions contain ions. The solution conducts electricity because the ions are free to migrate throughout the solution.

Strong electrolytes are substances that exist in solution, completely ionized. Ionic compounds and some molecular compounds called strong acids are strong electrolytes. For example, when solid sodium chloride dissolves in water, the ions dissociate completely:

$$NaCl(s) \dashrightarrow Na^{+}(aq) + Cl^{-}(aq)$$

Similarly, the strong acid, sulfuric acid, H_2SO_4, ionizes completely in aqueous solution:

$$H_2SO_4(l) \dashrightarrow H^{+}(aq) + HSO_4^{-}(aq)$$

Strong bases, like potassium hydroxide also dissociate completely:

$$KOH(s) \dashrightarrow K^{+}(aq) + OH^{-}(aq)$$

Your Turn 4.1

Write an equation for the ionization of gaseous hydrogen chloride in water. Write your answer in the space provided.

Table 4.1 lists the names and formulas of common strong acids and strong bases.

Table 4.1. The Names and Formulas of Common Strong Acids and Strong Bases.

Strong acids		**Strong bases**	
* Sulfuric acid	H_2SO_4	lithium hydroxide	$LiOH$
nitric acid	HNO_3	sodium hydroxide	$NaOH$
perchloric acid	$HClO_4$	potassium hydroxide	KOH
chloric acid	$HClO_3$	rubidium hydroxide	$RbOH$
hydrochloric acid	HCl	cesium hydroxide	$CsOH$
hydrobromic acid	HBr	**calcium hydroxide	$Ca(OH)_2$
hydroiodic acid	HI	**strontium hydroxide	$Sr(OH)_2$
		**barium hydroxide	$Ba(OH)_2$

* Sulfuric acid is a diprotic acid (It has two H^+ ions.) and only the first H^+ ion ionizes completely.

** $Ca(OH)_2$, $Sr(OH)_2$ and $Ba(OH)_2$, are dibasic (they each have two OH^- ions) and in each case, the second of the hydroxide groups is partially dissociated.

Common misconception: The terms, "ionize" and "dissociate," are often used interchangeably by many texts, even though their meanings differ in a subtle way. Both imply that ions exist in solution when a solute dissolves. When an ionic compound dissolves, dissociation of ions already present occurs. An electrolyte that is a covalent compound ionizes in solution because the ions are not present in the pure compound.

Weak electrolytes exist mostly as molecules in solution, with a small fraction in the form of ions. Molecular compounds called **weak acids** and **weak bases** are weak electrolytes. For example, acetic acid, $HC_2H_3O_2$ and ammonia, NH_3, are weak electrolytes. Only about one percent or less of each molecular compound's molecules ionize in aqueous solution. The percent dissociation is concentration dependent.

$$HC_2H_3O_2(l) \dashrightarrow H^+(aq) + C_2H_3O_2^-(aq)$$

$$NH_3(g) + H_2O(l) \dashrightarrow NH_4^+(aq) + OH^-(aq)$$

A non-electrolyte is a substance that does not form ions in solution and its solution does not conduct electricity. A non-electrolyte usually consists of a molecular compound, which when dissolved in water, usually consists of intact, un-ionized molecules. For example, when ethanol dissolves in water its molecules remain intact:

$$CH_3CH_2OH(l) \dashrightarrow CH_3CH_2OH(aq)$$

Your Turn 4.2

Classify each of these compounds as strong, weak, or nonelectrolytes. Explain your reasoning. Calcium chloride; ammonium sulfate; hydrocyanic acid, HCN; glucose, $C_6H_{12}O_6$*. Write your answer in the space provided.*

Section 4.2 Precipitation Reactions

A **precipitate** is an insoluble solid formed by a reaction in solution. For example, aqueous lead(II) nitrate reacts with aqueous sodium bromide to form solid lead(II) bromide and sodium nitrate. The complete equation for the reaction is:

$Pb(NO_3)_2(aq) + 2NaBr(aq) \dashrightarrow PbBr_2(s) + 2NaNO_3(aq)$

If the soluble strong electrolytes are shown as ions, a more accurate representation of the reaction is a complete ionic equation:

$$Pb^{2+}(aq) + 2NO_3^-(aq) + 2Na^+(aq) + 2Br^-(aq) \dashrightarrow$$
$$PbBr_2(s) + 2Na^+(aq) + 2NO_3^-(aq)$$

Spectator ions are ions that appear in identical form among both reactants and products of a complete ionic equation.

A **net ionic equation** omits spectator ions because they do not change from reactants to products:

$$Pb^{2+}(aq) + 2Br\text{-}(aq) \dashrightarrow PbBr_2(s)$$

Predicting Products of Precipitation Reactions

The AP exam often asks you to predict the products of precipitation reactions and to write their net ionic equations. Table 4.2 lists guidelines for ions that are expected to form precipitates.

Table 4.2. Guidelines for Predicting Precipitates

1. These cations rarely form precipitates:

Ammonium	sodium	potassium	(and all other alkali metal ions)
NH_4^+	Na^+	K^+	(Li^+, Rb^+, Cs^+)

There are no common exceptions.

2. These anions rarely form precipitates:

nitrate	acetate	chloride	bromide	iodide	sulfate
NO_3^-	$C_2H_3O_2^-$	Cl^-	Br^-	I^-	SO_4^{2-}

Common exceptions to Guideline 2:

$PbCl_2$	$AgCl$	Hg_2Cl_2			
$PbBr_2$	$AgBr$	Hg_2Br_2			
PbI_2	AgI	Hg_2I_2			
$PbSO_4$		Hg_2SO_4	$CaSO_4$	$BaSO_4$	$SrSO_4$

(Notice that the heavy alkaline earth sulfates and compounds of Pb^{2+}, Ag^+ and Hg_2^{2+} form precipitates.)

Table 4.2. continued

3. These ions commonly form precipitates:

hydroxide	carbonate	phosphate	sulfide
OH^-	CO_3^{2-}	PO_4^{3-}	S^{2-}

Common exceptions to Guideline 3:

ammonium	sodium	potassium	(and all other alkali metal ions)
NH_4^+	Na^+	K^+	(Li^+, Rb^+, Cs^+)

do not form precipitates

and $Ba(OH)_2$, BaS, CaS and SrS do not form precipitates.

4. Other common precipitates:

$PbCrO_4$	Ag_2CrO_4	$BaCrO_4$

To predict the products of precipitation reactions and to write their net ionic equations, we need to apply the solubility guidelines in Table 4.2 and the rules for writing net ionic equations in Table 4.3.

Table 4.3. General Rules for Writing Net Ionic Equations

1. Write all the reactants that are indicated to be solids (s), liquids (l), or gases (g).
2. Re-write the formulas of the aqueous reactants omitting the spectator ions (Usually: NH_4^+, Na^+, K^+, Cl^-, NO_3^-, SO_4^{2-})
3. Predict and write the product(s).
4. If necessary, use ions to balance the mass and charge. Then balance using coefficients. Check to see if the charges are balanced.

Examples:

Write the net ionic equation for each of the laboratory situations described below. Assume a reaction occurs in all cases. (Note: Although the Advanced Placement Exam does not require students to show the phases of reactants or products (g, l, s, aq), the phases are indicated in these examples for clarity.)

a. Aqueous solutions of silver nitrate and sodium phosphate are mixed.

Solution:

1. No solids, liquids or gases are indicated as reactants.

2. For precipitation reactions the spectator ions are usually: NH_4^+, Na^+, K^+, Cl^-, NO_3^-, SO_4^{2-}. *In this case, nitrate and sodium ions are spectator ions so they are omitted.*

$Ag^+(aq) + PO_4^{3-}(aq)$

3. Usually precipitates contain one of these anions: OH^-, CO_3^{2-}, PO_4^{3-}, S^{2-} *and/or one of these cations:* Pb^{2+}, Ag^+.

$Ag^+(aq) + PO_4^{3-}(aq) \longrightarrow Ag_3PO_4(s)$

4. The Ag^+ *ion requires a coefficient of 3 to balance the mass and charge:*
$3Ag^+(aq) + PO_4^{3-}(aq) \longrightarrow Ag_3PO_4(s)$

b. Aqueous copper(II) sulfate is added to solid sodium carbonate.

Solution:

1. $Na_2CO_3(s)$
Sodium carbonate is the only solid indicated.

2. $Na_2CO_3(s) + Cu^{2+}(aq)$
Sulfate ion is usually a spectator ion.

3. $Na_2CO_3(s) + Cu^{2+}(aq) \longrightarrow CuCO_3(s)$
Carbonate ions usually form precipitates.

4. $Na_2CO_3(s) + Cu^{2+}(aq) \longrightarrow CuCO_3(s) + 2Na^+(aq)$

Sodium ions are required on the right side of the equation to balance the mass and charge. Notice that in this case, Na^+ *is not a spectator ion because it changes form from solid to aqueous during the reaction.*

c. Hydrogen chloride gas is bubbled through an aqueous solution of lead(II) nitrate.

Solution:

1. $HCl(g)$
HCl is identified as a gas.

2. $HCl(g) + Pb^{2+}(aq)$
Nitrate is a spectator ion.

3. $HCl(g) + Pb^{2+}(aq) \longrightarrow PbCl_2(s)$

Lead(II) chloride is a common exception to the guideline that chlorides usually do not form precipitates.

4. $2HCl(g) + Pb^{2+}(aq) \dashrightarrow PbCl_2(s) + 2H^+(aq)$

H^+ ions on the right side of the equation balance the mass and charge.

Your Turn 4.3

Write a net ionic equation for the reaction of aqueous sodium carbonate with solid copper(II) sulfate. Write your answer in the space provided.

Acid-Base Reactions

Section 4.3

Acids are substances that ionize in aqueous solution to form H^+ ions.

Bases are substances that react with H^+ ions.

Strong acids and bases are strong electrolytes. They completely ionize in solution. Table 4.1 lists the names and formulas of common strong acids and bases.

Weak acids and bases are weak electrolytes. They ionize only slightly in water solution. Acids and bases not listed in Table 4.1 are weak.

A neutralization reaction occurs upon mixing a solution of an acid and a base. For example, a neutralization reaction between an acid and a metal hydroxide produces water and a salt:

Complete equation:

$Ca(OH)_2(s) + 2HNO_3(aq) \dashrightarrow 2H_2O(l) + Ca(NO_3)_2(aq)$

Net ionic equation: $Ca(OH)_2(s) + 2H^+(aq) \dashrightarrow 2H_2O(l) + Ca^{2+}(aq)$

To predict the products of acid-base reactions and to write their net ionic equations, we need to know that water is usually a product of neutralization reactions and apply the rules for writing net ionic equations in Table 4.3.

Examples:

Write the net ionic equation for each of the laboratory situations described below. Assume a reaction occurs in all cases.

a. Aqueous hydrochloric acid is mixed with a solution of sodium hydroxide.

Solution :

1. *Neither reactant is indicated as a solid, liquid or gas.*

2. *Both chloride and sodium ions are spectator ions.*
$H^+(aq) + OH^-(aq)$

3. *Water is a product of a neutralization reaction.*
$H^+(aq) + OH^-(aq) \dashrightarrow H_2O(l)$

b. *Gaseous hydrogen chloride is bubbled through a solution of sodium hydroxide.*

Solution:

1. *HCl is indicated as a gas.*
$HCl(g)$

2. *Sodium ion is a spectator.*
$HCl(g) + OH^-(aq)$

3. *Water is a product of a neutralization reaction.*
$HCl(g) + OH^-(aq) \dashrightarrow H_2O(l)$

4. *Chloride ion is needed to balance the mass and charge.*
$HCl(g) + OH^-(aq) \dashrightarrow H_2O(l) + Cl^-(aq)$

Notice that the reactants in questions a and b are the same but, because of the difference in phase of HCl, the net ionic equations are different.

c. *Solid potassium hydroxide is mixed with aqueous sulfurous acid.*

Solution:

1. *KOH is identified as a solid.*
$KOH(s)$

2. *Sulfurous acid is a weak acid (It does not appear on the list of strong acids in Table 4.1.) so it is shown in its molecular form.*
$KOH(s) + H_2SO_3 \dashrightarrow$

3. *Water is a product of a neutralization reaction.*
$KOH(s) + H_2SO_3 \dashrightarrow H_2O(l)$

4. *When the H^+ ion of the acid reacts with the base, free sulfite ion is a product and $2K^+$ are needed to balance the mass and charge.*
$2KOH(s) + H_2SO_3 \dashrightarrow 2H_2O(l) + 2K^+(aq) + SO_3^{2-}(aq)$

(Note: Because this problem does not specify the relative amounts of acid

and base added, it is also correct to assume a one to one mole ratio. The following equation is also acceptable:

$KOH(s) + H_2SO_3 \dashrightarrow H_2O(l) + K^+(aq) + HSO_3^-(aq)$

Your Turn 4.4

Write a net ionic equation for the reaction of aqueous potassium hydroxide with aqueous acetic acid. Write your answer in the space provided.

Acid-Base Reactions and Gas Formation

Ionic compounds containing carbonate, sulfite, or sulfide ions produce gases when they react with acids.

Metal carbonates and metal hydrogen carbonates react with acids to produce carbon dioxide gas and water.

Complete equation:

$NaHCO_3(aq) + HBr(aq) \dashrightarrow CO_2(g) + H_2O(l) + NaBr(aq)$

Net ionic equation:

$HCO_3^-(aq) + H^+(aq) \dashrightarrow CO_2(g) + H_2O(l)$

Complete equation:

$Na_2CO_3(s) + 2HCl(aq) \dashrightarrow CO_2(g) + H_2O(l) + 2NaCl(aq)$

Net ionic equation:

$Na_2CO_3(s) + 2H^+(aq) \dashrightarrow CO_2(g) + H_2O(l) + 2Na^+(aq)$

Metal sulfites and metal hydrogen sulfites react with acids to produce sulfur dioxide gas and water.

Complete equation:

$2KHSO_3(s) + H_2SO_4(aq) \dashrightarrow 2SO_2(g) + 2H_2O(l) + K_2SO_4(aq)$

Net ionic equation:

$2KHSO_3(s) + 2H^+ \dashrightarrow 2SO_2(g) + 2H_2O(l) + 2K^+(aq)$

(Notice that $H^+(aq)$ is the net ionic form of the strong acid, H_2SO_4.)

Complete equation:
$Li_2SO_3(aq) + 2HNO_3(aq) \dashrightarrow SO_2(g) + H_2O(l) + 2LiNO_3(aq)$

Net ionic equation:
$SO_3^{2-}(aq) + 2H^+(aq) \dashrightarrow SO_2(g) + H_2O(l)$

Metal sulfides react with acids to produce hydrogen sulfide gas and water

Complete equation:
$CaS(s) + 2HNO_3(aq) \dashrightarrow H_2S(g) + Ca(NO_3)_2(aq)$

Net ionic equation:
$CaS(s) + 2H^+(aq) \dashrightarrow H_2S(g) + Ca^{2+}(aq)$

To write the net ionic equations for acid-base reactions that produce gases, we need to apply the rules for writing net ionic equations in Table 4.3.

Examples:

Write the net ionic equation for each of the laboratory situations described below. Assume a reaction occurs in all cases.

a. Sulfuric acid is added to solid sodium sulfide.

Solution:

1. Sodium sulfide is identified as a solid.

$Na_2S(s)$

2. Sulfate ion is a spectator. The net ionic form of the strong acid, H_2SO_4, *is* $H^+(aq)$.

$Na_2S(s) + H^+(aq)$

3. The product of the reaction between an acid and a metal sulfide is $H_2S(g)$.

$Na_2S(s) + H^+(aq) \dashrightarrow H_2S(g)$

4. Two sodium ions written as products balance the mass and charge.

$Na_2S(s) + 2H^+(aq) \dashrightarrow H_2S(g) + 2Na^+(aq)$

b. Calcium hydrogen carbonate solid is added to an aqueous solution of acetic acid.

Solution:

1. Calcium hydrogen carbonate is identified as a solid.

$Ca(HCO_3)_2(s)$

2. Acetic acid is a weak acid because it is not listed among the strong acids in Table 4.1. It is written in molecular form.
$Ca(HCO_3)_2(s) + HC_2H_3O_2(aq)$

3. The products are carbon dioxide and water.
$Ca(HCO_3)_2(s) + HC_2H_3O_2(aq) \dashrightarrow CO_2(g) + H_2O(l)$

4. Calcium ions and acetate ions are needed to balance the mass and charge.
$Ca(HCO_3)_2(s) + 2HC_2H_3O_2(aq) \dashrightarrow 2CO_2(g) + 2H_2O(l) + Ca^{2+}(aq) + 2C_2H_3O_2^-(aq)$

c. Gaseous hydrogen chloride is bubbled through an aqueous solution of potassium sulfite.

Solution:

1. HCl is identified as a gas.
$HCl(g)$

2. Potassium ion is a spectator ion.
$HCl(g) + SO_3^{2-}(aq)$

3. The products are sulfur dioxide gas and water.
$HCl(g) + SO_3^{2-}(aq) \dashrightarrow SO_2(g) + H_2O(l)$

4. Two chloride ions are needed to balance the mass and charge.
$2HCl(g) + SO_3^{2-}(aq) \dashrightarrow SO_2(g) + H_2O(l) + 2Cl^-(aq)$

Your Turn 4.5

Write a net ionic equation for the reaction of aqueous nitric acid with solid lithium sulfite. Write your answer in the space provided.

Other Acid-Base Reactions
(For more information, see Section 7.6.)

Most nonmetal oxides are acidic. As they dissolve in water they react to produce acids. For example,

$SO_2(g) + H_2O(l) \dashrightarrow H_2SO_3(aq)$

$$P_4O_{10}(s) + 6H_2O(l) \dashrightarrow 4H_3PO_4(aq)$$

$$2NO_2(g) + H_2O(l) \dashrightarrow HNO_2(aq) + H^+(aq) + NO_3^-(aq)$$

Because they are acids, nonmetal oxides react with bases. For example,

Complete equation:
$$SO_2(g) + 2NaOH(aq) \dashrightarrow Na_2SO_3(aq) + H_2O(l)$$
Net ionic equation:
$$SO_2(g) + 2OH^-(aq) \dashrightarrow SO_3^{2-}(aq) + H_2O(l)$$

Most metal oxides are basic. They react to form bases when added to water. For example,

$$Li_2O(s) + H_2O(l) \dashrightarrow 2Li^+(aq) + 2OH^-(aq)$$

$$CaO(s) + H_2O(l) \dashrightarrow Ca^{2+}(aq) + 2OH^-(aq)$$

Because they are bases, metal oxides react with acids. For example,

Complete equation:
$$CaO(s) + 2HCl(aq) \dashrightarrow CaCl_2(aq) + H_2O(l)$$

Net ionic equation:
$$CaO(s) + 2H^+(aq) \dashrightarrow Ca^{2+}(aq) + H_2O(l)$$

Examples:

Write the net ionic equation for each of the laboratory situations described below. Assume a reaction occurs in all cases.

a. Sulfuric acid solution is added to solid aluminum oxide.

Solution:

1. Aluminum oxide is a solid
$Al_2O_3(s)$

2. Sulfate is a spectator ion.
$Al_2O_3(s) + H^+(aq)$

3. Aluminum oxide is a base and the product of a neutralization reaction is water.
$Al_2O_3(s) + H^+(aq) \dashrightarrow H_2O(l)$

4. Two aluminum ions are required to balance the mass and charge.
$Al_2O_3(s) + 6H^+(aq) \dashrightarrow 3H_2O(l) + 2Al^{3+}(aq)$

b. Carbon dioxide gas is bubbled through aqueous potassium hydroxide.

Solution:

1. $CO_2(g)$

2. Potassium ion is a spectator ion.

$CO_2(g) + OH^-(aq)$

3. The product, hydrogen carbonate ion, balances the equation.

$CO_2(g) + OH^-(aq) \dashrightarrow HCO_3^-(aq)$

(Think of this as an acid-base reaction and combine the reactants.)

c. Sulfur dioxide gas is passed over solid magnesium oxide.

Solution:

1. The phases of both reactants are designated.

$SO_2(g) + MgO(s)$

2. There are no aqueous reactants.

3. Again, this is an acid-base reaction so combine the reactants and the equation balances.

$SO_2(g) + MgO(s) \dashrightarrow MgSO_3(s)$

If the answers to either b or c are elusive, it might help to think of metal oxides as base anhydrides (literally, "without water".) Similarly metal oxides are acid anhydrides. In Problem b, add water to CO_2 and get H_2CO_3. Now the answer to the question is:

$$H_2CO_3(aq) + OH(aq) \dashrightarrow HCO_3^-(aq) + H_2O(l)$$

Now take away one water molecule from each side and the equation becomes the answer given previously:

$$CO_2(g) + OH^-(aq) \dashrightarrow HCO_3^-(aq)$$

Similarly, in Problem c, water added to SO_2 becomes H_2SO_3 and MgO(s) plus water is $Mg(OH)_2$. Now the equation is:

$$H_2SO_3(aq) + Mg(OH)_2(aq) \dashrightarrow 2H_2O(l) + MgSO_3(aq)$$

Removing two molecules of water from each side yields the final answer:

$$SO_2(g) + MgO(s) \dashrightarrow MgSO_3(s)$$

Oxidation-Reduction Reactions

Oxidation-reduction reactions (also called **redox** reactions) are reactions that transfer electrons between reactants.

For example, acids react with active metals to produce hydrogen gas.

An acid reacts with a metal to produce hydrogen gas:

Complete equation:

$2HNO_3(aq) + 2Na(s) \dashrightarrow H_2(g) + 2NaNO_3(aq)$

Net ionic equation:

$2H^+(aq) + 2Na(s) \dashrightarrow H_2(g) + 2Na^+(aq)$

In the reaction, electrons are transferred from sodium atoms to hydrogen ions. The sodium atoms lose one electron each to become sodium ions. The hydrogen ions gain one electron each to become a diatomic hydrogen molecule.

Oxidation is the loss of electrons. In this case sodium atoms lose electrons and are oxidized.

Reduction is the gain of electrons. In this case, hydrogen ions gain electrons and they are reduced.

Metals also react with metal ions to exchange electrons.

Complete equation:

$Mg(s) + Zn(NO_3)_2(aq) \dashrightarrow Mg(NO_3)_2(aq) + Zn(s)$

Net ionic equation:

$Mg(s) + Zn^{2+}(aq) \dashrightarrow Mg^{2+}(aq) + Zn(s)$

The metal is oxidized and the metal ion is reduced.

An activity series is a list of metals arranged in order of decreasing ease of oxidation. Any metal on the list can be oxidized by ions below it. Table 4.4 shows an activity series for metals in aqueous solution.

Table 4.4. An Activity Series. Metals at the top of the list are most likely to be oxidized (lose electrons).

Li(s)	-->	Li^+	(aq)	+	1 e-
K(s)	-->	K^+	(aq)	+	1 e-
Ca(s)	-->	Ca^{2+}	(aq)	+	2 e-
Na(s)	-->	Na^+	(aq)	+	1 e-
Mg(s)	-->	Mg^{2+}	(aq)	+	2 e-
Al(s)	-->	Al^{3+}	(aq)	+	3 e-
Zn(s)	-->	Zn^{2+}	(aq)	+	2 e-
Fe(s)	-->	Fe^{2+}	(aq)	+	2 e-

Table 4.4. continurd

Ni(s)	-->	$Ni2^{+}$ (aq)	+	2 e-
Pb(s)	-->	Pb^{2+} (aq)	+	2 e-
H_2(g)	-->	$2H^{+}$ (aq)	+	2 e-
Cu(s)	-->	Cu^{2+} (aq)	+	2 e-
Ag(s)	-->	Ag^{+} (aq)	+	1 e-
Au(s)	-->	Au^{3+} (aq)	+	3 e-

Oxidation state, also called **oxidation number**, is a positive or negative whole number assigned to an element in a chemical formula based on a set of formal rules. The oxidation state is used to track electron transfer in redox reactions. Table 4.5 lists the rules for assigning oxidation numbers to elements in chemical formulas.

Table 4.5. Simplified rules for determining oxidation numbers

1. The oxidation number of combined oxygen is usually 2-, except in the peroxide ion, O_2^{2-} where the oxidation number of oxygen is 1-.

   Examples: In H_2O and in H_2SO_4 the oxidation state of oxygen is 2-. In H_2O_2 and in BaO_2, O is 1-.

2. The oxidation number of combined hydrogen is usually 1+, except in the hydride ion, H^-, where it is 1-.

   Examples: In H_2O and in H_2SO_4 the oxidation state of hydrogen in 1+. In NaH and CaH_2 H is 1-.

3. The oxidation numbers of all individual atoms of a formula add to the charge on that formula. When in doubt, separate ionic compounds into common cation-anion pairs.

   Examples:

O_2	Na	K^+	Cl_2	Ca^{2+}	H_2SO_4	NO_3^-	$Mg_3(PO_4)_2$ = Mg^{2+}	PO_4^{3-}
0	0	1+	0	2+	1+ 6+ 2-	5+ 2-	2+ 5+ 2- 2+	5+ 2-

Your Turn 4.6

What is the oxidation number of each atom in potassium aluminum sulfate, $KAl(SO_4)_2$? Write your answer in the space provided.

Section 4.5

Concentrations of Solutions

A **solution** is a homogeneous mixture consisting of a solvent and one or more solutes.

Concentration is the amount solute dissolved in a given amount of solvent or solution.

Molar concentration, also called **molarity** (abbreviated M) is the number of moles of solute dissolved in a liter of solution.

Molarity = moles of solute/volume of solution in liters

For example, 1.5 M $Na_3PO_4(aq)$ means that every liter of solution contains 1.5 moles of sodium phosphate.

Because Na_3PO_4 is a strong electrolyte, its ions dissociate completely in aqueous solution:

$$Na_3PO_4(s) \dashrightarrow 3Na^+(aq) + PO_4^{3-}(aq)$$

A 1.00 M solution of Na_3PO_4 contains 3.00 M Na^+ ions and 1.00 M PO_4^{3-} ions.

Dilution, adding water to a solution decreases the molar concentration of each substance in the solution by a factor called the dilution factor. The dilution factor is the ratio of the original volume to the new volume.

Example:

Enough water is added to 500 mL of 1.00 M Na_3PO_4 to make the final volume 800 mL. What are the molar concentrations of each ion upon dilution?

Solution:

1.00 M Na_3PO_4 = 3.00 M Na^+ ions and 1.00 M PO_4^{3-} ions.

(3.00 M Na^+ ions)(500 mL/800 mL) = 1.88 M Na^+.

(1.00 M PO_4^{3-} ions)(500 mL/800 mL) = 0.625 M PO_4^{3-}.

Solution Stoichiometry

Section 4.6

Solution stochiometry involves calculations that relate moles of reactants and products to the volumes of solutions and their molar concentrations. Figure 4.1 illustrates the mole road for converting grams, moles, molarity and liters.

Figure 4.1. Converting moles, grams, molarity (M) and liters.

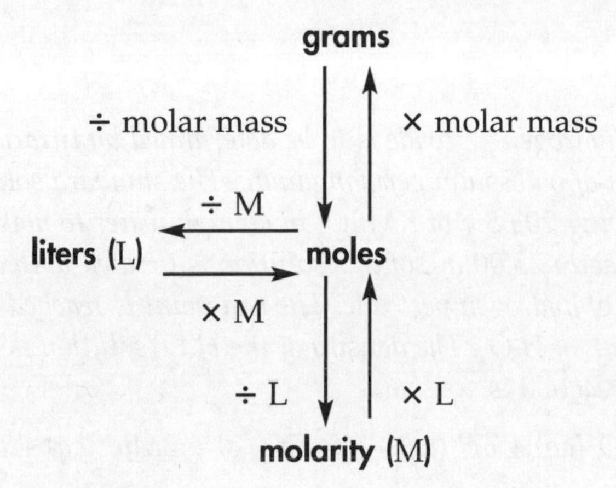

Example:

How many grams of solid sodium carbonate are in a sample if it takes 30.5 mL of 0.254 M hydrochloric acid to completely react with the sample?

Solution:

Write and balance a chemical equation, convert mL to L by dividing by 1000 and then follow the road map to the solution.

2 HCl(aq) + Na_2CO_3(s) --> CO_2(g) + H_2O(l) + 2NaCl(aq)

30.5 mL = .0305 L

? g Na_2CO_3(s)

x 0.254 M

x 106 g/mol

x 1mol Na_2CO_3

moles HCl

2 mol HCl

mol Na_2CO_3

(30.5mL)/(1000L/mL)

(30.5mL HCl_3)/(1000mL/L)(0.250mol/L)(1mol Na_2Cl_3/2mol HCl)(106g/mol)

A **titration** is an analytical technique used to determine the unknown concentration of a solution by reacting it with a solution of known concentration called a **standard solution**.

The **equivalence point** of the titration is the point at which the moles of substance dissolved in the unknown solution completely react with the moles of substance in the standard solution.

The **end point** is the point at which an indicator changes color. The end point is designed to coincide as closely as possible to the equivalence point. From the results of the experiment one can calculate the concentration of the unknown solution.

Example:

The percentage of hydrogen peroxide is to be determined by titration with a standard solution of potassium permanganate. The standard solution is prepared by dissolving 20.65 g of $KMnO_4$ *in enough water to make 500.0 mL of solution. Exactly 25.00 mL of this solution is titrated with an unknown solution of hydrogen peroxide. The end point is reached upon addition of 34.05 mL of* H_2O_2. *The density of the* H_2O_2 *solution is 1.05 g per milliliter. The reaction is:*

$$5H_2O_2(aq) + 2MnO_4^-(aq) + 6H^+(aq) \longrightarrow 5O_2(g) + 2Mn^{2+}(aq) + 8H_2O(l)$$

a. *What is the molar concentration of the standard* $KMnO_4$ *solution?*

b. *What is the molar concentration of the* H_2O_2 *solution?*

c. *How many grams of* H_2O_2 *are used in the experiment?*

d. *What is the percent of hydrogen peroxide in the solution?*

Solution:

a. *Molarity of* $KMnO_4$ *solution = moles* $KMnO_4$*/liters solution =*

$(20.65\ g/158.0\ g/mol)/0.5000\ L = 0.2614\ M\ KMnO_4$

b. *Follow the mole road from 25.00 mL of* $KMnO_4$ *solution to molarity of* H_2O_2 *solution:*

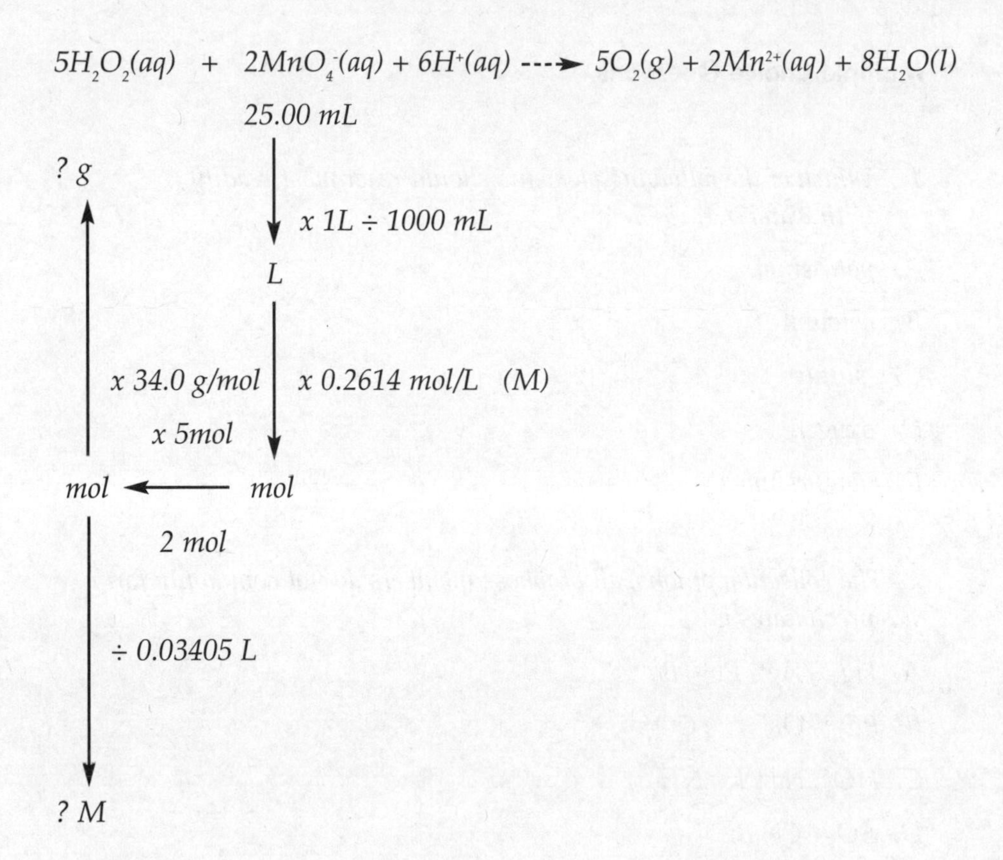

Molarity of H_2O_2 =
(25.00 mL/1000 mL/L)(0.2614 mol/L)(5 mol H_2O_2/2 mol MnO_4^-)/(0.03405 L) = 0.4798 M

c. On the road map, go from moles of H_2O_2 to grams of H_2O_2.

(25.00 /1000)(0.2614)(5/2) mol H_2O_2 (34.0 g/mol) = 0.555 g H_2O_2

d. % H_2O_2 = (g H_2O_2/g solution)(100)

g solution = 34.05 mL x 1.05 g/mL = 35.8 g solution

% H_2O_2 = (0.555g H_2O_2/35.8 g solution)(100) = 1.55% H_2O_2

Multiple Choice Questions

1. *Which of the following elements should react most readily with water?*

A) *potassium*

B) *calcium*

C) *sulfur*

D) *oxygen*

E) *magnesium*

2. *The collection of ions, all of whose members do not commonly form precipitates is*

A. Hg_2^{2+}, Ag^+, Pb^{2+}, Ba^{2+}

B. PO_4^{3-}, OH^-, S^{2-}, CO_3^{2-}

C. NO_3^-, Na^+, K^+, NH_4^+

D. SO_4^{2-}, Cl^-, Br^-, I^-

E. CrO_4^{2-}, Cu^{2+}, Fe^{2+}, SO_4^{2-}

3. *Which substance will **not** form a gas upon mixing with an aqueous acid?*

A) $NaHCO_3(s)$

B) $CaS(s)$

C) $Ca(s)$

D) $K_2SO_3(s)$

E) $Al_2O_3(s)$

4. *Which metal will **not** react with aqueous hydrochloric acid?*

A) *Fe*

B) *Al*

C) *Cu*

D) *K*

E) *Ni*

5. *How many milliliters of 0.40 M $FeBr_3$ solution would be necessary to precipitate all of the Ag^+ from 30 mL of a 0.40 M $AgNO_3$ solution?*

   $FeBr_3(aq) + 3AgNO_3(aq) \longrightarrow Fe(NO_3)_3(aq) + 3AgBr(s)$

A) 10 mL

B) 20 mL

C) 30 mL

D) 60 mL

E) 90 mL

6. *How many grams of baking soda, sodium hydrogen carbonate, are required to completely neutralize 1.00 L of of 6.00 M sulfuric acid that has been spilled on the floor?*

A) (1/6.00)(2/1)(84.0)

B) (6.00)(84.0)

C) (6.00)(2/1)/(84.0)

D) (6.00)(2/1)(84.0)

E) (6.00)(1/2)(84.0)

7. *It takes 37.50 mL of 0.152 M sodium chromate to titrate 25.00 mL of silver nitrate. What is the molarity of the silver nitrate solution?*

A) (2)(37.50)(0.152)(25.00)

B) (25.00)/(37.50)(0.152)(2)

C) (0.152)(25.00)/(37.50)

D) (37.50)(0.152)/(25.00)

E) (2)(37.50)(0.152)/(25.00)

8. *Which substance will react with water, at room temperature and pressure, to produce hydrogen?*

A) Mg

B) NaH

C) HN_3

D) NaOH

E) NH_3

9. *What is the net-ionic equation for the reaction of aqueous solutions of $CaCl_2$ and Na_2CO_3?*

A) $Ca^{2+}(aq) + 2Cl^{-}(aq) + 2Na^{+}(aq) + CO_3^{2-}(aq) \longrightarrow CaCO_3(s) + 2Na^{+}(aq) + 2Cl^{-}(aq)$

B) $Cl^{-}(aq) + Na^{+}(aq) \longrightarrow NaCl(s)$

C) $CaCl_2(aq) + CO_3^{2-}(aq) \longrightarrow CaCO_3(s) + 2Cl^{-}(aq)$

D) $Ca^{2+}(aq) + Na_2CO_3^{2-}(aq) \longrightarrow CaCO_3(s) + 2Na^{+}(aq)$

E) $Ca^{2+}(aq) + CO_3^{2-}(aq) \longrightarrow CaCO_3(s)$

10. *Magnesium burns in carbon dioxide to produce carbon and magnesium oxide. What is the ratio of carbon to magnesium oxide in the products?*

A) 1:1

B) 2:1

C) 1:2

D) 2:3

E) 3:2

Free Response Questions

1. *A student accidentally spills a 1.00 L bottle of concentrated 18.0 M sulfuric acid on the floor of the laboratory. She attempts to neutralize the spill by pouring a 5.00 kg box baking soda, sodium hydrogen carbonate, onto the acid.*

a. *Write a balanced net ionic equation for the reaction. Assume the concentrated sulfuric acid is 100% pure and not in aqueous solution.*

b. *What is the limiting reactant?*

c. *How many grams of sodium hydrogen carbonate are required to neutralize all the acid?*

d. *How many moles of excess reactant remain after all the limiting reactant has been consumed?*

e. *Would a floor consisting of bare concrete require more, less, or the same amount of baking soda to neutralize the spill? Explain.*

2. *Three unknown acid solutions are labeled A, B and C. One is sulfuric acid, one is hydrochloric acid, and the other is nitric acid. Each has a concentration of approximately one molar. Using only aqueous reagents of 0.20 M lead(II) nitrate and 0.50 M calcium nitrate write a short, con-*

cise experimental procedure, the results of which will be sufficient to identify each of the unknown acids. Tell what you would expect to see and what it means. Write net ionic equations to illustrate your answers.

Additional Practice in Chemistry the Central Science

For more practice answering questions in preparation for the Advanced Placement examination, try these problems in Chapter 4 of Chemistry the Central Science:

Additional Exercises: 4.91, 4.93, 4.94, 4.97, 4.98, 4.102, 4.103, 4.104.

Integrative Exercises: 4.106, 4.107, 4.108, 4.109, 4.110, 4.114.

Multiple Choice Answers and Explanations

1. *A. Alkali metals are highly reactive with water. Alkaline earth metals are less reactive. Nonmetals with the exception of white phosphorus do not react with water.*

   $2K(s) + 2H_2O(l) \dashrightarrow 2K^+(aq) + 2OH^-(aq) + H_2(g)$

2. *C.* NO_3^-*,* Na^+*,* K^+ *and* NH_4^+ *form no common precipitates.* PO_4^{3-}*,* OH^-*,* S^{2-}*,* CO_3^{2-} *form precipitates with most cations except* Na^+*,* K^+ *and* NH_4^+*.* Hg_2^{2+}*,* Ag^+*,* Pb^{2+} *commonly form precipitates with the halide ions,* Cl^-*,* Br^-*, and* I^-*. Cl, Br and I form precipitates with* Ag^+*,* Pb^{2+} *and* Hg_2^{2+}*.* SO_4^{2-} *forms precipitates with* Ag^+*,* Pb^{2+}*,* Sr^{2+} *and* Ba^{2+}*.* CrO_4^{2-} *forms precipitates with* Ag^+ *and* Pb^{2+-}*.*

3. *E. Upon reaction with acids:*

   carbonates produce carbon dioxide gas:

   $HCO_3^-(aq) + H^+(aq) \dashrightarrow CO_2(g) + H_2O(l)$

   sulfides produce hydrogen sulfide gas:

   $S^{2-}(aq) + 2H^+(aq) \dashrightarrow H_2S(g)$

   Active metals produce hydrogen gas:

   $Ca(s) + 2H^+(aq) \dashrightarrow H_2(g) + Ca^{2+}(aq)$

   Sulfites produce sulfur dioxide gas:

   $SO_3^{2-}(aq) + 2H^+(aq) \dashrightarrow SO_2(g) + H_2O(l)$

   Metal oxides are usually bases:

   $Al_2O_3(s) + 6H^+(aq) \dashrightarrow 2Al^{3+}(aq) + 3H_2O(l)$

4. C. *Metals above hydrogen on the activity series (or the table of standard reduction potentials given on the AP exam) are stronger reducing agents than hydrogen. They will react with H^+ in aqueous acids to give hydrogen gas and a corresponding metal ion.*

   $M(s) + 2H^+(aq) \dashrightarrow H_2(g) + M^{2+}(aq)$

   Metals below hydrogen on the same table will not react with H^+. Copper is the only metal listed that is below hydrogen on the activity series or the table of standard reduction potentials.

5. A. *$FeBr_3$ delivers three Br^- ions per mole of $FeBr_3$ so it would take 1/3 of the volume of $AgNO_3$ to completely precipitate the Ag+ ions. 1/3 X 30 mL = 10 mL. (Note: No calculators are permitted on the multiple choice section, and a calculator is not needed to solve this simple ratio.)*

   Using the mole road: mL $FeBr_3$ = (30 mL $AgNO_3$ /1000 mL/L)(0.40 mol/L)
   (1 mol $FeBr_3$/3 mol $AgNO_3$)(1 L/0.40 mol)(1000 mL/L) = 1/3 x 30 mL.

6. D. *The balanced equation and the corresponding mole road solution are:*

   $H_2SO_4(aq) + 2NaHCO_3(s) \dashrightarrow CO_2(g) + H_2O(l) + Na_2SO_4(aq)$

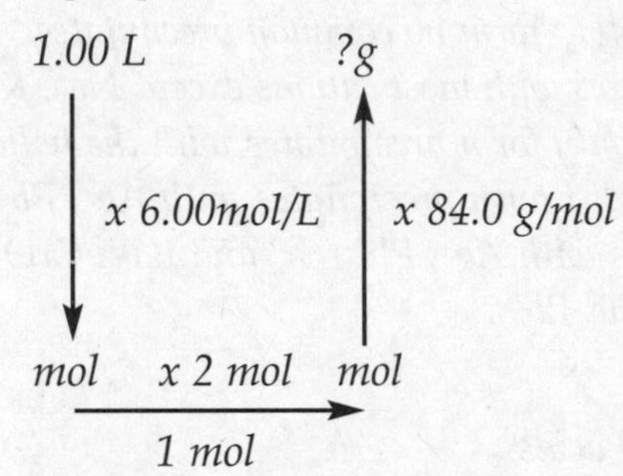

   g $NaHCO_3(s)$ = (1.00L)(6.00 mol/L)(2mol/1mol)(84.0 g/mol)

7. E. *The balanced equation and the mole road solution are:*

   $2AgNO_3(aq) + Na_2CrO_4(aq) \dashrightarrow Ag_2CrO_4(s) + 2NaNO_3(aq)$

   ?M *37.50 mL*

   ÷ 25.00 mL *x 0.152 mmol/mL*

   mmol x2mmol mmol

   1 mmol

   M $AgNO_3$ = (37.50 mL)(0.152 mmol/mL)(2 mmol /1 mmol)/(25.00 mL)

8. B. *Metal hydrides react with water to form metal hydroxides and hydrogen gas. Group 1 and 2 metals react with water to form hydrogen, but magnesium reacts only with steam.*

9. E. *Both solutions are aqueous, and the spectator ions are* Cl^- *and* Na^+ *so the reactants are written:* $Ca^{2+}(aq) + CO_3^{2-}(aq)$*. Carbonates commonly form precipitates so the product is* $CaCO_3(s)$*.*

10. C. *The balanced equation is:* $2Mg(s) + CO_2(g) \longrightarrow C(s) + 2MgO(s)$*.*

Free Response Answers

1a.

$$H_2SO_4(l) + 2NaHCO_3(s) \dashrightarrow 2CO_2(g) + 2H_2O(l) + 2Na^+(aq) + SO_4^{2-}(aq)$$

b. 1.00L 5.00 kg = 5000 g

?g

x 18.0mol/L x 84.0g/mol ÷ 84.0 g/mol

x 2 mol/1mol

mol → mol

÷ 1 mol ÷ 2 mol

1x18/1 = 18.0 5000/84.0/2 = 29.8

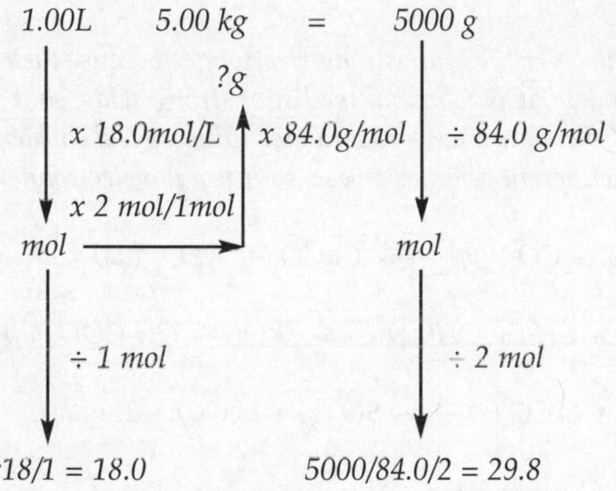

Sulfuric acid is limiting.

c. *x g* $NaHCO_3$ = (1.00 L H_2SO_4)x (18.0 mol/L)(2 mol $NaHCO_3$/1 mol H_2SO_4) x (84.0 g/mol) = 3020 g.

d. *x mol* $NaHCO_3$ *initially* = 5000 g ÷ 84.0 g/mol = 59.5 mol $NaHCO_3$ *initially.*

x mol $NaHCO_3$ *consumed* = 1.00 L x 18.0 mol/L x 2 mol/1mol = 36.0 mol $NaHCO_3$ *consumed.*

Moles $NaHCO_3$ *remaining* = 59.5 – 36.0 = 23.5 mol

e. *A concrete floor would require less baking soda to neutralize the spill because concrete contains significant amounts of calcium carbonate and other carbonates which would neutralize some the sulfuric acid.*

2. *Add one drop of each acid to one drop of each reagent and observe the results.*

   The acid that produces two precipitates is sulfuric acid.

   $Ca^{2+}(aq) + SO_4^{2-}(aq) \dashrightarrow CaSO_4(s)$

   $Pb^{2+}(aq) + SO_4^{2-}(aq) \dashrightarrow PbSO_4(s)$

   The acid that produces just one precipitate is hydrochloric acid.

   $Pb^{2+}(aq) + 2Cl^-(aq) \dashrightarrow PbCl_2(s)$

   The acid that does not produce a precipitate is nitric acid.

Your Turn Answers

4.1. $HCl(g) \dashrightarrow H^+(aq) + Cl^-(aq)$

4.2. *$CaCl_2$ and $(NH_4)_2SO_4$ are strong electrolytes because they are ionic compounds. HCN is not on the list of strong acids, so it is a weak acid and therefore a weak electrolyte. Glucose is a covalent molecule that is neither an acid nor a base, so it is a nonelectrolyte.*

4.3. $CuSO_4(s) + CO_3^{2-}(aq) \dashrightarrow CuCO_3(s) + SO_4^{2-}(aq)$

4.4. $OH^-(aq) + CH_3COOH(aq) \dashrightarrow H_2O(l) + CH_3COO^-(aq)$

4.5. $2H^+(aq) + Li_2SO_3(s) \dashrightarrow SO_2(g) + H_2O(l) + 2Li^+(aq)$

4.6. $K = 1+$, $Al = 3+$, $S = 6+$, $O = 2-$.

TOPIC

5

THERMOCHEMISTRY

Thermochemistry lays the groundwork for understanding energy as it relates, not only to chemical reactions and thermodynamics, but also to electron configurations, periodicity and chemical bonding. Advanced Placement exam questions usually combine the content of this chapter with the content of Chapter 19, Chemical Thermodynamics. Chapter 5 introduces the energy associated with chemical reactions and presents the concepts of enthalpy, calorimetry, Hess's Law and standard enthalpies of formation. Chapter 19 expands the treatment of energy and chemical reactions by introducing entropy and free energy concepts. Pay particular attention to these sections in Chemistry the Central Science.

The Nature of Energy

Section 5.1

Thermodynamics is the study of energy and its transformations.

Thermochemistry is the study of the relationships between energy changes involving heat and chemical reactions.

Energy is the capacity to do work or to transfer heat.

Heat is the energy transformed from one object to another because of a difference in temperature.

A **calorie** is an informal but still used unit for heat energy. One calorie is the amount of energy required to raise the temperature of one gram of water by one Celsius degree. The large Calorie (spelled with a capital C) is used to measure food energy. 1 Cal = 1000 cal.

A **joule** is the SI unit of energy. One calorie is equal to 4.184 joules.

1 cal = 4.184 J

Common misconception: Commonly the energy values of various foods are expressed as follows: for fats, 9 Cal/g; for carbohydrates, 4 Cal/g; for protein, 4 Cal/g. These caloric values are large calories and each refers to 1000 calories or 1 kcal.

Energy changes involve the transfer of heat between the **system**, that portion of the universe we single out for study, and the **surroundings**, everything else.

Section 5.2 The First Law of Thermodynamics

The **first law of thermodynamics,** also called the law of conservation of energy, states that energy is conserved. The energy of the universe is constant. Energy can neither be created nor destroyed, but it can be changed to other forms.

An **exothermic process** is a process that releases heat to the surroundings. Heat flows out of the system and into the surrounding. The temperature of the surroundings increases.

Common misconception: An exothermic process releases heat, but this does not mean that the system cools off. The word "release" means essentially that potential energy of the system is converted to heat energy so the surroundings increase in temperature during an exothermic process.

An **endothermic process** absorbs heat from the surroundings. Heat flows into the system from the surroundings. The temperature of the surroundings decreases.

Common misconception: An endothermic process absorbs heat, but this does not mean that the temperature of the system increases. The word "absorb" essentially means that heat energy from the surroundings is changed to potential energy of the system, and the temperature of the surroundings decreases.

Your Turn 5.1

Imagine an ice cube melting in your hand. Is the melting of ice endothermic or exothermic? Explain using the terms system and surroundings. Write your answer in the space provided.

Your Turn 5.2

Imagine warming your hands near a campfire. Is a campfire an endothermic or exothermic process? Explain. Write your answer in the space provided.

Enthalpy

Section 5.3

Enthalpy is the property of a system that accounts for heat flow between the system and the surroundings at constant pressure.

Enthalpy is a state function, a property of a system that depends only on the present state of the system, not on the path the system took to reach that state. A state is determined by specifying the system's temperature, pressure, location, and other conditions that define the system.

Enthalpy change, ΔH, is the heat absorbed during a physical or chemical process.

ΔH is positive for an **endothermic process**, one that absorbs heat.

ΔH is negative for an **exothermic process**, one that releases heat.

Enthalpies of Reaction

Section 5.4

Enthalpy changes are usually expressed in kJ or kJ/mol. For endothermic changes the heat absorbed is always a positive number because endothermic changes absorb heat. For an exothermic reaction, ΔH is negative because exothermic changes liberate (the opposite of absorb) heat.

Enthalpy change, ΔH, can be applied to either a physical or chemical change.

Common misconception: It's important to understand that the concept of enthalpy change, ΔH, is the same concept for a wide variety of processes, and often various subscripts for ΔH are used to denote specific types of change.

Table 5.1 lists examples of various types of enthalpy changes. Each subscript for ΔH is used to denote a specific kind of process.

Table 5.1. Enthalpy Changes for Various Types of Processes with Examples.

Symbol	Type of process	Definition of process with example
ΔH_{rxn}	Heat of reaction	Enthalpy change for any chemical reaction.
$AgNO_3(aq) + HCl(aq) \rightarrow AgCl(s) + HNO_3(aq)$		$\Delta H_{rxn} = -68$ kJ
ΔH_{comb}	Heat of combustion	Enthalpy change for a combustion reaction.
$CH_4(g) + 2O_2(g) \rightarrow CO_2(g) + 2H_2O(g)$		$\Delta H_{comb} = -802$ kJ
ΔH_{fus}	Heat of fusion	Heat change when a solid melts.
$H_2O(s) \rightarrow H_2O(l)$		$\Delta H_{fus} = +6.0$ kJ
ΔH_{vap}	Heat of vaporization	Heat change when a liquid vaporizes.
$H_2O(l) \rightarrow H_2O(g)$		$\Delta H_{vap} = +44$ kj
ΔH_{BDE}	Bond dissociation energy	Heat required to break a chemical bond.
$H_2(g) \rightarrow H(g) + H(g)$		$\Delta H_{BDE} = +436$ kJ
ΔH_f	Heat of formation	Heat change when one mole of a substance is formed from its elements.
$H_2(g) + \frac{1}{2}O_2(g) \rightarrow H_2O(l)$		$\Delta H_f = -285.83$ kJ

Table 5.1. continued

ΔH_{soln}	Heat of solution	Heat change when solute dissolves in a solvent.
$LiCl(s) \rightarrow Li^{+}(aq) + Cl^{-}(aq)$		$\Delta H_{soln} = -37.4$ kJ

The magnitude of the enthalpy change, ΔH, for any chemical or physical change is directly proportional to the amount of reactants and products involved in the change. For example, if one mole of methane burns in air to produce -802 kJ of energy, then two moles of methane will produce twice the energy.

$$CH_4(g) + 2O_2(g) \rightarrow CO_2(g) + 2H_2O(g) \qquad \Delta H_{comb} = -802 \text{ kJ}$$

$$2CH_4(g) + 4O_2(g) \rightarrow 2CO_2(g) + 4H_2O(g) \qquad \Delta H_{comb} = -1604 \text{ k}$$

Because it is directly proportional to the number of moles in a balanced equation, enthalpy change, ΔH, for any amount of reactant or product can be calculated using the mole road.

Example:

How many kJ of heat are absorbed when 25.0 g of methane burn in air?

Solution:

$$CH_4(g) + 2O_2(g) \rightarrow CO_2(g) + 2H_2O(g) \qquad \Delta H_{comb} = -802 \text{ kJ/mol}$$

25.0 g

÷16.0 g/mol

x -802 kJ/mol

mol ⟶ *?kJ*

$$x \text{ kJ} = (25.0 \text{ g } CH_4)/(16.0 \text{ g/mol}) \times (-802 \text{ kJ/mol}) = -1250 \text{ kJ}.$$

The sign of ΔH depends on the direction of the reaction. The magnitude of the enthalpy change for a forward reaction is equal to, but opposite in sign, for a reverse reaction. For example, if it takes +6 kJ to melt a mole of solid water, then one mole of liquid water releases -6 kJ of energy upon freezing.

$$H_2O(s) \rightarrow O(l) \qquad \Delta H = +6.0 \text{ kJ}$$

$$H_2O(l) \rightarrow H_2O(s) \qquad \Delta H = -6.0 \text{ kJ}$$

Common misconception: The sign of ΔH indicates the direction of energy flow, not a positive or negative value for energy. A positive sign indicates that heat is absorbed by the system. A negative sign means that heat is released by the system. Often the absolute value of ΔH is used in the context of sentences and its sign is implied. For example, when heat is said to be "absorbed", the sign of ΔH is positive. When heat is "liberated", the sign is negative.

Your Turn 5.3

What scientific law requires that the magnitude of the heat change for forward and reverse processes be the same with opposite signs? Explain. Write your answer in the space provided.

It is always true that phase transitions from solid to liquid and liquid to gas always absorb heat. They are endothermic. Phase transitions from gas to liquid and from liquid to solid always liberate heat. They are exothermic.

$$H_2O(l) \rightarrow H_2O(g) \quad \Delta H_{vap} = +44 \text{ kJ}$$

$$H_2O(g) \rightarrow H_2O(l) \quad \Delta H_{vap} = -44 \text{ kJ}$$

Your Turn 5.4

Why are phase changes from solid to liquid and from liquid to gas alway endothermic? Is the process of sublimation (changing directly from a solid to a gas) endothermic or exothermic? Explain. Write your answer in the space provided.

Section 5.5

Calorimetry

Calorimetry is the measurement of heat flow. A calorimeter is an apparatus that measures heat flow.

Heat capacity, C, is the heat required to raise the temperature of an object by 1 K. The units of C are J/K.

Common misconception: Remember that one Kelvin is the same size as one degree Celsius (even though they are 273 degrees apart on their respective scales). Kelvin and °C are often used interchangeably in various units.

Molar heat capacity, C_{molar} is the amount of heat absorbed by one mole of a substance when it experiences a one degree temperature change. The units of C_{molar} are J/mol K.

Specific heat capacity is the heat capacity of one gram of a substance. The units are J/g K.

The specific heat of water is worth remembering: 1 cal/g K = 4.184 J/g K.

Your Turn 5.5

Explain why the value of the specific heat of water is 1 cal/g K or 4.184 J/g K. Write your answer in the space provided.

Heat capacities, temperature, energy, moles and grams are related by an analogy to the mole road as shown in Figure 5.1.

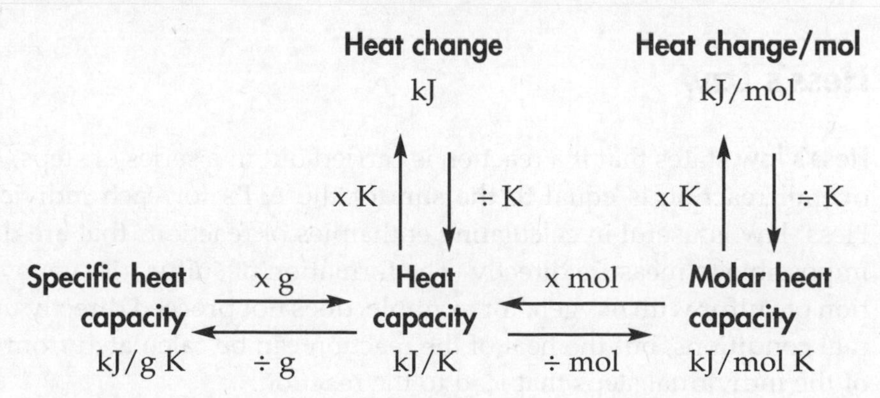

Figure 5.1. Mole road analogy for heat capacity calculations.

Calorimetry is often used to measure enthalpy changes in chemical reactions. For example, the heat of combustion of ethanol, CH_3CH_2OH, can be measured by placing a measured quantity of ethanol and excess oxygen in a bomb calorimeter having a known heat capacity. The ethanol is ignited, the temperature change of the calorimeter is observed and the heat released by the combustion reaction is calculated from the heat capacity of the calorimeter.

Example:

What is the molar heat of combustion of liquid ethanol if the combustion of 9.03 grams of ethanol causes a calorimeter to increase in temperature by 3.54 K? The heat capacity of the calorimeter is 75.8 kJ/K

Solution:

The units for molar heat of combustion are kJ/mol. Our task is to use the mole road to change the given values of grams and temperature into moles and kJ and then divide kJ by moles.

$CH_3CH_2OH(l) + 3O_2(g) \dashrightarrow 2CO_2(g) + 3H_2O(g) + 3.54\ K$

9.03 g ↓ ÷ 46.0 g/mol → mol

3.54 K ↓ x 75.8 kJ/K → kJ

x kJ/mol = (75.8 kJ/K)(3.54 K) ÷ [(9.03 g)/(46.0 g/mol)] = -1370 kJ/mol

Keep in mind that the negative sign means that the reaction is exothermic, that heat is released to the environment.

Section 5.6 Hess's Law

Hess's law states that if a reaction is carried out in a series of steps, ΔH of the overall reaction is equal to the sum of the ΔH's for each individual step. Hess' law is useful in calculating enthalpies of reactions that are difficult or impossible to measure directly. The formation of sulfur trioxide by the reaction of sulfur with oxygen, for example, does not proceed directly under normal conditions, but the heat of the reaction can be calculated from the heats of the individual steps that lead to the reaction.

Example:

Calculate ΔH for the following reaction,

$2S(s) + 3O_2(g)\ 2SO_3(g)$

from the enthalpies or these related reactions.

Reaction	ΔH_{rxn}
$S(s) + O_2(g) \longrightarrow SO_2(g)$	-296.9 kJ
$2SO_2(g) + O_2(g) \longrightarrow 2SO_3(g)$	-196.6 kJ

Solution:

The given reactions in can be manipulated like algebraic quantities to yield the reaction in question. Their enthalpy values are mathematically manipulated in the same way.

Reaction	ΔH_{rxn}
$2[S(s) + O_2(g) \longrightarrow SO_2(g)]$	2(-296.9 kJ)
$+2SO_2(g) + O_2(g) \longrightarrow 2SO_3(g)$	-196.6 kJ
$2S(s) + 3O_2(g) \longrightarrow 2SO_3(g)$	-790.4 kJ

Hess's law says that the enthalpy of the reaction is calculated by manipulating the enthalpy quantities in the same way as we manipulated the corresponding reactions.

Common misconception: Be sure to manipulate the ΔH values in the same way you manipulate the reactions. That is, if you multiply a reaction by 2, you must multiply its ΔH value by 2. If you reverse a reaction, you must change the sign of its ΔH value. Always be careful of signs.

Figure 5.2 shows an enthalpy diagram illustrating Hess's Law.

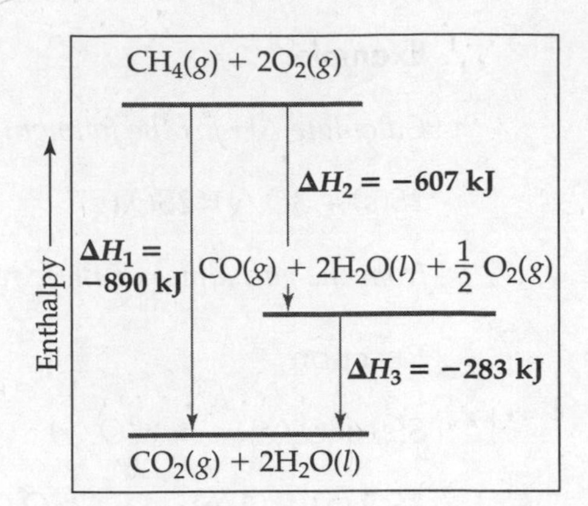

Figure 5.2. An enthalpy diagram illustrating Hess's Law. The quantity of heat generated in any chemical reaction is independent on the pathway of the reaction.

Section 5.7 Enthalpies of Formation

A **formation reaction** is a reaction that produces one mole of a substance from its elements in their most stable thermodynamic state. For example, the formation reaction for gaseous hydrogen iodide, HI, includes only one mole of HI as the sole product, and the elements hydrogen and iodine in their most stable forms as the only reactants:

$$\frac{1}{2} H_2(g) + \frac{1}{2} I_2(s) \longrightarrow HI(g) \qquad \Delta H = +25.94 \text{ KJ}$$

Gaseous diatomic hydrogen and solid diatomic iodine are the most stable thermodynamic states of those elements at 25°C.

The standard heat of formation, $\Delta H°_f$, is the heat absorbed when one mole of a substance is formed from its elements in their standard states at 25°C and 1 atmosphere pressure. A heat of formation is ΔH for a formation reaction. For example, the standard enthalpy of formation, $\Delta H°_f$, of HI is the enthalpy of the formation reaction for HI. Its value is + 25.94 kJ. Table 5.2 gives the standard heats of formation of some selected substances. Appendix C of Chemistry the Central Science provides a much more complete table of $\Delta H°_f$ values. Keep in mind that each substance listed is the sole product of a formation reaction and that the corresponding $\Delta H°_f$ value listed is the enthalpy of that formation reaction under standard conditions.

Table 5.2. Standard Enthalpies of Formation, $\Delta H°_f$ for selected substances.

Substance	$\Delta H°_f$ (kJ/mol)	Substance	$\Delta H°_f$ (kJ/mol)
$AlCl_3(s)$	-705.6	$MnO_4^-(aq)$	-541.4
$Br_2(g)$	+30.71	$NH_3(aq)$	-80.29
$Br_2(l)$	0	$NH_3(g)$	-46.19

Table 5.2. continued

Ca(g)	+179.3	O(g)	+ 247.5
Ca(s)	0	O_2(g)	0
C(s, diamond)	+1.88	O_3(g)	+142.3
C(s, graphite)	0	H_2O(g)	-241.82
H(g)	+217.97	H_2O(l)	-285.83
H_2(g)	0		

Your Turn 5.6

Write the thermochemical equation associated with the standard heat of formation of $AlCl_3$(s) listed in Table 5.2. Write your answer in the space provided.

Notice that the substances in Table 5.2 include elements, compounds and ions. Notice also that the heat of formation of any element in its most stable thermodynamic state is zero. For example, the ΔH°_f values for solid monatomic calcium, liquid diatomic bromine, and gaseous diatomic oxygen are all zero because these are the most stable forms of these elements at 25°C and one atmosphere pressure.

Your Turn 5.7

Based on the definition of a formation reaction, explain why the standard heat of formation of an element in its most stable thermodynamic state is zero. Write a chemical equation to illustrate your answer. Write your concise answer in the space below.

Hess's law allows us to use a relatively small number of measurements to calculate ΔH for a vast number of reactions. Using data from Table 5.2 or from Appendix C of Chemistry the Central Science, the standard enthalpy of many reactions can be calculated from the enthalpies of the reactants and products.

The standard enthalpy of any reaction, ΔH°_{rxn}, is equal to the sum of the standard enthalpies of formation of products minus the sum of standard enthalpies of formation of reactants. The values for n and m in the following equation represent the coefficients of the balanced equation.

$$\Delta H^\circ_{rxn} = \Sigma n\Delta H^\circ_{f\ products} - \Sigma m\Delta H^\circ_{f\ reactants}$$

Example:

Calculate the standard enthalpy change for the combustion of one mole of liquid ethanol.

Solution:

Write and balance the equation. Use Appendix C in Chemistry the Central Science to determine the standard enthalpies of formation for reactants and products. Apply the summation equation. Be sure to multiply each ΔH°_f value by the corresponding coefficient that balances the equation.

$$CH_3CH_2OH(l) + 3O_2(g) \longrightarrow 2CO_2(g) + 3H_2O(g)$$

From appendix C: ΔH°_f = -238.6 kJ 0 kJ -393.5 kJ -241.82 kJ

$$\Delta H^\circ_{rxn} = \Sigma\, n\, \Delta H^\circ_{f\ products} - \Sigma\, m\, \Delta H^\circ_{f\ reactants}$$

$$\Delta H^\circ_{rxn} = 2(-393.5\ kJ) + 3(-241.82\ kJ) - (-238.6\ kJ) - 3(0\ kJ)$$

$$\Delta H^\circ_{rxn} = -1273.9\ kJ$$

Common misconception: The hardest part of this type of problem is the arithmetic. Especially be careful to manipulate the + and – signs correctly.

Examine Table 5.2. again. Notice that there is a difference between the ΔH°_f values for liquid water and gaseous water. The difference represents the molar heat of vaporization of water. The equation is:

$$H_2O(l) \longrightarrow H_2O(g)$$

$$\Delta H^\circ_f = -285.83\ kJ \qquad -241.82\ kJ$$

$$\Delta H^\circ_{rxn} = \Delta H^\circ_{vap} = \Delta H^\circ_f\ H_2O(g) - \Delta H^\circ_f\ H_2O(l)$$

$$\Delta H^\circ_{vap} = -241.82\ kJ - (-285.83\ kJ) = 44.01\ kJ$$

Your Turn 5.8

What is, a) the enthalpy of sublimation of solid calcium? b) The heat of solution gaseous ammonia? c) The bond dissociation energy of hydrogen gas? d) The heat change when gaseous bromine condenses to a liquid? Where can you find the data needed to solve these problems? Write chemical equations to illustrate your answers. Write your answers in the space provided.

Multiple Choice Questions

1. *The standard enthalpy of formation ($\Delta H^o f$) for potassium chloride is the enthalpy change for the reaction*

A) $K(g) + \frac{1}{2} Cl_2(g) \dashrightarrow KCl(g)$

B) $K^+(g) + Cl^-(g) \dashrightarrow KCl(s)$

C) $2K(s) + Cl_2(g) \dashrightarrow 2KCl(s)$

D) $K(s) + \frac{1}{2} Cl_2(g) \dashrightarrow KCl(s)$

E) $K^+(g) + Cl^-(g) \dashrightarrow KCl(s)$

2. *For which of these processes is the value of ΔH expected to be negative?*

I. *The temperature increases when calcium chloride dissolves in water.*

II. *Steam condenses to liquid water.*

III. *Water freezes.*

IV. *Dry ice sublimes.*

A) *IV only*

B) *I, II and III*

C) *I only*

D) *II and III only*

E) *I and II only*

3. *Which is expected to not have a ΔH°_f value of zero?*

A) $F_2(g)$

B) $Br_2(g)$

C) $I_2(s)$

D) *C(s, graphite)*

E) $N_2(g)$

4. *For which of the following equations is the change in enthalpy at 25°C and 1 atm equal to ΔH°_f of $CH_3OH(l)$?*

A) $CH_3OH(l) + 3/2\ O_2(g) \dashrightarrow CO_2(g) + 2H_2O(l)$

B) $CH_3OH(l) + 3/2\ O_2(g) \dashrightarrow CO_2(g) + 2H_2O(g)$

C) $2CH_3OH(l) + 3O_2(g) \dashrightarrow 2CO_2(g) + 4H_2O(l)$

D) $CH_3OH(l)$ ---► $C(s) + 2H_2O(l)$

E) $C(s) + 2H_2(g) + ½ O_2(g)$ ---► $CH_3OH(l)$

5. *Which change will result in an increase in enthalpy of the system?*

A) *burning a candle*

B) *freezing water*

C) *evaporating alcohol*

D) *dropping a ball*

E) *condensing steam*

6. *The standard enthalpy of formation of Cl(g) is +242 kJ/mol. What is the dissociation energy of a Cl-Cl bond?*

A) *+242 kJ/mol*

B) *-242 kJ/mol*

C) *+484 kJ/mol*

D) *+121 kJ/mol*

E) *-121 kJ/mol*

7. *For which process is the sign of ΔH negative?*

A) *Photosynthesis*

B) $CO_2(g)$---► $C(s) + O_2(g)$

C) $N_2(g)$ ---► $2N(g)$

D) $NaOH(s)$ ---► $Na^+(aq) + OH^-(aq)$ + *heat*

E) H_2O + *electricity* ---► $½ H_2(g) + O_2(g)$

8. *Given the following data, what is the heat of formation of methane gas?*

1.	$CH_4(g) + 2O_2(g)$ ---► $CO_2(g) + 2H_2O(g)$	$\Delta H = -803$ kJ
2.	$H_2(g) + ½ O_2(g)$ ---► $H_2O(g)$	$\Delta H = -242$ kJ
3.	$C(s) + O_2(g)$ ---► $CO_2(g)$	$\Delta H = -394$ kJ
4.	$C(s) + ½ O_2(g)$---►$CO(g)$	$\Delta H = -111$ kJ

A) *- 803 kJ/mol*

B) *-75 kJ/mol*

C) *+167 kJ/mol*

D) +208 kJ/mol

E) -1439 kJ/mol

9. *The standard heat of formation of gaseous sulfur trioxide is -396 kJ/mol. What is the enthalpy of reaction represented by the following balanced equation?*

$2SO_3(g) \dashrightarrow 2S(s) + 3O_2(g)$

A) - 396 kJ

B) +396 kJ

C) +792 kJ

D) -792 kJ

E) +198 kJ

10. *Given only the following data, what can be said about the following reaction?*

$3H_2(g) + N_2(g) \dashrightarrow 2NH_3(g) \quad \Delta H = -92 \text{ kJ}$

A) *The enthalpy of products is greater than the enthalpy of reactants.*

B) *The total bond energies of products are greater than the total bond energies of reactants.*

C) *The reaction is very fast.*

D) *Nitrogen and hydrogen have very stable bonds compared to the bonds of ammonia.*

E) *The reaction is endothermic.*

Free Response Questions

1. *The heat of combustion of gaseous butane is -2658 kJ/mol and the heat of combustion of liquid butane is -2635 kJ/mol when, in both cases, all products are gases.*

a. *Write a balanced chemical equation for the combustion of gaseous butane.*

b. *How many grams of gaseous butane combust when 1550 kJ of heat are produced?*

c. *What is the magnitude and sign of the molar heat of vaporization of butane? Explain your reasoning using Hess's law. Is your sign for the heat of vaporization realistic? Explain.*

2. *When 15.00 g of propane are burned in air to produce all gaseous products, 730.0 kJ of heat are produced.*

a. *Calculate the molar heat of combustion of propane.*

b. *When 15.00 g of propane are combusted in air to produce gaseous carbon dioxide and liquid water, 790.0 kJ of heat are produced. Explain why the amount of heat available from the combustion of propane depends on the phase of the products.*

c. *Calculate the heat of vaporization of water in units of kJ/g.*

Additional Practice in Chemistry the Central Science

For more practice answering questions in preparation for the Advanced Placement examination try these problems in Chapter 5 of Chemistry the Central Science:

Additional Exercises: 5.96, 5.97, 5.99, 5.100, 5.101, 5.103, 5.104.

Integrative Exercises: 5.108, 5.110, 5.111, 5.113, 5.114, 5.115.

Multiple Choice Answers and Explanations

1. *D. The enthalpy of formation of a substance is the enthalpy change for the formation of one mole of that substance from its elements in their most stable thermodynamic forms at 25°C and one atmosphere pressure. Solid potassium and gaseous diatomic chlorine are the most stable forms of those elements.*

2. *B. A negative ΔH value is characteristic of an exothermic process, one that releases heat to its surroundings. Exothermic processes increase the temperature of their surroundings. Phase changes from gas to liquid and from liquid to solid are all exothermic processes. The opposite processes are endothermic Sublimation is the process by which a solid changes directly into a gas, also an endothermic process.*

3. *B. By definition, the standard heats of formation of elements in their*

most stable thermodynamic state at 25°C and one atmosphere pressure is zero. Bromine is a liquid at 25°C and one atm. Solid graphite is the most stable form of carbon.

4. E. *Although the direct formation of methanol from its elements is improbable, the heat of formation of methanol is defined as the enthalpy change for this reaction. Answers A, B and C are all combustion reactions of methanol and answer D is a decomposition reaction.*

5. C. *An increase in enthalpy is associated with an endothermic process, one that absorbs energy from its surroundings. Phase changes from solid to liquid or gas, and from liquid to gas are always endothermic. Combustion reactions are always exothermic as are phase changes from gas to liquid and liquid to solid.*

6. C. *The bond dissociation of chorine gas is the enthalpy change for this reaction:* $Cl_2(g) \longrightarrow 2Cl(g)$. *Bond dissociation energies always have positive values because bond breaking is always endothermic, so the sign of the heat change is always positive. The formation reaction for Cl(g) is:* $\frac{1}{2}Cl_2(g) \longrightarrow Cl(g)$. *Notice that the formation equation is exactly half the reaction in question. So the bond dissociation energy is twice the enthalpy of formation.*

7. D. *ΔH is negative for exothermic processes and positive for endothermic processes. Exothermic processes release heat to their surroundings. Photosynthesis involves absorbing energy from sunlight and is endothermic. Combustion reactions are always exothermic and answer B is the reverse of a combustion reaction so it is endothermic. Bond breaking is always endothermic and bond making is always exothermic. The process of forming nitrogen atoms from nitrogen molecules requires bond breaking. The electrolysis of water absorbs electrical energy and is endothermic.*

8. B. *The heat of formation of methane gas is the heat change for the formation reaction:* $C(s) + 2H_2(g) \longrightarrow CH_4(g)$. *One way to solve the problem is to notice that Equation 1 represents the combustion of methane. The other given equations are the formation reactions for the reactants and products of the combustion reaction. (Equation 4 is extraneous.) Apply the corresponding heats to the summation equation and solve for the unknown quantity, the heat of formation of methane.*

$$\Delta H^\circ_{rxn} = \Sigma n\Delta H^\circ_{f\,products} - \Sigma m\Delta H^\circ_{f\,reactants}$$

$$\Delta H^\circ_{comb} = 2\Delta H^\circ_f H_2O + \Delta H^\circ_f CO_2 - 2\Delta H^\circ_f O_2 - \Delta H^\circ_f CH_4$$

$-803 = 2(-242) + (-394) - O - \Delta H^o_f CH_4$

$\Delta H^o_f CH_4 = -75$ kJ/mol

Another approach is to use Hess's law and mathematically manipulate the given equations to yield the formation reaction for methane. Apply the same mathematical manipulations to the given heats to find the heat of the formation reaction. (This approach is the essence of the summation equation used in the first approach.)

1. $-[CH_4(g) + 2O_2(g) \longrightarrow CO_2(g) + 2H_2O(g)]$ $\Delta H = -803$ kJ x -1 = + 803
2. $+2[H_2(g) + ½\, O_2(g) \longrightarrow H_2O(g)]$ $\Delta H = -242$ kJ x +2 = -484
3. $+[C(s) + O_2(g) \longrightarrow CO_2(g)]$ $\Delta H = -394$ kJ = -394

$C(s) + 2H_2(g) \longrightarrow CH_4(g)$ 75 kJ/mol

9. *C. The standard heat of formation of gaseous sulfur trioxide is heat change for the formation reaction:* $S(s) + 3/2\, O_2(g) \longrightarrow SO_3(g)$. *Recognize that the reaction in question is double the formation reaction and the reverse of it. The heat change for the reaction in question is:*

   $-2(-396) = +792$ kJ.

10. *B. Bond making is always exothermic and bond breaking is always endothermic. The reaction is exothermic, one that releases energy to its surroundings, because the sign of the heat change is negative. The bonds formed by the products are more stable than those broken in the reactants leaving a net energy released to the surroundings. The higher the bond energy the more stable the bond.*

Free Response Answers

1. *i. Butane is the alkane hydrocarbon containing four carbon atoms. Alkanes have all single carbon-carbon bonds and have the general formula,* C_nH_{2n+2}. *Water and carbon dioxide are the products of the combustion of hydrocarbons.*

   $C_4H_{10}(g) + 13/2\, O_2(g) \longrightarrow 4CO_2(g) + 5H_2O(g)$

   (Double coefficients are acceptable.)

ii. *The amount of heat in a thermochemical equation is proportional to the amount of reactants and products. Use the mole road to convert 1550 kJ to grams.*

$C_4H_{10}(g) + 13/2\ O_2(g) \longrightarrow 4CO_2(g) + 5H_2O(g) + 1550\ kJ$

?g

↑ x 58.0 g/mol

÷ 2658 kJ/mol $C_4H_{10}(g)$

mol ← 1550 kJ

$x\ g = (1550\ kJ)(58.0\ g/mol)/(2658\ kJ/mol\ C_4H_{10}(g)) = 33.8\ g\ C_4H_{10}(g)$

iii. *The molar heat of vaporization of butane is the difference between the heat of combustion of liquid vs. that of gaseous butane. Subtracting the equation for the combustion of gaseous butane from the equation for the combustion of liquid butane yields the equation for the vaporization of butane. Subtracting the corresponding heats yields the heat of vaporization of butane. Vaporization processes are always positive.*

$C_4H_{10}(l) + 13/2\ O_2(g) \longrightarrow 4CO_2(g) + 5H_2O(g) \quad \Delta H = -2635\ kJ/mol$

$-[C_4H_{10}(g) + 13/2\ O_2(g) \longrightarrow 4CO_2(g) + 5H_2O(g)] \quad \Delta H = -(-2658\ kJ/mol)$

$C_4H_{10}(l) \longrightarrow C_4H_{10}(g) \quad \Delta H = +23\ kJ/mol$

b. i. *Use the mole road and the balanced chemical equation to calculate the number of kJ burned in one mole of propane. Propane is the alkane hydrocarbon having three carbon atoms.*

$C_3H_8(l) + 5O_2(g) \longrightarrow 3CO_2(g) + 4H_2O(g) \quad \Delta H = ?$

1.000 mol

↓ x 44.00 g/mol

x -730.0 kJ/15.00 g

g → kJ

$x\ kJ = 1.000\ mol(44.0\ g/mol)(-730\ kJ/15.0\ g) = -2141\ kJ/mol$

ii. *Upon combustion of the same amount of butane, more heat is liberated when liquid water forms than when gaseous water forms because heat is released when water vapor condenses. That is, water vapor possesses more enthalpy (potential energy) than liquid water and when gas changes to liquid, the heat is released.*

iii. *Apply the same mole road approach to calculate the molar heat of combustion of propane when liquid water forms. The difference is the heat of vaporization of four moles of water because the balanced equation contains four moles of water. Divide by four and by the molar mass of water to obtain the heat of vaporization of water in kJ/g.*

$$x\ kJ = 1.000\ mol(44.0\ g/mol)(-792.9\ kJ/15.0\ g) = -2326\ kJ/mol$$

$$C_3H_8(l) + 5O_2(g) \dashrightarrow 3CO_2(g) + 4H_2O(g) \qquad \Delta H = -2141\ kJ/mol$$

$$-[C_3H_8(l) + 5O_2(g) \dashrightarrow 3CO_2(g) + 4H_2O(l)] \quad \Delta H = -(-2326\ kJ/mol)$$

$$4H_2O(l) \dashrightarrow 4H_2O(l) \qquad \Delta H = +185\ kJ/\ 4\ mol$$

$$\Delta H = +46.25\ kJ/\ mol$$

$$x\ kJ/g = (42.0\ kJ/mol)/(18.0\ g/mol) = 2.57\ kJ/g$$

Your Turn Answers

5.1. *Melting of ice (the system) is endothermic as evidenced by the cooling of your hand (the surroundings) as the ice melts. Endothermic processes cool the surroundings by taking heat away from the surroundings, and converting it to potential energy of the system.*

5.2. *A campfire (the system) is an exothermic process. It releases heat to your hands (the surroundings). Potential energy of the system is converted to heat energy, and your hands feel the heat coming from the campfire.*

5.3. *The law of conservation of energy says that the energy of the universe is constant. Energy is neither created nor destroyed so the amount of energy that flows into a forward process will be the same as that flowing out of a reverse process. The value for ΔH will be the same for both processes but will have opposite signs. The sign of ΔH indicates the direction of the process.*

5.4. *It requires energy to melt a solid or to vaporize a liquid to overcome the strong forces of attraction between particles that exist in solids and liquids. This energy is released when gases change to liquids, and liquids change to solids. Sublimation is always an endothermic process because it requires energy to overcome strong forces of attraction in solids and liquids.*

5.5. *The value of the specific heat of water derives from the original and modern definitions of a calorie. One calorie is the amount of heat necessary to change one gram of water by one Celsius (or Kelvin) degree. One calorie is 4.184 J.*

5.6. $Al(s) + 3/2\ Cl_2(g) \dashrightarrow AlCl_3(s)$

5.7. *The formation of a stable form of an element from that element represents no net change so there is no heat change.*

$Ca(s) \dashrightarrow Ca(s) \quad \Delta H_{rxn} = 0$

5.8.

a) $Ca(s) \dashrightarrow Ca(g) \quad \Delta H^\circ_f = +179.3\ kJ - 0\ kJ = +179.3\ kJ/mol$

b) $NH_3(g) \dashrightarrow NH_3(aq) \quad \Delta H^\circ_{soln} = -80.29\ kJ - (-46.19\ kJ) = -34.10 kJ/mol$

c) $H_2(g) \dashrightarrow 2H(g) \quad \Delta H^\circ_{BDE} = 2(+217.97\ kJ) - 0\ kJ) = +435.94 kJ/mol$

b) $Br_2(g) \dashrightarrow Br_2(l) \quad \Delta H^\circ_{cond} = 0\ kJ - (+30.70\ kJ) = -30.70\ kJ/mol$

Table 5.2 or Appendix C in Chemistry the Central Science has the thermodynamic data to calculate the enthalpies of hundreds of processes.

TOPIC 6

ELECTRONIC STRUCTURE OF ATOMS

The wave nature of light and ideas from quantum theory lead to an understanding of atomic structure and electron configurations of atoms. Electronic structure is closely connected to periodic relationships, ionic and covalent bonding, and to chemical reactivity. Pay particular attention to using the periodic table as a guide to write ground state electron configurations of atoms. Know how to mathematically manipulate the properties of waves: frequency, wavelength, and energies of photons. Also know how quantum numbers relate to electron configurations and how the Bohr model explained the atomic emission spectrum of hydrogen.

The Wave Nature of Light

Section 6.1

Electromagnetic radiation (also called radiant energy or light) is a form of energy having both wave and particle characteristics. It propagates through a vacuum at the speed of light, 3.00×10^8 m/s.

Wavelength, λ, is the distance between two adjacent peaks of the wave.

Frequency, ν, is the number of wavelengths (or cycles) that pass a given point in a second.

The **electromagnetic spectrum** includes all the wavelengths of radiant energy from short gamma rays to long radio waves. (See Figure 6.1.)

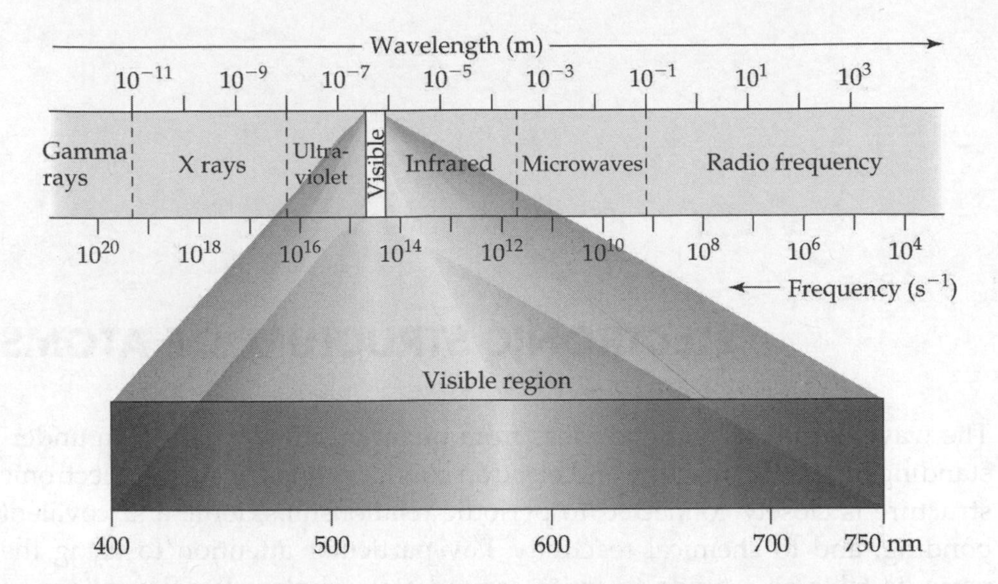

Figure 6.1. The electromagnetic spectrum. As wavelength decreases, frequency, and energy increase.

The **visible spectrum** is that part of the electromagnetic spectrum that is visible to the eye, generally with wavelengths ranging between about 400 and 700 nm.

Section 6.2

Quantized Energy and Photons

A **quantum** (also called a **photon**) is a specific particle of light energy that can be emitted or absorbed as electromagnetic radiation. The energy of a photon is described by the equation, $E = h\nu$. All energy is quantized. That is, matter is allowed to emit or absorb energy only in discrete amounts, whole number multiples of $h\nu$.

The **speed of light**, c, in a vacuum is 3.00×10^8 m/s.

Table 6.1 shows how the wavelength, frequency and energy of a single photon of light are related mathematically.

Table 6.1. Mathematical relationships regarding electromagnetic radiation.

$\nu = c/\lambda$	$E = h\nu$	$E = hc/\lambda$
λ = wavelength in nm ν = frequency in 1/s or hertz. 1 Hz = 1 s^{-1}	E = energy of a single photon in joules	
c = speed of light = 3.00×10^8 m/s = 3.00×10^{17} nm/s 1 nm = 10^{-9} m.	h = Planck's constant = 6.63×10^{-34} J-s	

Example:

What is the frequency, the energy of a single photon, and the energy of a mole of photons of light having a wavelength of 555 nm?

Solution:

First calculate the frequency from the wavelength using $\nu = c/\lambda$. Use your answer to calculate the energy of a single photon using $E = h\nu$. Finally, multiply the resulting energy by Avogadro's number of photons per mole to obtain the energy of a mole of photons.

$\nu = c/\lambda = (3.00 \times 10^{17}\ nm/s)/(555\ nm) = 5.41 \times 10^{14}\ 1/s$

$E = h\nu = (6.63 \times 10^{-34}\ Js)(5.41 \times 10^{14}\ 1/s) = 3.58 \times 10^{-19}\ J$

$x\ J/mol = (3.58 \times 10^{-19}\ J/photon)(6.02 \times 10^{23}\ photons/mol) =$

$216{,}000\ J/mol = 216\ kJ/mol$

Common misconception: When calculating frequency from wavelength, be sure the units for the speed of light match the units for wavelength. For instance, when wavelength is expressed in nanometers it's convenient to use the value of 3.00×10^{17} nm/s for the speed of light.

Your Turn 6.1

Explain why the units for frequency are reciprocal seconds, 1/s. Write your answer in the space provided.

Section 6.3

Line Spectra and the Bohr Model

An **atomic emission spectrum** (or **line spectrum**) is a pattern of discrete lines of different wavelengths when the light energy emitted from energized atoms is passed through a prism or diffraction grating. Each element produces a characteristic and identifiable pattern. Figure 6.1 shows the atomic emission spectrum of hydrogen.

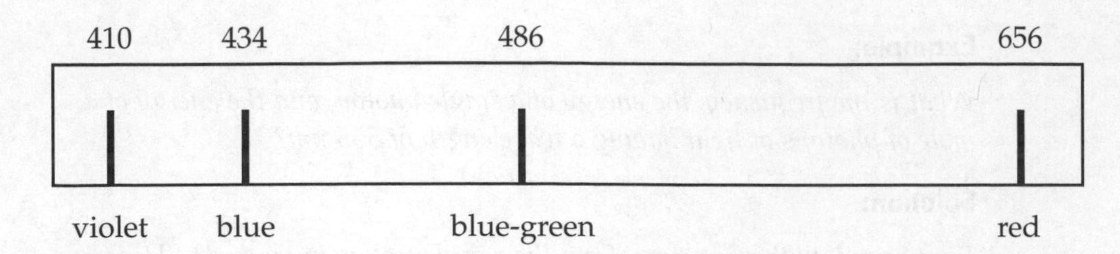

Figure 6.2. The atomic emission spectrum of hydrogen showing the colors and wavelengths of the visible lines.

The Bohr model of the atom (after Danish physicist Neils Bohr) explains the origin of the lines of the atomic emission spectrum of hydrogen. Adopting the idea that energies are quantized, Bohr proposed that electrons move in circular, fixed energy orbits around the nucleus. He postulated that each circular orbit corresponds to an "allowed" stable energy state. The ground state is the lowest energy state. Excited states are states of higher energy than the ground state. Energy, in the form of a photon, is emitted or absorbed by an electron only when it changes from one allowed energy state to another. The lines of the atomic emission spectrum of hydrogen result when an electron falls from a higher allowed state to a lower allowed state. The increment between each allowed state is proportional to Planck's constant, the speed of light and the Rydberg constant, R_H. (See Figure 6.2.)

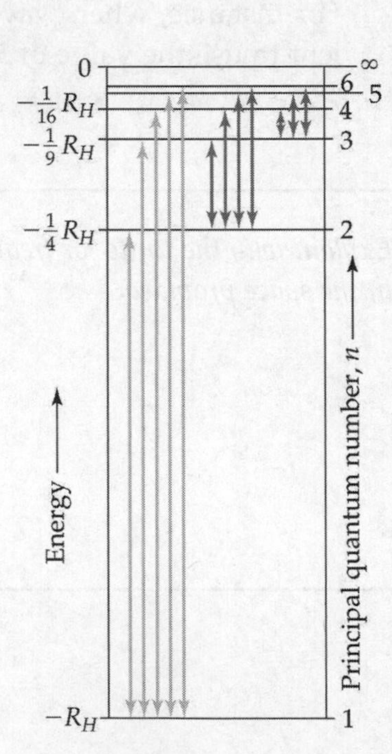

Figure 6.2. Energy levels in the hydrogen atom from the Bohr model. The four transitions with n=2 as the lower state, represent the four colored lines in the atomic emission spectrum of hydrogen.

The energy transition from allowed states n=3 to n=2 in Figure 6.2 gives rise to the red line in the atomic emission spectrum of hydrogen. The three transitions immediately to the right of the n=3 to n=2 transition correspond to the other three colored lines seen in the hydrogen spectrum.

Your Turn 6.2

The energy transitions shown in Figure 6.2 correspond to which colors and wavelengths of the visible lines of the hydrogen spectrum? Which line corresponds to the largest energy change? Explain. Write your answer in the space provided.

The Wave Behavior of Matter

Section 6.4

Like light, electrons have characteristics of both waves and particles. Because a wave extends in space, its location is not precisely defined.

The **uncertainty principle**, applied to electrons in an atom, states that it is inherently impossible to simultaneously determine the exact position and momentum of an electron. The best that can be done is to calculate a probability of finding an electron in a certain region of space.

Quantum Mechanics and Atomic Orbitals

Section 6.5

The **quantum mechanical model** of the atom is a mathematical model that incorporates both wave and particle characteristics of electrons in atoms.

Quantum numbers arise from the quantum mechanical model of the atom and describe properties of electrons and orbitals. Each electron in an atom has assigned to it a series of four quantum numbers. Table 6.2 lists the orbital or electron characteristic of each quantum number and the possible values each can have.

Table 6.2. Characteristics Associated with Quantum Numbers

Quantum Number	**Principal**	**Azimuthal**	**Magnetic**	**Spin**
Symbol	n	l	m_l	m_s
Possible Values	1,2,3,4…∞	0,1,2,3…n-1 s,p,d,f	$-l$,…0…$+l$	± ½
Electron/Orbital Characteristic	Size	Shape	Orientation	Magnetic spin

Letters are often used in place of the numerical values for the azimuthal quantum number, *l*. The letters s, p, d, and f correspond to *l* values of 0, 1, 2 and 3, respectively. For example, in Figure 6.3 an s orbital has a value of *l* equal to 0 and a p orbital has a value of *l* equal to 1.

Sublevels within an atom are traditionally designated using the first two quantum numbers, n and *l*. For example, sublevel designations of 1s, 3p, 4d and 5f have these values of n and *l*:

1s: (1,0); 3p: (3,1); 4d: (4,2); 5f: (5,3).

Table 6.3 shows the relationship among the values of quantum numbers and the orbital designations.

Table 6.3. Summary of Quantum Numbers and orbital designations for the first four energy levels.

Principal n	Azimuthal l	Magnetic m_l	Spin m_s	Sublevel Designation	Number of Orbitals
1	0	0	± ½	1s	1
2	0	0	± ½	2s	1
	1	-1,0,+1	± ½	2p	3
3	0	0	± ½	3s	1
	1	-1,0,+1	± ½	3p	3
	2	-2,-1,0,+1,+2	± ½	3d	5
4	0	0	± ½	4s	1
	1	-1,0,+1	± ½	4p	3
	2	-2,-1,0,+1,+2	± ½	4d	5
	3	-3,-2,-1,0,+1,+2,+3	± ½	4f	7

The **Pauli exclusion principle** states that no two electrons in an atom can have the same set of four quantum numbers.

Notice that the exclusion principle limits the number of electrons that can occupy any orbital to two. For example, the quantum numbers for electrons occupying a 1s orbital, in the order of n, *l*, m_l, and m_s, are (1,0,0, + ½) or (1,0,0, -½). Two electrons in a 3s sublevel will have quantum numbers (3,0,0,+½) and (3,0,0,-½).

A sublevel designated 3p can have six electrons, each having a different set of quantum numbers: (3,1,0,+½), (3,1,0,-½), (3,1,-1,+½), (3,1,-1,-½), (3,1,+1,+½) and (3,1,+1,-½).

Your Turn 6.3

How many electrons can occupy a 3d sublevel? What set of quantum numbers will each have? Write your answer in the space provided.

Section 6.6

Representations of Orbitals

An **orbital** (or the square of a **wave function**) is a calculated probability of finding an electron of a given energy in a region of space. Figure 6.3 shows the electron probability distributions of various orbitals.

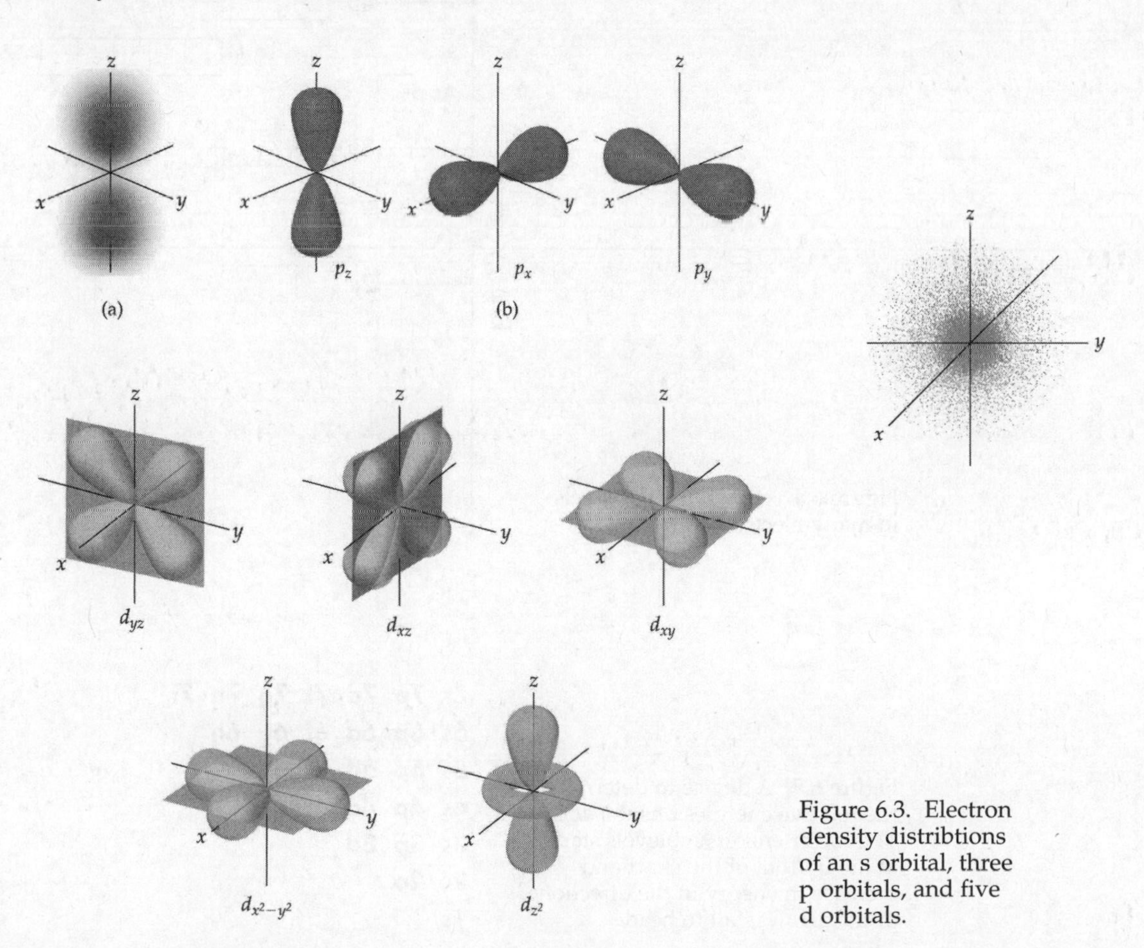

Figure 6.3. Electron density distribtions of an s orbital, three p orbitals, and five d orbitals.

Common misconception: An orbital is not the same as an orbit. An orbital is not a defined path of an electron. Rather it is a three-dimensional probability density distribution where an electron is likely to be found in the space surrounding the nucleus of an atom.

Section 6.7

Many Electron Atoms

In many-electron atoms (atoms other than hydrogen), electron-electron repulsions cause different sublevels to have different energies. Figure 6.4 shows the relative energy levels in many-electron atoms. Figure 6.5 is a convenient device for determining the relative energies, from lowest to highest of each sublevel of a many electron atom.

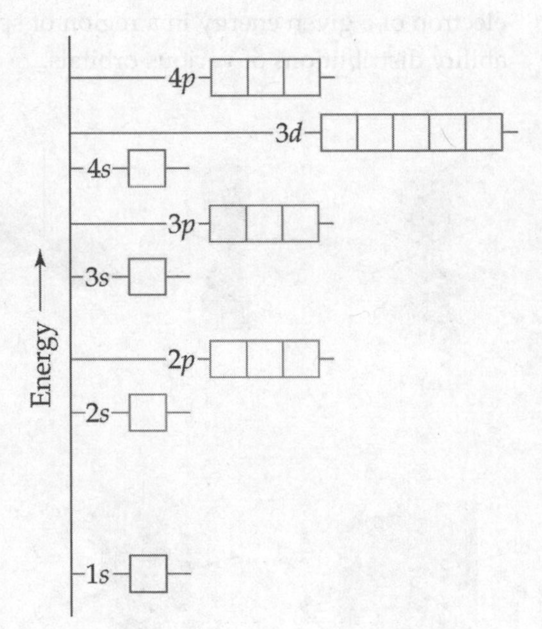

Figure 6.4. Orbital energy levels in many-electron atoms.

7s 7p 7d 7f 7g 7h 7i
6s 6p 6d 6f 6g 6h
5s 5p 5d 5f 5g
4s 4p 4d 4f
3s 3p 3d
2s 2p
1s

Figure 6.5. A device to determine the relative energies of sublevels. The lower energy sublevels start at the bottom of the chart and increase in energy in the direction of the arrows, tail to head.

Your Turn 6.4

List the first sixteen energy sublevels shown in Figure 6.5 in order of increasing energy. Write your answer in the space provided.

Electron Configurations

Section 6.8

An **electron configuration** is a distribution of electrons among various orbitals of an atom. Table 6.4 shows the electron configurations of some light elements.

Table 6.4 Electron Configurations of Several Lighter Elements

Element	Total Electrons	Orbital Diagram 1s	2s	2p			3s	Electron Configuration
Li	3	↑↓	↑					$1s^2 2s^1$
Be	4	↑↓	↑↓					$1s^2 2s^2$
B	5	↑↓	↑↓	↑				$1s^2 2s^2 2p^1$
C	6	↑↓	↑↓	↑	↑			$1s^2 2s^2 2p^2$
N	7	↑↓	↑↓	↑	↑	↑		$1s^2 2s^2 2p^3$
Ne	10	↑↓	↑↓	↑↓	↑↓	↑↓		$1s^2 2s^2 2p^6$
Na	11	↑↓	↑↓	↑↓	↑↓	↑↓	↑	$1s^2 2s^2 2p^6 3s^1$

Paramagnetic is a term referring to an atom having one or more un-paired electrons. Table 6.4 shows that lithium, boron, carbon, nitrogen, and sodium are all paramagnetic.

Diamagnetic means that all electrons in an atom are paired.

Your Turn 6.5

Which of the elements listed in Table 6.4 are diamagnetic? Explain. Write your answer in the space provided.

Figures 6.4 or 6.5 can be used to write the ground state electron configurations of most atoms.

Rules for writing ground state electron configurations for atoms using Figures 6.4 or 6.5.

1. Fill the lowest energy level first. Electrons in the same orbital must have opposite spins (different spin quantum numbers). The total number of electrons to use is the atomic number of the element.

2. Place no more than two electrons per orbital to satisfy the Pauli exclusion principle.

3. Do not pair electrons in degenerate (same energy) orbitals until each orbital has one electron of the same spin. (This is called Hund's Rule.)

4. Write the electron configuration by using sublevel designations and superscripts to designate the number of electrons in each sublevel.

Example:

Write the electron configuration for iron.

Solution:

The atomic number of iron is 26. Fill 26 electrons in the orbital diagram like this:

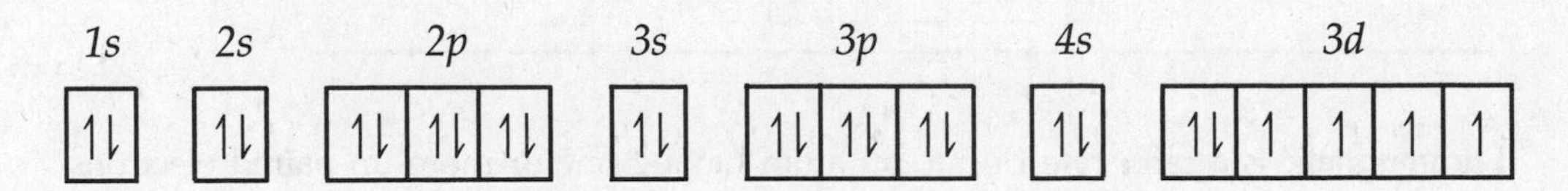

Write the corresponding ground state electron configuration like this:

$1s^2 2s^2 2p^6 3s^2 3p^6 4s^2 3d^6$

An **excited state configuration** has higher energy than the ground state configuration. Excited state configurations have one or more electrons occupying higher energy levels than would be predicted from Figures 6.4 or 6.5. For example, one of many possible excited state configurations for iron arises when an electron moves from a 4s sublevel to a 3d sublevel. The excited state orbital diagram looks like this:

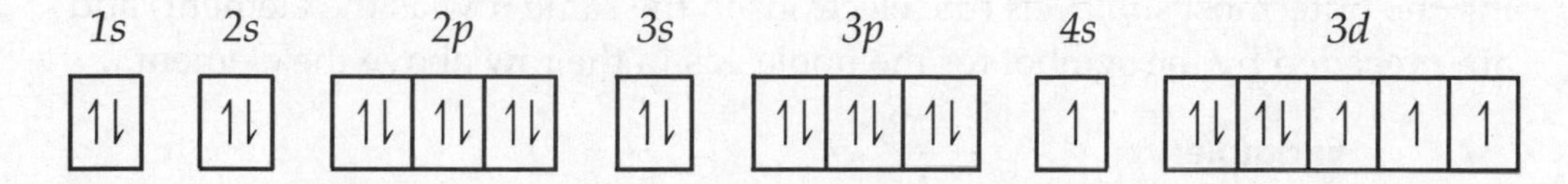

$1s^22s^22p^63s^23p^64s^13d^7$ is an excited state electron configuration.

Electron Configurations and the Periodic Table

Section 6.9

The periodic table is organized so that elements with similar electron configurations are arranged in columns as shown in Figure 6.6. Electron configurations of most elements can be determined by their location on the periodic table using Figure 6.6.

s^1	s^2	d^1	d^2	d^3	s^1d^5	d^5	d^6	d^7	d^8	s^1d^{10}	d^{10}	p^1	p^2	p^3	p^4	p^5	p^6
1s																	
2s	→											2p	→				
3s	→											3p	→				
4s	→	3d	→		Cr					Cu		4p	→				
5s	→	4d	→		Mo					Ag		5p	→				
6s	→	5d	→				X			Au		6p	→				
7s	→	6d	→														
			4f	→													
			5f	→													

Figure 6.6. Elements on the periodic table are largely arranged according to electron configurations.

Example:

Write the electron configuration of the element marked by the "X"

Solution:

Element "X" is located in the row labeled 5d and the column labeled d^6 so there are six electrons in the 5d sublevel, the highest occupied sublevel.

The electron configuration for element X ends in $5d^6$. Write the **complete electron configuration** *by starting at the top left and working left to right and top to bottom until the element X is reached:*
$1s^2 2s^2 2p^6 3s^2 3p^6 4s^2 3d^{10} 4p^6 5s^2 4d^{10} 5p^6 6s^2 4f^{14} 5d^6$.

A **condensed electron configuration** for an element shows only electrons occupying the outermost sublevels (the electrons in the same row as the element) and are preceded by the symbol for the noble gas in the row above the element.

Example:

Write the condensed electron configuration for element X.

Solution:

Element X has these sublevels occupied in its row: $6s^2 4f^{14} 5d^6$ and the noble gas in the row above is xenon, Xe. The condensed configuration is, (Xe) $6s^2 4f^{14} 5d^6$.

Your Turn 6.6

In a condensed electron configuration, what does the symbol (Xe) represent? Write your answer in the space provided.

The Group 6 atoms, Cr & Mo, are unusual because the ground state outermost electrons in these atoms are arranged s^1d^5 rather than s^2d^4. There is an unusual stability associated with a precisely half filled d orbital. All of the group 11 atoms, Cu, Ag, and Au similarly exhibit s^1d^{10} ground state configurations rather than s^2d^9 because completely filled d orbitals are especially stable.

Common misconception: The seemingly anomalous ground state electron configurations of Cr, Mo, Cu, Ag and Au are often mistaken for excited state configurations. These configurations arise because the s and d orbitals are very close in energy and precisely half-filled degenerate d orbitals (as in chromium) and completely full d orbitals (as in copper) are more stable than other partially filled configurations.

Your Turn 6.7

Scan Chapter 6 of Chemistry the Central Science and state in one sentence each, the major contribution(s) to the modern structure of the atom made by each of the following scientists: Max Planck, Neils Bohr, Louis de Broglie, Werner Heisenberg, Erwin Schrodinger. Write your answer in the space provided.

Multiple Choice Questions

1. *How many electrons in the ground state of a copper atom have quantum numbers $n = 3$ and $l = 2$?*

A) 2

B) 6

C) 8

D) 10

E) 18

2. *What is the wavelength of light that has a frequency of 6.0×10^{14} Hz?*

A) 2000 nm

B) 500 nm

C) 200 nm

D) 2.0×10^{6} nm

E) 5.0×10^{-7} nm

3. *What is the maximum number of electrons that can occupy the 5f sublevel?*

A) 2

B) 5

C) 10

D) 14

E) 18

4. *What is the maximum number of orbitals in a 4d sublevel?*

A) 2

B) 5

C) 10

D) 14

E) 18

5. X: $1s^2 2s^2 2p^3$ Y: $1s^2 2s^1$

Atoms X and Y have the ground state electronic configuration shown above. The formula for the compound most likely formed from X and Y is

(A) YX

(B) Y_2X

(C) Y_3X

(D) YX_3

(E) Y_2X

6. *A blue line in the atomic emission spectrum of hydrogen has a wavelength of 434 nm. What is the energy of this light per mole of photons?*

A) $(10^6)(6.63)(3.00)(6.02)/(434)$ kJ/mol

B) $(10^3)(6.63)(3.00)(6.02)/(434)$ kJ/mol

C) $(10^6)(6.63)(3.00)(6.02)/(434)$ J/mol

D) $(10^3)(6.63)(3.00)(6.02)/(434)$ J/mol

E) $(10^3)(434)(6.02)/(6.63)(3.00)$ kJ/mol

7. *The wavelength of electromagnetic radiation is longer when*

A) its energy is small and its frequency is large.

B) its energy is small and its frequency is small.

C) its energy is large and its frequency is large.

D) its energy is large and its frequency is small.

E) its energy is large and its amplitude is high.

8. *The outermost electron in a ground state potassium atom can be described by which of the following sets of four quantum numbers?*

(A) 4, 0, 0, ½

(B) 4, 1, 0, ½

(C) 4, 1, 1, ½

(D) 5, 0, 0, ½

(E) 5, 1, 0, ½

9. *Gaseous atoms of which of the following elements are paramagnetic in their ground states?*

I. Na   II. Mg   III. Al   IV. P

A) I, II, III, IV.

B) I, II, III only.

C) I, III, IV only.

D) II only.

E) III, IV only.

10. Which set of quantum numbers is ***not*** *allowed?*

(A) (2, 2, 1, ½)

(B) (3, 2, 0, -½)

(C) (4, 3, -3, ½)

(D) (5, 4, 4, ½)

(E) (6, 2, -1, ½)

Free Response Questions

1. *A line having a wavelength of 656 nm exists in the atomic emission spectrum of hydrogen.*

a. *For the line calculate the following values and specify their units:*

i. frequency

ii. energy of a photon

iii. energy of a mole of photons

b. *What color is the line? Explain your reasoning.*

c. *Discuss the origin of the line in terms of the Bohr theory of the atom. Specify any energy transitions that are applicable.*

2. *Molecules of oxygen are converted to atomic oxygen in the upper atmosphere by absorbing photons having wavelengths of 240 nm and shorter.*

a. *Write the electron configuration of oxygen and tell why atomic oxygen is diamagnetic or paramagnetic.*

b. *Write the electron configuration of the oxide ion. Assign a set of four quantum numbers to each of the electrons in the oxide ion. Correlate the sets to the electron configuration.*

c. *Calculate the energy equivalent of a photon of wavelength 240 nm in units of kJ/mol.*

Additional Practice in Chemistry the Central Science

For more practice answering questions in preparation for the Advanced Placement examination try these problems in Chapter 6 of Chemistry the Central Science:

Additional Exercises: 6.78, 6.89, 6.90, 6.92, 6.96, 6.97.

Integrative Exercises: 6.99, 6.100, 6.101, 6.102, 6.104.

Multiple Choice Answers and Explanations

1. *D. The quantum numbers n=3 and l = 2 refer to the 3d orbital which has 10 electrons in the ground state configuration of copper. Copper's ground state outer electron configuration is anomalous because it is a s^1d^{10} rather than an s^2d^9 as would be predicted from its location in Group 11 on the periodic table. Completely full d orbitals are more stable than partially filled d orbitals.*

2. *B. Wavelength and frequency of light are related by the equation, $\nu = c/\lambda$ where ν is the frequency in Hz (or 1/s), λ is the wavelength and c is the speed of light. Rearranging the equation and solving:*
   $\lambda = c/\nu$
   $\lambda = (3.00 \times 10^{17} nm/s)/(6.0 \times 10^{14} 1/s) = 500$ nm.

3. *D. Any f sublevel contains 7 orbitals, each of which can contain 2 electrons for a total of 14 electrons maximum. The number of orbitals and maximum number of electrons in the other sublevels are: s, 1, 2; p, 3, 6; and d, 5, 10. The key to this question is to interpret a 5f sublevel as a set of seven orbitals.*

4. *B. Any d sublevel contains a set of five orbitals. The key to this question is understand that it asks for the number of orbitals, not electrons.*

5. *C. Atom X is nitrogen and atom Y is lithium. Lithium will lose one electron to form a 1+ ion and nitrogen will gain three electrons to form a 3- ion.*

6. *B. The energy of a single photon is given by the following equation: $E = hc/\lambda$. Solving for E in J/photon and multiplying by 6.02×10^{23} photons/mol:*

E = (6.63 x 10^{-34} Js)(3.00 x 10^{17} nm/s)(6.02 x 10^{23} photons/mol)/(434 nm) = (6.63)(3.00)(6.02)(10^{-34})(10^{17})(10^{23})/(434) J/mol = (6.63)(3.00)(6.02)(10^{6})/(434) J/mol = (6.63)(3.00)(6.02)(10^{3})/(434) kJ/mol

Be careful to convert joules to kilojoules to match the units of the answer.

7. *B. Wavelength is inversely proportional to both energy and frequency as seen by the equations, √ = c/λ and E = hc/λ. Small energies and frequencies are associated with large wavelengths.*

8. *A. The electron configuration of potassium is (Ar)$4s^1$. The outer electron configuration consists of the electrons in orbitals higher than the noble gas core, in this case, sublevel 4s, which corresponds to quantum numbers n = 4 and l = 0. Answers B and C refer to 4p sublevels and answers D and E refer to 5s and 5p sublevels, respectively.*

9. *C. Paramagnetic is a term referring to an atom having one or more un-paired electrons. Diamagentic means that all electrons in an atom are paired. In orbital diagram form, the electron configurations of each of the elements listed are:*

Na:	*(Ne)*	*$3s^1$*	
		1⇂	
Mg:	*(Ne)*	*$3s^2$*	
		1⇂	
Al:	*(Ne)*	*$3s^2$*	*$3p^1$*
		1⇂	1 — —
P:	*(Ne)*	*$3s^2$*	*$3p^1$*
		1⇂	1 1 1

All the electrons in the neon core are paired and the 3s electrons of magnesium is paired. Na, Al and P all have one or more unpaired electrons.

10. *A. The quantum numbers are listed in order of n, l, m_l and m_s. The value of l must be at least one fewer than the value of n. The possible values of each quantum number follows:*

n = 1,2,3,4,5,6,7... ∞

l = 0, 1, 2...n-1.

ml = -l...0...+l

ms = ±½

Free Response Answers

a. i. *Wavelength and frequency are related by the equation,* $\nu = c/\lambda$
Substituting:

$\nu = c/\lambda$

$\nu = (3.00 \times 10^{17}\ nm/s)/656\ nm = 4.57 \times 10^{14}\ 1/s.$

ii. *The energy of a single photon is related to its frequency by the equation,* $E = h\nu$.

$E = h\nu$

$E = (6.63 \times 10^{-34}\ Js)(4.57 \times 10^{14}\ 1/s) = 3.03 \times 10^{-19}\ J/photon.$

iii. *A mole of photons will have the energy in Answer a ii times Avogadro's number of photons:*
$x\ J/mol = (3.03 \times 10^{-19}\ J/photon(6.02 \times 10^{23}\ photons/mol) = 182{,}000\ J/mol = 182\ kJ/mol.$

b. *The 656 nm line is red. The visible spectrum ranges from a short wavelength of about 400 nm for violet to a long wavelength of about 650 nm for red.*

c. *The red 656 nm line originates when an electron gives off a photon of corresponding energy upon making a transition from the n=3 level to the n=2 level in the Bohr model.*

2a. $1s^2 2s^2 2p^4$
Atomic oxygen is paramagnetic because two of its 2p electrons are unpaired.

b. *The oxide ion has two more electrons than the oxygen atom. The electron configuration of the oxide ion,* O^{2-}, *is* $1s^2 2s^2 2p^6$. *The quantum numbers for electrons in the oxide ion correspond to these sublevels:*

1s: (1,0,0, + ½) and (1,0,0, - ½)

2s: (1,0,0, + ½) and (1,0,0, - ½)

2p: (2,1,0, + ½), (2,1,0, - ½) (2,1,-1, + ½), (2,1,-1, - ½), (2,1,1, + ½), and (2,1,1, - ½)

c. *For a photon of wavelength 240 nm, λ = 240 nm. Convert 240 nm to kJ/mol:*

$E = hc/\lambda = (6.63 \times 10^{-34}$ Js)(3.00×10^{17} nm/s)/240 nm = 8.28×10^{-19} J/photon

(8.28×10^{-19} J/photon(6.02×10^{23} photons/mol)(1kJ/1000J) = 499 kJ/mol

Your Turn Answers

6.1. *The unit for frequency is 1/s because frequency of light is measured in complete wavelengths per second or, more generally, complete cycles of anything per second.*

6.2. *Blue-green line, n=4 to n=2; blue line, n=5 to n=2; violet line, n=6 to n=2.*
The violet line represents the largest energy change because violet light is the most energetic visible light.

6.3. *Ten electrons can occupy a 3d sublevel having designations (3,2,-2,+½), (3,2,-1,+½), (3,2,0,+½), (3,2,+1,+½), (3,2,+2,+½), (3,2,-2,-½), (3,2,-1,-½), (3,2,0,-½), (3,2,+1,-½) and (3,2,+2,-½).*

6.4. *1s, 2s, 2p, 3s, 3p, 4s, 3d, 4p, 5s, 4d, 5p, 6s, 4f, 5d, 6p, 7s.*

6.5. *Table 6.4 shows that Be and Ne are diamagnetic because all the electrons in their respective configurations are paired.*

6.6. *The symbol (Xe) represents the electron configuration of Xe: $1s^2 2s^2 2p^6 3s^2 3p^6 4s^2 3d^{10} 4p^6 5s^2 4d^{10} 5p^6$.*

6.7. *Max Planck first enunciated the quantum theory that says that energy can be released or absorbed by atoms only in discrete chunks called quanta.*

Neils Bohr proposed a model for the atom that explains the atomic emission spectrum of hydrogen as arising from photons of light given off when electrons make quantum transitions within atoms.

Louis de Broglie proposed that matter, specifically electrons, show both wave and particle characteristics.

Werner Heisenberg proposed the principle of uncertainty, the idea that it is impossible to simultaneously determine the exact position and momentum of an electron.

Erwin Schrodinger developed the wave mechanical model of the atom based on a mathematical wave equation.

TOPIC

7

PERIODIC PROPERTIES OF THE ELEMENTS

Effective nuclear charge and its direct affect on how the periodic properties of elements vary across the periodic table is, by far, the most important concept to master in Chapter 7. Students should recognize and explain the role that electron configurations play in determining periodic properties. They should know the trends in atomic radius, ionic radius, ionization energy, and electron affinity and be able to explain them by applying the concepts of effective nuclear charge and the shielding effect. Students should also be able to identify and explain the anomalies in the trends of first ionization energy and electron affinity. Finally, students should be able to write balanced equations for the common reactions of Group 1 and 2 metals. Focus especially on these sections:

- 7.2 **Effective Nuclear Charge**
- 7.3 **Sizes of Atoms and Ions**
- 7.4 **Ionization Energy**
- 7.5 **Electron Affinities**
- 7.6 **Metals, Nonmetals and Metalloids**
- 7.7 **Group Trends for the Active Metals**
- 7.8 **Group Trends for Selected Nonmetals**

The **force of attraction** between two charged particles is given by the equation, $F = kQ_1Q_2/d^2$ where Q_1 and Q_2 are the charges of the particles and d is the distance between them. The equation tells us that, in general, electrons close to the nucleus will be held with a greater force than those that are more distant from the nucleus. Also, higher positive nuclear charges will draw electrons closer to the nucleus and hold them with greater force.

Valence electrons are electrons in the outermost orbitals of atoms, those farthest from the nucleus.

Core electrons are inner electrons and include noble gas core electrons and electrons in completely full "d" orbitals.

Section 7.2 Effective Nuclear Charge

Effective nuclear charge is the net positive charge experienced by an electron in an atom. This charge is not the full nuclear charge because core electrons closer to the nucleus **shield** (cancel) part of the positive nuclear charge. As a result, valence electrons experience less than the full nuclear charge.

The **shielding effect** is the reduction of the full nuclear charge experienced by an outer electron as a result of screening by inner core electrons. Valence electrons shield one another ineffectively. The shielding effect arises from core electrons within the atom screening part of the positive nuclear charge.

Effective nuclear charge increases from left to right across any row (period) of the periodic table. Although the number of shielding core electrons remains constant in any row, the nuclear charge increases and thus, the effective nuclear charge increases.
A simple relationship between effective nuclear charge, Z_{eff}, and the number of protons in the nucleus, Z, is:

$$Z_{eff} = Z - S$$

S is the screening constant. The value of S, for comparison purposes, can be taken as approximately the total number of core electrons in the atom. The effective nuclear charge increases left to right within any period of the periodic table. Compare electron configurations of sodium, magnesium, and aluminum and their corresponding effective nuclear charges.

$_{11}$Na: $1s^2 2s^2 2p^6 3s^1$ or $(Ne)3s^1$. $Z_{eff} = Z - S = 11 - 10 = +1$.

$_{12}$Mg: $1s^2 2s^2 2p^6 3s^2$ or $(Ne)3s^2$. $Z_{eff} = Z - S = 12 - 10 = +2$.

$_{13}$Al: $1s^2 2s^2 2p^6 3s^2 3p^1$ or $(Ne)3s^2 3p^1$. $Z_{eff} = Z - S = 13 - 10 = +3$.

All three elements contain ten noble gas core electrons, which shield ten protons in the nucleus. However, the outer electrons in each element experience a different effective nuclear charge because they have a different nuclear charge. As a consequence, the valence electrons of aluminum are drawn more closely to the nucleus and are held more tightly than the valence electron on sodium. (Remember, the valence electrons typically do not shield one another effectively.) The shielding effect gives rise to periodic properties of elements as the effective nuclear charge varies across the periodic table. Those properties include, atomic and ionic size, ionization energy and electron affinity.

The effective nuclear charge remains fairly constant going down a group of the periodic table. This is because the number of valence electrons within a group is constant while an increasing number of protons balances an increasing number of shielding noble gas core electrons. For example, compare the electron configurations of sodium, potassium, and rubidium.

$_{11}$Na: $1s^22s^22p^63s^1$ or $(Ne)3s^1$. $Z_{eff} = 11 - 10 = +1$.

$_{19}$K: $1s^22s^22p^63s^23p^64s^1$ or $(Ar)4s^1$. $Z_{eff} = 19 - 18 = +1$.

$_{37}$Rb: $1s^22s^22p^63s^23p^64s^23d^{10}4p^65s^1$ or $(Kr)5s^1$. $Z_{eff} = 37 - 36 = +1$.

Your Turn 7.1

What is the approximate effective nuclear charge of scandium? Would you expect that its valence electrons are held more or less tightly than those of potassium? Use scandium's electron configuration to explain your answer. Place your answer in the space provided.

Sizes of Atoms and Ions

Section 7.3

Atomic radius is an estimate of the size of an atom. According to the quantum mechanical model, atoms do not have sharply defined boundaries because of the inherent uncertain distributions of electrons within atomic orbitals. Definite sizes are not obtainable and although atomic size is defined in different ways based on the distances between atoms in various situations, comparisons of the relative size of atoms is instructive.

Within a group, atomic size increases from top to bottom of the periodic table. Within a period, atomic size increases from right to left. Atoms get smaller left to right because shielding remains roughly constant as nuclear charge increases. Atoms get larger top to bottom because outer electrons have higher principal quantum numbers and thus, greater probabilities of being farther from the nucleus.

Common misconception: Don't confuse a trend with an explanation for a trend. For example, if asked why the atomic radius of argon is smaller than that of chlorine, its not acceptable to say that argon is farther to the right than is chlorine on the periodic table. That's a statement of the trend, not an answer to the question. The correct explanation is that argon's higher effective nuclear charge draws its valence electrons closer to the nucleus. Both atoms have ten core electrons screening their respective nuclei but argon has one more proton than does chlorine giving argon the higher effective nuclear charge.

Cations are always smaller than their parent atoms because the electrons lost upon formation of a cation vacate the outermost orbitals, decreasing the size of the ion. Additionally, there are fewer electron-electron repulsions.

Anions are always larger than their parent atoms because additional electrons cause increased electron-electron repulsions causing the electrons to spread out more in space.

For ions carrying the same charge, the trend in size is the same as that for neutral atoms. The size of both cations an anions increases down a group and increases from right to left along a period.

Figure 7.1 shows the relative sizes of some parent atoms and their ions.

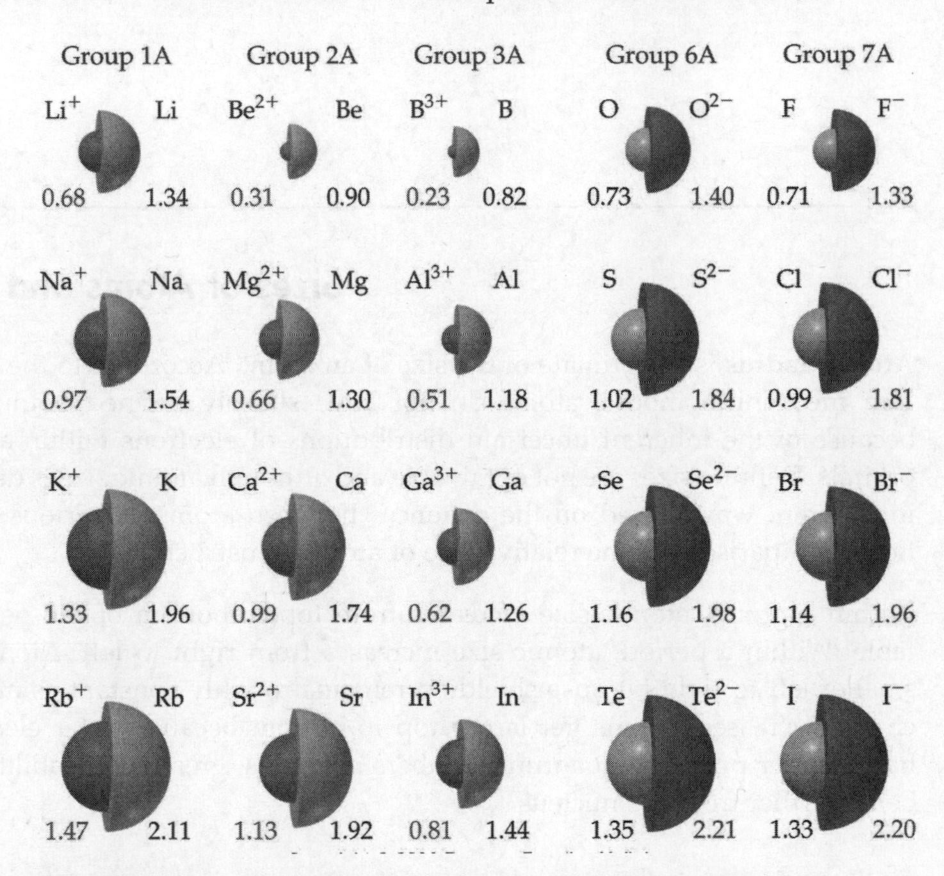

Figure 7.1. The relative sizes of atoms and ions in angstroms, Å. 1 Å = 10^{-10} m.

Section 7.4

Ionization Energy

The first ionization energy (potential), I_1, of an atom is the energy required to remove the outermost electron from the ground state of a gaseous atom. For example, the first ionization energy of sodium is +495 kJ per mole:

$$495 \text{ kJ} + \text{Na} \longrightarrow \text{Na}^+ + e^-$$

The **second ionization energy**, I_2, is the energy needed to remove the second electron, and so forth, for successive removal of electrons. For all atoms, $I_1 < I_2 < I_3$, and so forth, because with each successive removal, an electron is pulled away from an increasingly more positive ion.

Figure 7.2 shows that first ionization energy (potential) increases from left to right and along a period, and from bottom to top within a group on the periodic table. In general, the smaller atoms have the higher first ionization potentials. Outer electrons in smaller atoms are closer to the nucleus, less shielded and held more tightly.

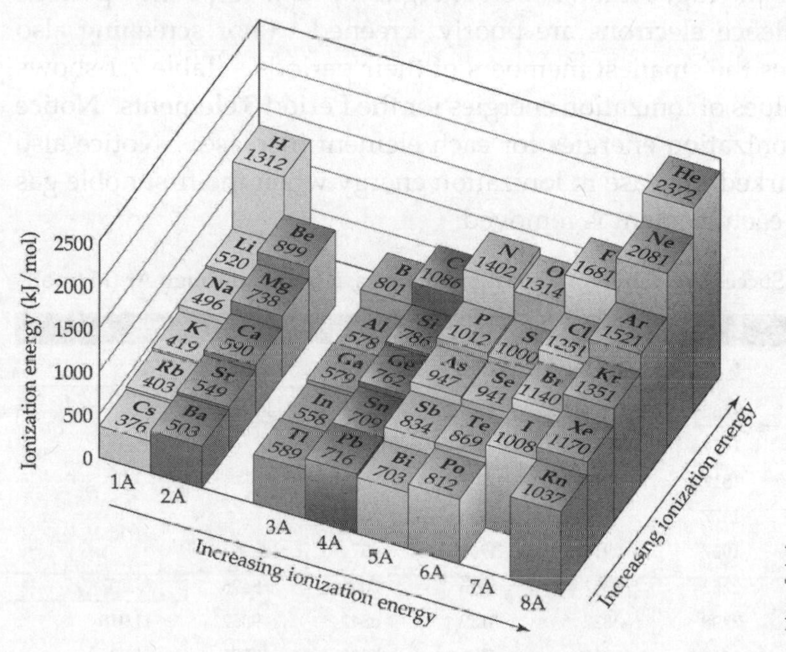

Figure 7.2. Trends in first ionization energy.

Subtle anomalies exist in the trend for first ionization energies between atoms of Groups 2 and 13 and between atoms of groups 15 and 16.

First ionization energies decrease from Be to B and from Mg to Al because electrons in filled s or d orbitals provide limited screening for electrons in the p subshells.

First ionization energies also decrease from N to O, P to S, As to Se due to repulsion of paired electrons in the p^4 configuration of the Group 16 atoms.

Your Turn 7.2

Arrange the Period 3 elements in order of increasing first ionization energy, lowest to highest. Justify your answer noting any anomalies. Place your answer in the space provided.

Noble gases have the highest ionization energies of their respective periods because their valence electrons are poorly screened. Poor screening also makes noble gases the smallest members of their periods. Table 7.1 shows the successive values of ionization energies for the Period 3 elements. Notice that successive ionization energies for each element increase. Notice also that there is a marked increase in ionization energy when the first noble gas core electron for each element is removed.

Table 7.1. Successive Values of Ionization Energies, I, for Na through Ar (kJ/mol)

Element	I_1	I_2	I_3	I_4	I_5	I_6	I_7
Na	495	4562			(inner-shell electrons)		
Mg	738	1451	7733				
Al	578	1817	2745	11,577			
Si	786	1577	3232	4356	16,091		
P	1012	1907	2914	4964	6274	21,267	
S	1000	2252	3357	4556	7004	8496	27,107
Cl	1251	2298	3822	5159	6542	9362	11,018
Ar	1521	2666	3931	5771	7238	8781	11,995

Section 7.5

Electron Affinities

Electron affinity, ΔH_{ea}, is the energy change when an electron is added to a gaseous atom.

$$F(g) + e^- \dashrightarrow F^-(g) \quad \Delta H_{ea} = -328 \text{ kJ/mol}$$

When most atoms attract electrons, energy is released so the sign of ΔH_{ea} is negative. Electron affinity increases from left to right along a period and from bottom to top within a group. Smaller atoms are less shielded and attract electrons more readily. Fluorine is an anomaly because it has a smaller electron affinity than chlorine. Fluorine's smaller size causes electron-electron repulsion making it less likely to attract an electron than chlorine. Figure 7.3 shows the electron affinities of the representative elements. Notice that the elements in groups 2A, 5A and 8A are anomalies having very low electron affinities.

1A	2A	3A	4A	5A	6A	7A	8A
H −73							He > 0
Li −60	Be > 0	B −27	C −122	N > 0	O −141	F −328	Ne > 0
Na −53	Mg > 0	Al −43	Si −134	P −72	S −200	Cl −349	Ar > 0
K −48	Ca −2	Ga −30	Ge −119	As −78	Se −195	Br −325	Kr > 0
Rb −47	Sr −5	In −30	Sn −107	Sb −103	Te −190	I −295	Xe > 0

Figure 7.3

Common misconception: It is important to recognize the difference between ionization energy and electron affinity. Ionization energy measures the ease with which an atom loses an electron, whereas electron affinity measures the ease with which an atom gains an electron. Ionization energies are usually endothermic (with positive energies) and electron affinities are usually exothermic (with negative energies). Keep in mind, that the more negative the electron affinity, the larger its relative value. The sign of the energy value tells what direction the energy flows, not how large the energy is.

Tables 7.2 and 7.3 summarize the trends of important periodic properties and their anomalies with explanations.

Table 7.2. Related Properties and Trends on the Periodic Table.

Periodic Property	**Trend: increases across the table**	**Explanation for trends**
Atomic size Ionic size	← (right to left), ↓ (top to bottom)	As effective nuclear charge increases left to right, outer elec trons are held more closely and more strongly to the nucleus. Top to bottom, outer electrons are in larger atomic orbitals and farther from the nucleus. As size decreases, attraction for electrons increases, increasing electron affinity and first ionization energy.
First ionization energy, I_1	→ (left to right), ↑ (bottom to top)	
Electron affinity	→ (left to right), ↑ (bottom to top)	

Table 7.3. Periodic "Anomalies" and Explanations

Periodic Property	Periodic "anomalies"	Explanation
Atomic size	None	—
First ionization energy, I_1	B < Be, Al < Mg,	Partial shielding by s valence electrons decreases Z_{eff}.
First ionization energy, I_1	O < N, S < P, Se < As, Te < Sb	Electron-electron repulsion in p^4 configurations of group 16 atoms.
Electron affinity	F < Cl Groups 2, 15 and 18 are very low.	The small size of fluorine contributes to strong electron-electron repulsion.

Section 7.6 Metals, Nonmetals, and Metalloids

Metals tend to have low inization energies and lose electrons readily. Metals tend to form positive ions.

Metals exhibit luster, conduct heat and electricity, are malleable (can be hammered into thin sheets), and are ductile (can be drawn into wires).
Metallic character is a reflection of the extent to which elements exhibit the physical and chemical properties of metals. In general, metallic character increases right to left along a period and top to bottom within a group.

Most metal hydrides, oxides and nitrides are basic:

$$CaH_2(s) + 2H_2O(l) \dashrightarrow Ca^{2+}(aq) + 2OH^-(aq) + 2H_2(g)$$

$$Li_2O(s) + H_2O(l) \dashrightarrow 2Li^+(aq) + 2OH^-(aq)$$

$$Mg_3N_2(s) + 6H_2O(l) \dashrightarrow 3Mg^{2+}(aq) + 6OH^-(aq) + 2NH_3(aq)$$

Nonmetals tend to have high electron affinities and gain electrons readily. Nonmetals tend to form negative ions. They do not exhibit luster and are poor conductors of heat and electricity.

Compounds composed entirely of nonmetals are molecular substances.

Most nonmetal oxides are acidic. They react with water to produce acids. Usually the nonmetal retains its oxidation number.

$SO_2(g) + H_2O(l) \dashrightarrow H_2SO_3(aq)$

$SO_3(g) + H_2O(l) \dashrightarrow H^+(aq) + HSO_4^-(aq)$

$CO_2(g) + H_2O(l) \dashrightarrow H_2CO_3(aq)$

$Cl_2O_7(s) + H_2O(l) \dashrightarrow 2H^+(aq) + 2ClO_4^-(aq)$

$P_2O_5(s) + 3H_2O(l) \dashrightarrow 2H_3PO_4(aq)$

$Cl_2O_3(s) + H_2O(l) \dashrightarrow 2HClO_2(aq)$

Notice that strong acids are shown in ionic form and weak acids are shown in molecular form. When nitrogen dioxide reacts with water both nitrous and nitric acid is formed.

$2NO_2(g) + H_2O(l) \dashrightarrow HNO_2(aq) + H^+(aq) + NO_3^-(aq)$

Metalloids have properties intermediate between those of metals and those of nonmetals.

Group Trends of the Active Metals

Section 7.7

The alkali metals (Group 1) are soft, metallic solids. They all have s^1 valence electron configurations and lose one electron to form +1 cations. Because ionization energy decreases down the group, the alkali metals become more reactive.

All alkali metals react with water to produce hydrogen gas:

$2Na(s) + 2H_2O(l) \dashrightarrow 2Na^+(aq) + 2OH^-(aq) + H_2(g)$

All form hydrides:

$2Li(s) + H_2(g) \dashrightarrow 2LiH(s)$

All alkali metals react with most nonmetals:

$2K(s) + S(s) \dashrightarrow K_2S(s)$ $2Rb(s) + Cl_2(g) \dashrightarrow 2RbCl(s)$

$6Li(s) + N_2(g) \dashrightarrow 2Li_3N(s)$ $4Li(s) + O_2(g) \dashrightarrow 2Li_2O(s)$

Na, K, Rb and Cs form peroxides:

$2Na(s) + O_2(g) \dashrightarrow Na_2O_2(s)$

K, Rb and Cs form superoxides:

$K(s) + O_2(g) \dashrightarrow KO_2(s)$

Common misconception: Know the difference between the oxide ion, O^{2-}, the peroxide ion, O_2^{2-} and the superoxide ion, O_2^-.

Each alkali metal emits a characteristic color when placed in a flame:

Li (crimson-red), Na (yellow) and K (violet).

Your Turn 7.3

Write and balance a chemical equation to describe what happens when solid potassium is added to water. Classify the reaction as acid-base, redox, precipitation or complex ion formation. Describe what you would observe when the reaction takes place. Write your answer in the space provided.

The alkaline earth metals (Group 2), all have s^2 valence electron configurations and lose two electrons to form 2+ cations. They all have low ionization energies but each is higher than the corresponding alkali metal of the same period. The alkaline earth metals are less reactive than the alkali metals and become more reactive as they progress down a group.

Calcium and the elements below it react with liquid water:

$$Ca(s) + 2H_2O(l) \dashrightarrow H_2(g) + Ca^{2+}(aq) + 2OH^-(aq)$$

Magnesium does not react with liquid water but reacts with steam:

$$Mg(s) + 2H_2O(g) \dashrightarrow H_2(g) + MgO(s)$$

Mg and Ca form oxides:

$$2Ca(s) + O_2(g) \dashrightarrow 2CaO(s)$$

Barium forms a peroxide:

$$Ba(s) + O_2(g) \dashrightarrow BaO_2\ (s)$$

All react with nonmetals:

$$Sr(s) + H_2(g) \dashrightarrow SrH_2(s)$$

$$3Mg(s) + N_2(g) \dashrightarrow Mg_3N_2(s)$$

$$Ba(s) + Cl_2(g) \dashrightarrow BaCl_2(s)$$

The characteristic flame colors are:

Ca (brick red), Sr (crimson-red), and Ba (green-yellow).

Your Turn 7.4

Write and balance a chemical equation to describe what happens when solid calcium oxide is added to water. Classify the reaction as acid-base, redox, precipitation, or complex ion formation. A drop of phenolphthalein would turn the resulting solution what color? Why? Write your answer in the space provided.

Section 7.8

Group Trends for Selected Nonmetals

Hydrogen's $1s^1$ electron configuration places it in Group 1 above the alkali metals. However, hydrogen has many unique properties and does not belong to any particular group. The ionization energy of hydrogen is extremely high (1312 kJ/mol) because of the complete lack of shielding of its single electron. Its high ionization energy, its ability to gain an electron to form the hydride ion, H^-, and its tendency to share electrons to form covalent bonds are the basis for classifying hydrogen as a nonmetal.

Allotropes are different forms of the same element in the same state. For example, several members of the carbon, nitrogen, and oxygen families can exist as allotropes as shown in Table 7.4.

Table 7.4. Allotropes are different forms of the same element in the same physical state.

Element	**Allotropes**
Carbon	C(s, graphite). C(s, diamond)
Phosphorus	P_4(s, white). P(s, red)
Oxygen	Dioxygen, O_2(g). Ozone, O_3(g)
Sulfur	S(s), S_8(s)

The halogens (Group 17) exist as diatomic molecules. Fluorine, F_2(g), and chlorine, Cl_2(g) are gases. Bromine, Br_2(l) is a liquid at room temperature, and Iodine, I_2(s) is a solid which readily sublimes at room temperature. All share the s^2p^5 valence electron configuration and each gains one electron to form a -1 anion. Artists use fluorine, in the form of hydrofluoric acid, HF, to etch glass. Chlorine is commonly used as a disinfectant in drinking water and swimming pools:

$$Cl_2(g) + H_2O(l) \dashrightarrow H^+(aq) + Cl^-(aq) + HOCl(aq)$$

Common misconception: The above reaction is a redox reaction called a disproportionation where chlorine is both oxidized and reduced. Notice that the strong acid, HCl is written in ionized form.

The noble gases (Group 18) are all nonmetallic monatomic gases at room temperature. They all have completely filled s and p sublevels and form a limited number of compounds: XeF_2, XeF_4, XeF_6, KrF_2 and HArF.

Multiple Choice Questions

1. *The first five ionization energies, in kJ/mol, for a particular element are shown below.*

I	I_2	I_3	I_4	I_5
786	1577	3232	4356	16,091

The element is likely to form ionic compounds in which its charge is

A) 1+

B) 2+

C) 3+

D) 4+

E) 5+

2. *Which list of elements is arranged in order of increasing atomic size (largest last)?*

A) Be, Mg, Ca, Sr, Ba

B) Ba, Sr, Ca, Mg, Be

C) Be, Ca, Ba, Mg, Sr

D) Be, Ba, Ca, Mg, Sr

E) Ba, Sr, Ca, Be, Mg

3. *Which atom in the ground state is paramagnetic?*

A) He

B) Be

C) Ba

D) C

E) Ne

4. *Which sequence is arranged in order of increasing ionization energies, lowest to highest?*

A) Be, B, C, N, O

B) B, Be, C, O, N

C) Be, B, C, O, N

D) B, Be, C, N, O

E) O, N, C, B, Be

5. *Which is not true of nonmetals?*

A) Most of their oxides are acidic.

B) They are poor conductors of heat.

C) They are poor conductors of electricity.

D) Most tend to lose electron readily.

E) Many are gases at room temperature.

The following responses are to be used to answer questions 6-10. Use an answer once, more than once or not at all.

A. Mg & Al

B. As & Se

C. Cl & F

D. Cr & Mo

E. Cu & Ag

6. *Show a reversal in the trend for first ionization energy because of electron-electron repulsion.*

7. *Show a reversal in the trend for first ionization energy because of screening by full orbitals.*

8. *Show a reversal in the trend for electron affinity because of electron-electron repulsion.*

9. *Exhibit an anomaly in outer electron configuration because full d orbitals are especially stable.*

10 *Exhibit an anomaly in outer electron configuration because half full d orbitals are especially stable.*

Free Response Questions

1. *Use details of the modern atomic theory and periodicity to explain why*

a. *atomic radii become larger as the atomic number within a family gets larger.*

b. *atomic radii become smaller as the atomic number within a period gets larger.*

c. *the radius of an oxide ion is larger than the radius of a oxygen atom.*

d. *the first ionization energy of aluminum is smaller than the first ionization energy of magnesium.*

e. *the third ionization energy of an element is always larger than its second ionization energy.*

2. *Write the formulas to show the reactants and products for the following laboratory situations. Balance the resulting chemical equation and answer the question about each. Be sure to write substances as ions if they are extensively ionized in solution.*

a. *Solid lithium nitride is added to water.*
What would be the effect on wetted pH paper held over the vessel?

b. *Solid calcium is placed in water.*
What would you observe?

c. *Solid strontium is heated in the presence of bromine gas.*
Would you expect the reaction to be endothermic or exothermic? Explain

d. *Sulfur dioxide gas is bubbled through water.*
Is this a redox reaction? Explain.

e. *Magnesium is heated in dry air.*
Write the equation that represents another possible reaction for this mixture.

f. *Sodium hydride is mixed with water.*
Write the chemical equation for the spontaneous ignition of one of the products with air.

Additional Practice in Chemistry the Central Science

For more practice answering questions in preparation for the Advanced Placement examination try these problems in Chapter 7 of Chemistry the Central Science:

Additional Exercises: 7.81, 7.82, 7.85, 7.91, 7.92, 7.94.
Integrative Exercises: 7.105, 7.107, 7.108, 7.110.

Multiple Choice Answers and Explanations

1. *D. The marked difference between ionization energies, I_4 and I_5 indicates that the fifth electron to be ionized is a poorly screened core electron, probably a noble gas core electron. Thus, the four valence elec*

tons will be lost relatively easily but the fifth electron will not, resulting in a 4+ ion.

2. A. *From top to bottom of any group the valence electrons of atoms reside in increasingly larger orbitals having higher principal quantum numbers. The outer electrons are farther from the nucleus as the atoms proceed down a group and their size increases.*

3. D. *Paramagnetic means that one or more electrons are unpaired. The electron configuration of carbon contains two unpaired electrons. The electron configurations of all the other elements listed show that all the electrons are paired, a condition termed diamagnetic.*

4. B. *Answer A reflects the general trend that ionization energy increases left to right along any period to parallel the general increase in effective nuclear charge. However, B has a lower ionization energy than Be because the $2s^2$ electrons of B partially shield the $2p^1$ electron. Also the electron-electron repulsion of the two paired electrons in the $2p^4$ configuration of oxygen makes the ionization energy of oxygen slightly lower than that of nitrogen.*

5. D. *Nonmetals tend to have high effective nuclear charges, which give them high electron affinities and high ionization energies so they tend to gain electrons readily. Metals have low effective nuclear charges making them lose electrons readily due to their relatively low ionization energies.*

6. B. *Selenium has two paired electrons in its $4p^4$ sublevel. They repel each other giving Se a lower ionization energy than arsenic which has only three electrons in the 4p sublevel, all of which are unpaired.*

7. A. *The $3s^2$ electrons of aluminum partially shields its $3p^1$ electron giving aluminum a lower than expected first ionization energy.*

8. C. *Although fluorine is a smaller atom than chlorine with its valence electrons drawn more closely to the nucleus, it's so small that significant electron-electron repulsion occurs upon attracting an extra electron to the small 2p sublevel.*

9. E. *Among transition metals, the outer s and d orbitals are very close in energy and Cu, Ag and Au all exhibit s^1d^{10} electron configurations because completely full d orbitals are especially stable, even more stable than s orbitals.*

10. D. *Five unpaired electrons in an exactly half-filled d sublevel in both*

Cr and Mo make the ground state configurations of s^1d^5 more stable than the s^2d^4 configurations that would be predicted by their relative positions on the periodic table.

Free Response Answers

1. a. *Atomic radii become larger as the atomic number within a family gets larger because as we go down a group, the outer electrons have higher principal quantum numbers, occupy larger and larger orbitals and have greater probability of being farther from the nucleus.*

b. *Because the number of shielding electrons remains constant along any period, the effective nuclear charge increases as the atomic number increases. An increasing effective nuclear charge tends to draw valence electrons closer to the nucleus thus decreasing the size of the atoms.*

c. *A 2- oxide ion has two more electrons in the valence level than does a neutral oxygen atom. and these added electrons increase electron-electron repulsion expanding the region they occupy in the oxide ion.*

d. *The $3s^2$ electrons configuration of aluminum partially shield from the nucleus the outermost $3p^1$ aluminum electron making that electron more loosely held than the outermost $3s^2$ electrons of magnesium which do not shield each other. The result is that aluminum has a smaller first ionization energy than magnesium.*

e. *The third ionization energy of an element is always larger than its second ionization energy because the third electron is pulled away from a 2+ ion whereas the second electron is pulled away from a 1+ ion. The higher nuclear charge of the 2+ ions holds the remaining electrons with more force.*

2 a. $Li_3N(s) + 3H_2O(l) \dashrightarrow 3Li^+(aq) + 3OH^-(aq) + NH_3(g)$

Wetted pH paper held over the vessel would indicate that a basic gas, ammonia, was liberated from the solution.

b. $Ca(s) + 2H_2O(l) \dashrightarrow Ca^{2+}(aq) + 2OH^-(aq) + H_2(g)$

Bubbles of hydrogen would emerge from the container.

c. $Sr(s) + Br_2(g) \dashrightarrow SrBr_2(s)$

The reaction would be highly exothermic owing to the fact that strontium is an active metal and bromine is a highly reactive nonmetal. The product is a very stable salt.

d. $SO_2(g) + H_2O(l) \dashrightarrow H_2SO_3(aq)$

This is not a redox reaction because all elements in reactants and products retain their oxidation numbers. This is an acid-base reaction.

e. $2Mg(s) + O_2(g) \dashrightarrow 2MgO(s)$

Besides oxygen, air also contains nitrogen, and magnesium reacts with nitrogen to form magnesium nitride:

$3Mg(s) + N_2(g) \dashrightarrow Mg_3N_2(s)$

f. $NaH(s) + H_2O(l) \dashrightarrow Na^+(aq) + OH^-(aq) + H_2(g)$

Hydrogen is explosive in the presence of oxygen in air:

$2H_2(g) + O_2(g) \dashrightarrow 2H_2O(g)$

Note: Depending on the temperature and pressure, water will form as a liquid or a gas.

Your Turn Answers

7.1. $_{21}Sc$: $1s^22s^22p^63s^23p^64s^23d^1$ or $(Ar)4s^23d^1$ $Z_{eff} = 21 - 18 = +3$
Scandium's eighteen noble gas core electrons shield eighteen of its 21 protons giving it an effective nuclear charge of about +3. As a result of its increased Z_{eff} *scandium's valence electrons are held closer to the nucleus and more tightly than those of potassium.*

7.2. *First ionization energies:* $Na < Al < Mg < Si < S < P < Cl < Ar$.
Generally first ionization energies increase form left to right along a period because the effective nuclear charge increases drawing valence electrons more tightly to the nucleus. Aluminum has a lower ionization energy than magnesium because its s^2 *valence electrons partially screen the nucleus making the effective nuclear charge smaller. Sulfur has a lower ionization energy than phosphorus because electron-electron repulsion in sulfur's* p^4 *valence configuration aids the removal of a p electron.*

7.3. $2K(s) + 2H_2O(l) \dashrightarrow 2K^+(aq) + 2OH^-(aq) + H_2(g)$
The chemical equation represents a redox reaction. Bubbles of hydrogen are visible as the potassium dissolves. The hydrogen might catch fire.

7.4. $CaO(s) + H_2O(l) \dashrightarrow Ca^{2+}(aq) + 2OH^-(aq)$
The chemical equation represents an acid-base reaction. Phenolphthalein would be pink owing to the presence of hydroxide ions.

TOPIC 8

BASIC CONCEPTS OF CHEMICAL BONDING

Although each section contains important information, many of the topics of this chapter may have been learned in first year chemistry and need only be reviewed. Students should be able to write both octet and non-octet Lewis structures for atoms, ions and molecules, determine bond polarity from electronegativity values, draw resonance structures, correlate bond multiplicity to bond length and strength, and calculate enthalpies of reactions from bond dissociation energies.

Chemical Bonds, Lewis Symbols and the Octet Rule

Section 8.1

A **chemical bond** is a strong force of attraction that holds two atoms together. Chemical bonding involves the valence electrons of atoms.

An **ionic bond** is an attractive force between ions of opposite charges, often a metal cation and a nonmetal anion.

A **covalent bond** results from sharing electrons between two atoms, usually nonmetal atoms.

A **metallic bond** occurs when the nuclei of a collection of metals atoms simultaneously attract their collective electrons.

Your Turn 8.1

Classify the following compounds as ionic or covalent. On what do you base your answer? Which compounds contain both ionic and covalent bonds? KBr, SO_2, H_2SO_4, CH_3COOH, Na_3PO_4, $CaCO_3$. *Write your answers in the space provided.*

A **Lewis symbol** shows the valence electrons as dots around the symbol for the element as seen in Table 8.1. The number of valence electrons for a representative element is the same as its group number on the periodic table.

Element	Electron Configuration	Lewis Symbol
Li	[He]$2s^1$	Li·
Be	[He]$2s^2$	·Be·
B	[He]$2s^2 2p^1$	·Ḃ·
C	[He]$2s^2 2p^2$	·Ċ·
N	[He]$2s^2 2p^3$	·Ṅ:
O	[He]$2s^2 2p^4$	:Ȯ:
F	[He]$2s^2 2p^5$	·F̈:
Ne	[He]$2s^2 2p^6$	:N̈e:

Table 8.1 Lewis Symbols

The **octet rule** states that representative elements tend to gain, lose, or share electrons until they are surrounded by eight valence electrons (an octet). An octet of electrons gives an atom a noble gas configuration with completely full s and p sublevels. Noble gas configurations are stable because they are poorly shielded from the nucleus. Hydrogen provides an exception to the octet rule because it requires only two electrons to attain a noble gas configuration.

Common misconception: When an atom gains or loses electrons to attain an octet of valence electrons, it does not become a noble gas. For example, chloride ion, Cl^-, has the same electron configuration of argon, but it is not argon. Chloride ion is isoelectronic (the same number of electrons) with argon but chloride has one fewer proton.

Your Turn 8.2

Place the following chemical species into isoelectronic groups. N^{3-}, K^+, Ca^{2+}, O^{2-}, F^-, *Ne*, Br^-, *Kr*, Sc^{3+}, Na^+, Al^{3+}, Se^{2-}, Mg^{2+}. *Write your answer in the space provided.*

Ionic Bonding

Section 8.2

Ions form when electrons transfer from an atom of low ionization energy (usually a metal) to an atom of high electron affinity (usually a nonmetal). The electrostatic attraction between two oppositely charge ions constitutes an ionic bond.

A **lattice** is a stable, ordered, solid three-dimensional array of ions associated with ionic compounds. Figure 8.1 shows a lattice of sodium and chloride ions.

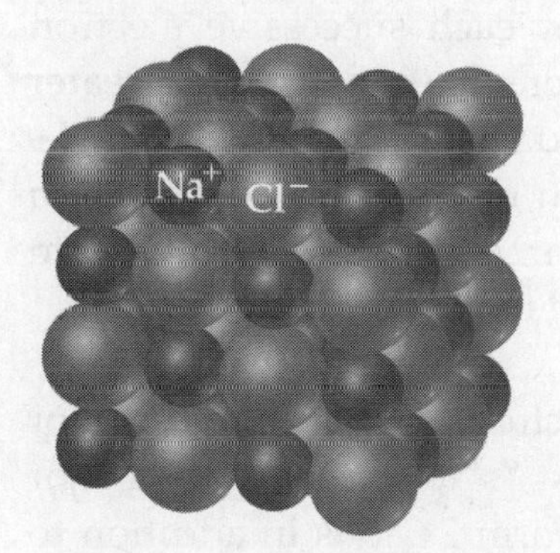

Figure 8.1. The crystal structure of sodium chloride.

Lattice energy, $\Delta H_{lattice}$ is the energy required to completely separate a mole of a solid ionic compound into its gaseous ions. For example, the lattice energy of potassium fluoride is 808 kJ/mol.

$$KF(s) \rightarrow K^+(g) + F^-(g) \qquad \Delta H_{lattice} = +808 \text{ kJ/mol}$$

Table 8.2 lists the lattice energies of some common ionic compounds. Lattice energy increases with smaller, more highly charged ions. The potential energy of two interacting charged particles is given by the equation, $E = KQ_1Q_2/d$ where Q_1 and Q_2 are the charges on the ions and d is the distance between them.

Table 8.2 Lattice Energies for Some Ionic Compounds

Compound	Lattice Energy (kJ/mol)	Compound	Lattice Energy (kJ/mol)
LiF	1030	$MgCl_2$	2326
LiCl	834	$SrCl_2$	2127
LiI	730		
NaF	910	MgO	3795
NaCl	788	CaO	3414
NaBr	732	SrO	3217
NaI	682		
KF	808	ScN	7547
KCl	701		
KBr	671		
CsCl	657		
CsI	600		

Ionic compounds tend to have high lattice energies, which tend to make ionic compounds hard and brittle with relatively high melting points.

The charge of a monatomic ion is determined by the number of valence electrons of the parent atom. Representative metals in Groups 1, 2 or 3 will typically lose all of their valence electrons when forming monatomic cations. Because ionization energies increase rapidly for each successive electron removed, metals tend not to form monatomic ions having charges greater than 3+. (Common exceptions are tin, lead and titanium, which form 4+ cations.) Similarly, nonmetals gain only enough electrons to complete an octet. Because of electron-electron repulsion, nonmetals rarely form monatomic anions having charges more negative than 3-.

Transition metals lose their valence level "s" electrons first and then one or two "d" orbital electrons. Having two outermost "s" electrons is the major reason why many transition metals commonly form 2+ ions in addition to ions of other charges. For example, the electron configurations of Fe, Fe^{2+} and Fe^{3+} are shown below:

Fe: (Ar) $3d^6 4s^2$

Fe^{2+}: (Ar) $3d^6 4s^0$ or (Ar) $3d^6$

Fe^{3+}: (Ar) $3d^5 4s^0$ or (Ar) $3d^5$

Common misconception: When writing electron configurations for transition metals, the outer "s" levels are filled before the outer "d" levels. However, when writing electron configurations for transition metal ions, remove the outer "s" electrons before removing the "d" electrons. (Note that often electron configurations are written so that the outer "s" sublevel comes before outer "d" sublevel.)

Lead and tin commonly form monatomic cations having 2+ and 4+ charges. For example the electron configurations of Pb, Pb^{2+} and Pb^{4+} are shown:

Pb: (Xe) $4f^{14}5d^{10}6s^{2}6p^{2}$

Pb^{2+}: (Xe) $4f^{14}5d^{10}6s^{2}6p^{0}$ or (Xe) $4f^{14}5d^{10}6s^{2}$

Pb^{4+}: (Xe) $4f^{14}5d^{10}6s^{0}6p^{0}$ or (Xe) $4f^{14}5d^{10}$

Your Turn 8.3

Write the electron configurations of Cr^{3+} and Sn^{4+}. Write your answer in the space provided.

Covalent Bonding

Section 8.3

A **covalent bond** is formed between two atoms that share one or more pairs of electrons.

Lewis structures for covalent molecules often show shared pairs of electrons (bonding pairs) as lines and unshared electron pairs (nonbonding pairs) as lines or dots. For example, the Lewis structure for carbon dioxide is represented as:

$$\ddot{\underset{\cdot\cdot}{O}} = C = \ddot{\underset{\cdot\cdot}{O}} \quad \text{or} \quad \overline{\underline{O}} = C = \overline{\underline{O}}$$

Section 8.4

Bond Polarity and Electronegativity

A **nonpolar covalent bond** is one in which the electrons are shared equally between two atoms.

A **polar covalent bond** is a bond in which one atom attracts the pair of electrons more strongly than the other atom.

Electronegativity is the relative ability of an atom to attract a pair of electrons. Electronegativity generally increases from left to right along a period and from bottom to top within a group as shown in Figure 8.2. The trend in increasing electronegativity parallels increasing effective nuclear charge from left to right along a period. Increasing Z_{eff} increases an atom's tendency to attract electrons. Within a group, smaller atoms attract electrons better because force of attractions increases with decreasing distance from the nucleus.

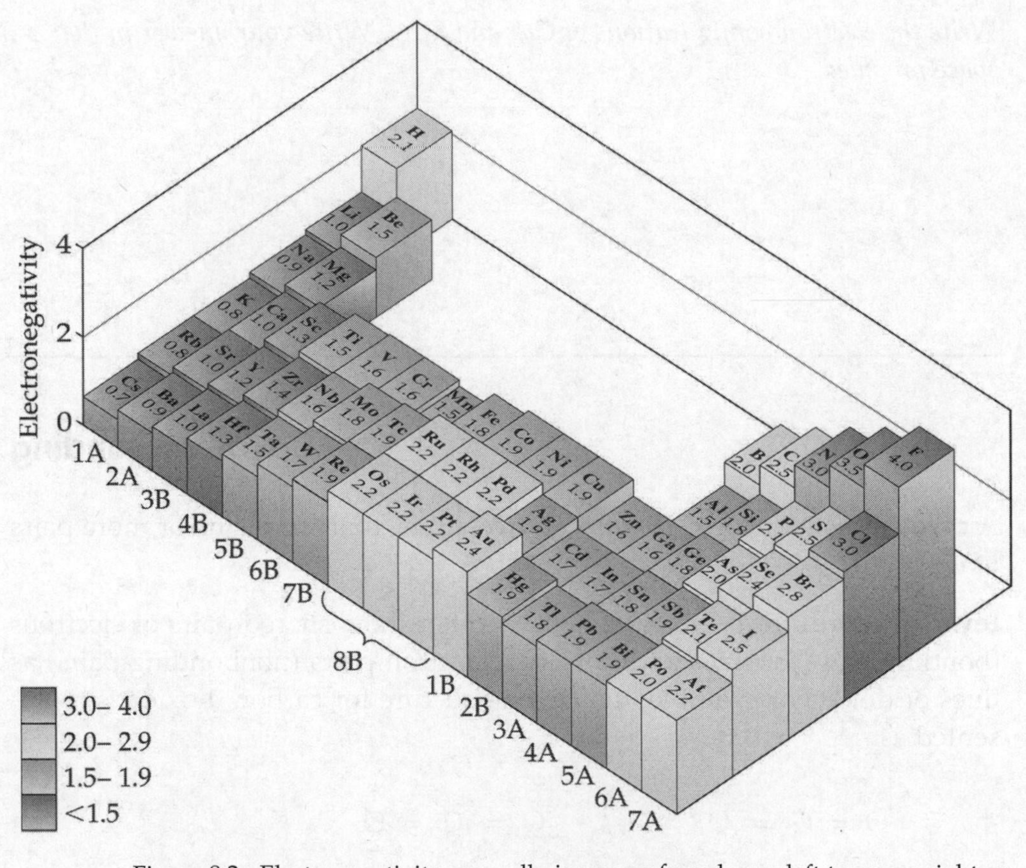

Figure 8.2. Electronegativity generally increases from lower left to upper right on the periodic table.

Common misconception: Noble gases have relatively high electronegativities. But becasue they don't make many covalent bonds, their electronegativities are not usually considered.

Fluorine has the highest electronegativity of all the reactive atoms and francium has the lowest.

The electronegativity difference between two bonded atoms is used to estimate the polarity of a covalent bond. The greater the electronegativity difference between two atoms the more polar the bond. In general, if the

electronegativity difference is:	the bond is:	example	(difference)
0-0.4	nonpolar covalent	Cl-Cl	(0.0)
0.4-1.0	polar covalent	H-Cl	(0.9)
1.0-2.0	very polar covalent	H-F	(1.9)
>2.0	ionic	NaCl	(2.1)

A **dipole** is a molecule with one end having a slight positive charge and the other end having a slight negative charge due to differences in electronegativity. For example, HF is a dipole and is often represented as:

$$\overset{\delta+}{H} \rightarrow \overset{\delta-}{F}$$

where the Greek letter delta, δ, means "partial" and the arrow represents a polar bond and points toward the higher electron concentration, the more electronegative element.

A **dipole moment** is a quantitative measure of the magnitude of a dipole. The higher the dipole moment, the more polar is the molecule.

Drawing Lewis Structures

Section 8.5

Lewis structures use valence electrons to aid our understanding of the bonding in molecules and ions.

Rules for drawing Lewis Structures of Molecules and Ions

1. Total the valence electrons of all bonded atoms.

2. Use one pair of electrons to bond each outer atom to the central atom (usually the atom in least abundance).

3. Complete the octets around all of the outer atoms.

4. Place any remaining electrons on the central atom.

5. If there are not enough electrons to give the central atom an octet, make multiple bonds.

Common misconception: Hydrogen can never be a central atom because it needs only two electrons to attain a noble gas configuration. In oxyacids and oxyanions, hydrogen bonds to oxygen, not to the central atom.

Example:

Write the Lewis structure for the nitrate ion, NO_3^-.

Solution:

1. The groups numbers of nitrogen (5) and oxygen (6) tell us how many valence electrons each atom contributes to the Lewis structure. The negative charge means to add one extra electron. The total number of valence electrons is 5 + 6 + 6 + 6 + 1 = 24 electrons.

2. Nitrogen is the central atom:

```
      O
      |
  O - N - O
```

3. Each oxygen needs three more pairs to complete its octet:

```
      ..
    : O :
 ..   |   ..
: O - N - O :
  ..      ..
```

4. Nitrogen still does not have an octet and we have used all 24 electrons.

5. Make a double bond:

```
    : O :
 ..   ||  ..
: O - N - O :
  ..      ..
```

Resonance Structures

Section 8.6

Resonance structures are two or more Lewis structures that are equally good representations of the bonding in a molecule or ion. Resonance structures usually differ only in the positions of multiple bonds or single, unpaired electrons. For example, the nitrate ion is represented by three equally good Lewis structures differing only in the placement of the double bond:

```
      : O :                     ..                       ..
        ||                    : O :                    : O :
   ..        ..                 |        ..       ..     |      ..
: O – N – O :  ◄--►  : O = N – O :  ◄--►  : O – N = O :
   ..        ..         ..        ..        ..
```

Common misconception: Once atoms and nonbonding pairs of electrons are placed, it is not correct to move an atom, or a nonbonding pair, to represent a different resonance structure.

The **formal charge** of an atom is the number of valence electrons in an isolated atom minus the number of electrons assigned to the atom in the Lewis structure. Formal charge is used to determine which of a variety of Lewis structures is most suitable to represent bonding in a molecule of ion. In general, the most suitable Lewis structure is the one whose atoms bear formal charges closest to zero. For example, three resonance structures of carbon dioxide are possible and the formal charges are assigned:

a	b	c
O = C = O :	O—C≡O	O≡C— O :
0 0 0	-1 0 +1	+1 0 -1

To calculate the formal charge on an atom in a Lewis structure:

	O	= C =	O :
+ nonbonding electrons assigned to an atom	+4	0	+4
+ ½ bonding electrons assigned to an atom	+4/2	+8/2	+4/2
- valence electrons in isolated atom	- 6	-4	-6
= formal charge	0	0	0

Lewis structure **a** for carbon dioxide is the most suitable representation of bonding because its atoms all bear zero formal charge.

Section 8.7

Exceptions to the Octet Rule

There are many common exceptions to the octet rule for writing Lewis structures. These exceptions are summarized as follows:

1. When the species contains an odd number of electrons, one atom will have only seven electrons around it.

   Examples: NO, NO_2.

2. When the species contains a central atom from Groups 2 or 3, the number of electrons around the central atom will be twice the group number.

   Examples: BeI_2, BCl_3, GaF_3.

3. When a central atom has more than four other atoms bonded to it, it will have more than eight electrons around it. This "expanded octet" usually occurs with larger nonmetal central atoms in Period 3 and beyond because of the availability of empty "d" orbitals for bonding. Also, the larger the central atom, the larger is the number of atoms that can surround it. The size of the surrounding atoms is also important. Expanded octets occur most often when large central atoms bond to small, electronegative atoms like F, Cl and O.

   Examples: PCl_5, SF_6, AsF_6^-.

4. An expanded octet may also occur for many species where: (total valence electrons of isolated central atom) ÷ (number of atoms attached to the central atom) > 8.

   Examples: SF_4, BrI_3, I_3^-, XeF_4, XeF_2, ICl_4^-

   (Note: Correct octet structures exist for species like PCl_3, AsF_3 SO_2 and SCl_2.)

Your Turn 8.4

PCl_5 and SF_6 exist but their analogs NCl_5 and OF_6 do not. Explain. Write your answer in the space provided.

Strengths of Covalent Bonds

Section 8.8

Bond enthalpy or bond dissociation energy is the enthalpy change, ΔH_{BDE}, for the breaking of bonds in one mole of a gaseous substance. Table 8.3 shows the average bond enthalpies of various bonds.

Common misconception: Keep in mind that bond enthalpies are always positive values because they represent breaking of bonds. Bond breaking is always endothermic and bond making is always exothermic.

Table 8.3 Average Bond Enthalpies (kJ/mol)

Single Bonds							
C—H	413	N—H	391	O—H	463	F—F	155
C—C	348	N—N	163	O—O	146		
C—N	293	N—O	201	O—F	190	Cl—F	253
C—O	358	N—F	272	O—Cl	203	Cl—Cl	242
C—F	485	N—Cl	200	O—I	234		
C—Cl	328	N—Br	243			Br—F	237
C—Br	276			S—H	339	Br—Cl	218
C—I	240	H—H	436	S—F	327	Br—Br	193
C—S	259	H—F	567	S—Cl	253		
		H—Cl	431	S—Br	218	I—Cl	208
Si—H	323	H—Br	366	S—S	266	I—Br	175
Si—Si	226	H—I	299			I—I	151
Si—C	301						
Si—O	368						
Si—Cl	464						
Multiple Bonds							
C=C	614	N=N	418	O_2	495		
C≡C	839	N≡N	941				
C=N	615	N=O	607	S=O	523		
C≡N	891			S=S	418		
C=O	799						
C≡O	1072						

Common misconception: The bond enthalpy of a multiple bond is not an integer multiple of the bond enthalpy of a single bond. For example, the bond enthalpy of a C=C double bond is not double that of a C—C single bond. Nor is the bond enthalpy of a C≡C triple bond three times that of a C—C single bond. Each type of bond has its own enthalpy value.

The enthalpy change of any reaction, ΔH_{rxn}, is estimated as the sum of the bond enthalpies of the broken bonds minus the sum of the bond enthalpies of the bonds formed:

$$\Delta H_{rxn} = \Sigma(\text{bond enthalpies of broken bonds}) - \Sigma(\text{bond enthalpies of bonds formed})$$

Example:

Use average bond enthalpies to estimate the enthalpy change of the following reaction:

$2\,H_2O \rightarrow 2\,H_2 + O_2$

Solution:

Draw Lewis structures of all the reactants and products, identify all the bonds that are broken and formed and apply the bond enthalpies from Table 8.3 to the summation equation.

$2\,H_2O \rightarrow 2\,H_2 + O_2 = H\text{-}O\text{-}H + H\text{-}O\text{-}H \rightarrow H\text{-}H + H\text{-}H + O{=}O$

The following bond enthalpies apply: ***H-O*** *= +463 KJ/mol,* ***H-H*** *= +436 KJ/mol, and of* ***O=O*** *= +495 KJ/mol.*

$\Delta H_{rxn} = 4(+463) - 2(436) - 495 = +485$ *KJ/mol*

Multiple bonds are shorter and stronger than single bonds. Table 8.4 shows that bond length decreases from single to double to triple bonds involving the same atoms. The more electrons involved in bonding, the more closely and tightly the atoms are held together.

Table 8.4 Average Bond Lengths for Some Single, Double, and Triple Bonds

Bond	Bond Length (Å)	Bond	Bond Length (Å)
C—C	1.54	N—N	1.47
C=C	1.34	N=N	1.24
C≡C	1.20	N≡N	1.10
C—N	1.43	N—O	1.36
C=N	1.38	N=O	1.22
C≡N	1.16		
		O—O	1.48
C—O	1.43	O=O	1.21
C=O	1.23		
C≡O	1.13		

Multiple Choice Questions

1. *Which pair of atoms should form the most polar bond?*

A) F and B

B) C and O

C) F and O

D) N and F

E) B and N

2. *Which pair of ions should form the ionic lattice with the highest energy?*

A) Na^+ and Br^-

B) Li^+ and F^-

C) Cs^+ and F^-

D) Li^+ and O^{2-}

E) K^+ and F^-

3. *The electrostatic force of attraction is greatest in which compound?*

A) BaO

B) MgO

C) CaS

D) MgS

E) CaO

4. *Which molecule has the weakest bond?*

A) CO

B) O_2

C) NO

D) N_2

E) Cl_2

5. *For which species are octet resonance structures necessary and sufficient to describe the bonding satisfactorily?*

A) BCl_3

B) SO_2

C) CO_2

D) BeF_2

E) SO_4^{2-}

6. How are the bonding pairs arranged in the best Lewis structure for ozone?

A) O—O—O

B) O=O—O

C) O≡O—O

D) O≡O=O

E) O=O=O

7. Which species has the shortest bond length?

A) CN^-

B) O_2

C) SO_2

D) SO_3

E) CO_2

8. Which species requires the least amount of energy to remove an electron from the outermost energy level?

A) Na^+

B) He

C) F^-

D) O^{2-}

E) Mg^{2+}

9. Which species has a valid non-octet Lewis structure?

A) $GeCl_4$

B) SiF_4

C) NH_4^+

D) $SeCl_4$

E) CO_3^{2-}

10. The Lewis structure for SeS_2 has a total of

A) 2 bonding pairs and 7 nonbonding pairs.

B) 2 bonding pairs and 6 nonbonding pairs.

C) 3 bonding pairs and 6 nonbonding pairs.

D) 4 bonding pairs and 5 nonbonding pairs.

E) 5 bonding pairs and 4 nonbonding pairs.

Free Response Questions

1. *Consider the following chemical species: the nitrogen molecule, the nitrite ion and the nitrate ion.*

a. *Write the chemical formulas for each of the species and identify the oxidation number of the nitrogen atom in each formula.*

b. *Draw Lewis structures for each of the species. Where appropriate, draw resonance structures for each.*

c. *List the chemical species in order of increasing N-O bond length, the formula with the shortest bond first. Justify your answer.*

d. *Write a balanced net ionic equation for the reaction of nitrogen dioxide with water. Comment on the molecular and/or ionic species that are formed.*

e. *Draw each of the resonance structures for nitrogen monoxide and assign formal charges to each atom of each structure.*

2. *Carbon dioxide gas is bubbled into water.*

a. *Write and balance a chemical equation to describe the process.*

b. *Draw the Lewis structures of the reactants and products. Include any valid resonance structures.*

c. *Given the following bond enthalpies, estimate the enthalpy of the reaction.*

Bond	*Bond enthalpy (kJ/mol)*
H—H	*436*
H—O	*463*
O—O	*146*
C—O	*358*
C═O	*799*
C≡O	*1072*

d. *Are the C—O bonds in the reactant stronger or weaker than those in the product? Explain. Is your explanation consistent with the sign of the enthalpy change you estimated? Explain.*

e. Excess aqueous sodium hydroxide is added to the solution. Write and balance a net ionic equation for the resulting reaction.

Additional Practice in Chemistry the Central Science

For more practice answering questions in preparation for the Advanced Placement examination try these problems in Chapter 8 of Chemistry the Central Science:

Additional Exercises: 8.74, 8.75, 8.80, 8.81, 8.83, 8.84, 8.86, 8.87, 8.88, 8.90.

Integrative Exercises: 8.94, 8.97, 8.99, 8.100, 8.101, 8.102, 8.106.

Multiple Choice Answers and Explanations

1. *A. Bond polarity increases with an increasing difference in electronegativity of the bonded atoms. Generally, electronegativity increases from left to right along a period and from bottom to top within a group. Electronegativity differences are generally largest for those elements that are farther apart on the periodic table.*

2. *D. Lattice energy is largest for ions of small size and large charge. Charge generally has a greater effect on lattice energy than does size. Lithium ion is the smallest cation listed and oxide has the largest negative charge.*

3. *B. A strong electrostatic force of attraction between two ions is dependent on high charge and small size. All of the responses are compounds having 2+ cations and 2- anions. The size of cations and anions increase top to bottom within a group. Magnesium ion is the smallest cation represented and oxide is the smallest anion.*

4. *E. The strength of covalent bonds increases with increasing multiplicity and decreasing length. Triple bonds are stronger than double bonds, which are stronger than single bonds. Lewis structures show that CO and N_2 each have a triple bond, O_2 and NO have double bonds and Cl_2 has a single bond.*

5. *B. Valid resonance structures for SO_2 are* |O═S—Ö| *and* |Ö—S═O|. *The Lewis structure for CO_2 is* |O═C═Ö| *where the formal charge of each atom is zero. Two possible resonance forms can be eliminated because the formal charges on the oxygens are +1 and -1:* |O≡C—Ö| *and* |Ö—C≡O|.

6. *B. The Lewis structure for ozone is* |O=O-Ö|.

7. *A. Triple bonds are shorter and stronger than double bonds which are shorter and stronger than single bonds. The Lewis structure for CN^- has a triple bond. Double bonds exist in O_2, SO_2, CO_2 and SO_3.*

8. *D. All five are isoelectronic. Each of the species is shielded by two inner core electrons. Oxide ion has the fewest number of protons and the smallest effective nuclear charge.*

9. *D. The Lewis structure of each of the species includes four single bonds to the central atom. In $SeCl_4$ the central atom has an additional nonbonding pair of electrons. Because of their small size and the lack of availability of "d" orbitals for bonding, central atoms from Period 2 cannot form expanded octets. Atoms from Period 3 and beyond can form expanded octets especially if they bond to small highly electronegative outer atoms like F and O.*

10. *C. The Lewis structure for SeS_2 is* $|\underline{S}{=}\underline{Se}{-}\overline{\underline{S}}|$.

Free Response Answers

1a. *nitrogen = N_2, O.N. = 0; nitrite ion = NO_2^-, O.N. = 4+; nitrate ion = NO_3^-, O.N. = 5+.*

b.

```
:N≡N:

  ..   ..           ..    ..
: O − N = O :  ←→  : O = N − O :
  ..   ..                 ..    ..

     : O :             : O :             : O :
  ..  ||  ..             |    ..     ..    |    ..
: O − N − O :  ←→  : O = N − O :  ←→  : O − N = O :
  ..      ..        ..      ..     ..
```

c.

$N_2 < NO_2^- < NO_3^-$

Multiple bonds are shorter and stronger than single bonds. The Lewis structure for N_2 shows that it has a triple bond. The resonance structures for nitrite show that each N-O bond has some double bond character. The resonance structures for nitrate show that each N-O bond has less double bond character than does nitrite.

d. $2\ NO_2(g) + H_2O(l) \dashrightarrow H^+(aq) + NO_3^-(aq) + HNO_2(aq)$
Nitric acid is a strong electrolyte so it is written in ionic form.
Nitrous acid is a weak electrolyte so it is written in molecular form.

e.

$$\overset{..}{:N} = \overset{..}{O}\cdot \longleftrightarrow \cdot\overset{..}{N} = \overset{..}{O}:$$
$$+1 \quad -1 \qquad\qquad 0 \quad 0$$

Formal charge of nitrogen = 4 nonbonding electrons + 2 bonding electrons – 5 valence electrons = 1+.

Formal charge of oxygen = 3 nonbonding electrons + 2 bonding electrons – 6 valence electrons = 1-.

Formal charge of nitrogen = 3 nonbonding electrons + 2 bonding electrons – 5 valence electrons = 0.

Formal charge of oxygen = 4 nonbonding electrons + 2 bonding electrons – 6 valence electrons = 0.

2a. $CO_2(g) + H_2O(l) \dashrightarrow H_2CO_3(aq)$ *Carbonic acid is a weak acid so it is written in molecular form.)*

b.

$$H-\ddot{\underset{..}{O}}-H$$

$$:O=C=O:$$

$$H-\underset{..}{\ddot{O}}-\overset{\overset{:O:}{||}}{C}-\underset{..}{\ddot{O}}-H$$

c. *Two H-O bonds are broken but two more H-O bonds are formed for a net gain or loss of energy of about zero. One C=O bond is broken and two C-O bonds are formed.*
$\Delta H_{rxn} = +799 - (2)(+358) = +83$ *kJ/mol*

d. *The C-O bonds in CO_2 are stronger than they are in H_2CO_3. Multiple bonds are stronger and shorter than single bonds. The Lewis structures show that both C-O bonds are double bonds in CO_2, whereas in H_2CO_3, the resonance structures show then the C-O bonds have more single bond character. Yes, the enthalpy change is* *positive indicating that the reaction is endothermic. The bond energy of one C═O bond is greater than than the total bond energies of two C—O bonds which means that it requires more energy to break the bond in the reactant than is delivered by the bonds forming in the product.*

e. $H_2CO_3(aq) + 2OH^-(aq) \dashrightarrow 2H_2O(l) + CO_3^{2-}(aq)$

Your Turn Answers

8.1. *KBr, Na_3PO_4 and $CaCO_3$ are ionic compounds. They contain metals and non-metals.*

SO_2, H_2SO_4 and CH_3COOH are covalent compounds. They contain only nonmetals.

Na_3PO_4 and $CaCO_3$ have both ionic and covalent bonds. Phosphate and carbonate ions form ionic bonds with their cations but also have covalent bonds within each ion.

8.2. *N^{3-}, O^{2-}, F^-, Ne, Na^+, Mg^{2+} Al^{3+} all contain ten electrons.*

Se^2, Br, Kr, K^+, Ca^{2+}, Sc^{3+} all contain 18 electrons.

8.3. *Cr^{3+}: (Ar) $3d^3 4s^0$ or (Ar) $3d^3$; Sn^{4+}: (Kr) $4d^{10} 5s^0 5p^0$ or (Kr) $4d^{10}$*

8.4. *P and S can form expanded octets because they are much larger than N and O. Additionally, besides having valence s and p orbitals available for bonding, P and S have empty 3d orbitals that can be used for bonding.*

TOPIC

9

MOLECULAR GEOMETRY AND BONDING THEORY

Students should be able to apply the VSEPR theory to both octet and non-octet Lewis structures of ions and molecules, explain molecular shape and bond angles in terms of both bonding and non-bonding pairs of electrons, determine polarity of molecules and orbital hybridization from their geometries and distinguish between sigma and pi bonds. Students should understand the difference between electron domain geometry and molecular geometry. Pay particular attention to the content in these sections:

9.2 **The VSEPR Model**

9.3 **Molecular Shape and Bond Polarity**

9.5 **Hybrid Orbitals**

9.6 **Multiple Bonds**

The VSEPR Model

Section 9.2

The **valence-shell electron-pair repulsion** (VSEPR) model is a way to use Lewis structures to determine the geometries of molecules. It's based on the natural repulsive forces that electron pairs exhibit within a molecule.

The **geometry** of a molecule refers to the arrangement of its atoms in three-dimensional space.

A **bond angle** is an angle made by the lines joining the nuclei of atoms in a molecule.

An **electron domain** is a region around an atom in which electrons will most likely be found. An electron domain is produced by a nonbonding pair, a single bond, a double bond, or a triple bond.

An **electron domain** geometry is the three-dimensional arrangement of electron domains around the central atom of a molecule.

A **molecular geometry** is the arrangement of only the atoms in the molecule.

Tables 9.1 and 9.2 show the common electron domain and molecular geometries of atoms.

Table 9.1 Electron domain geometries and molecular shapes for molecules with two, three, and four electron domains around the central atom.

Number of Electron Domains	Electron-Domain Geometry	Bonding Domains	Nonbonding Domains	Molecular Geometry	Example
2	Linear	2	0	Linear	CO_2
3	Trigonal planar	3	0	Trigonal planar	BF_3
		2	1	Bent	NO_2^-
4	Tetrahedral	4	0	Tetrahedral	CH_4
		3	1	Trigonal pyramidal	NH_3
		2	2	Bent	H_2O

Table 9.2 Electron domain geometries and molecular shapes for molecules with five and six electron domains around the central atom.

Total Electron Domains	Electron-Domain Geometry	Bonding Domains	Nonbonding Domains	Molecular Geometry	Example
5	Trigonal bipyramidal	5	0	Trigonal bipyramidal	PCl_5
		4	1	Seesaw	SF_4
		3	2	T-shaped	ClF_3
		2	3	Linear	XeF_2
6	Octahedral	6	0	Octahedral	SF_6
		5	1	Square pyramidal	BrF_5
		4	2	Square planar	XeF_4

Rules for determining geometry from VSEPR theory:

1. Draw the Lewis structure.

2. Count the number of electron domains around the central atom. Count each nonbonding pair, single bond, double bond and triple bond as one domain.

3. Arrange the domains of electrons as far away in 3-D space as possible.

Examples are given in Table 9.3.

Table 9.3. Examples of geometries obtained from the VSEPR model. Notice that the existence of nonbonding electrons affect the name of the molecular geometry.

Electron domains, non-bonding pairs	**Electron domain geometry**	**Molecular geometry**	**Bond angles**	**Orbital hybridiz-ation**	**Examples**
2, 0	Linear	Linear	180°	sp	BeI_2, CO_2, HCN, OCN^-
3, 0	Trigonal planar	Trigonal planar	120°	sp^2	BF_3, SO_3, PCl_3, NO_3,CO_3^{2-}
3, 1	Trigonal planar	Non-linear	$<120^{\circ}$	sp^2	SO_2, O_3, NO_2^-
4,0	Tetrahedral	Tetrahedral	109.5°	sp^3	CH_4, SO_4^{2-}, ClO_4^-, NH_4^+
4, 1	Tetrahedral	Trigonal pyramidal	$<109^{\circ}$	sp^3	NH_3, SO_3^{2-}, ClO_3^{2-}
4,2	Tetrahedral	Non-linear	$<109^{\circ}$	sp^3	H_2O, OF_2,
5,0	Trigonal bipyramidal	Trigonal bipyramidal	90° 120°	dsp^3	PCl_5
5,1	Trigonal bipyramidal	See-saw	$<90^{\circ}$ $<180^{\circ}$	dsp^3	SF_4
5,2	Trigonal bipyramidal	T-shape	$<90^{\circ}$ $<120^{\circ}$	dsp^3	ClF_3
5,3	Trigonal bipyramidal	Linear	180°	dsp^3	I_3^-, XeF_2
6,0	Octahedral	Octahedral	90°	d^2sp^3	SF_6
6,1	Octahedral	Square pymamidal	$<90^{\circ}$ $<180^{\circ}$	d^2sp^3	BrF_5
6,2	Octahedral	Square planar	90° 180°	d^2sp^3	XeF_4

Common misconception: The vital and subtle difference between an electron domain geometry and a molecular geometry is the presence of one or more nonbonding pairs of electrons around the central atom. When no nonbonding pair is present, the electron domain geometry has the same name as the molecular geometry. When one or more nonbonding pairs exist on the central atom, the name changes for the molecular geometry to describe only the locations of the atoms and not the nonbonding pair(s).

Common misconception: Except for the 109.5° angles in tetrahedral species, most bond angles need not be memorized. Most bond angles can be deduced from the electron domain geometries using basic geometrical principles.

Any nonbonding pairs on a five electron domain geometry occupy the radial (or equatorial) positions, not the axial positions as seen in Figure 9.1. Because nonbonding pairs have greater repulsion, the radial positions offer fewer 90° interactions with other domains. In the axial position, a nonbonding pair experiences three 90° interactions. By contrast, the equatorial position offers only two 90° interactions.

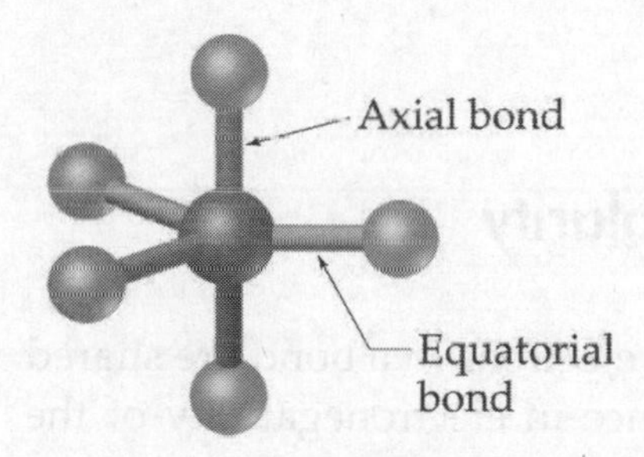

Figure 9.1. The trigonal bipyramidal geometry showing equatorial and axial bonds.

Your Turn 9.1

Examine Tables 9.1 and 9.2 and deduce the bond angles for the following electron domain geometries: linear, trigonal planar, trigonal bipyramidal, octahedral? Write your answers in the space provided.

Electron domains for non-bonding electron pairs and for double bonds exert greater repulsive forces on adjacent domains and thus tend to compress the bond angles. Table 9.4 lists several examples of molecules having slightly smaller bond angles than predicted by VSEPR.

Table 9.4. Molecules having smaller than normal bond angles due to nonbonding electron pairs or double bonds.

Molecule	Electron domain geometry	Molecular geometry	Actual bond angle
NH_3	Tetrahedral	Trigonal pyramid	107°
H_2O	Tetrahedral	Non-linear	104.5°
SF_4	Trigonal bipyramidal	See-saw	116°
$Cl_2C{=}O$	Trigonal planar	Trigonal planar	111.4° (Cl-C-Cl)
SO_2	Trigonal planar	Non-linear	116°

Section 9.3 Molecular Shape and Molecular Polarity

Bond polarity is a measure of how equally the electrons in a bond are shared between two atoms. The greater the difference in electronegativity of the bonded atoms, the more polar the bond.

Dipole moment is a quantitative measure of the charge separation in a molecule.

A **bond dipole** is the dipole moment for each bond in a molecule.

An **overall dipole moment** of a molecule is the vector sum of all of its bond dipoles. For a molecule that consists of more than two atoms, the dipole moment depends on both the polarities of the individual bonds and the geometry of the molecule. Figure 9.2 illustrates how bond polarity and geometry affects the polarity of a molecule.

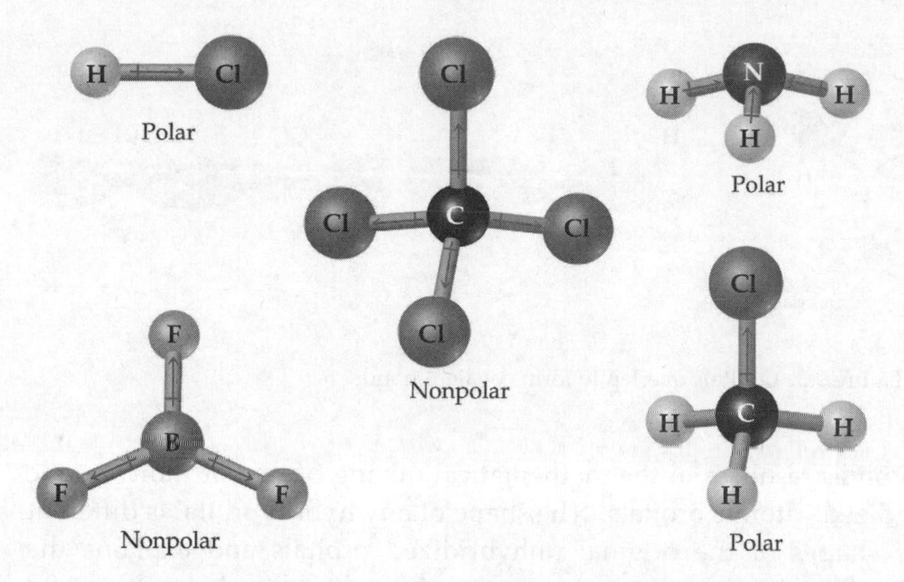

Figure 9.2. Geometry affects whether a molecule containing polar bonds will be polar.

Your Turn 9.2

Explain why the molecules shown in Figure 9.2 are polar or nonpolar. Write your answer in the space provided.

Hybrid Orbitals

Section 9.5

Valence-bond theory defines a covalent bond as an overlap of orbitals allowing two electrons of opposite spin to share a common region of space between the nuclei. Figure 9.3 illustrates bonding by orbital overlap in the molecules H_2, HCl and Cl_2.

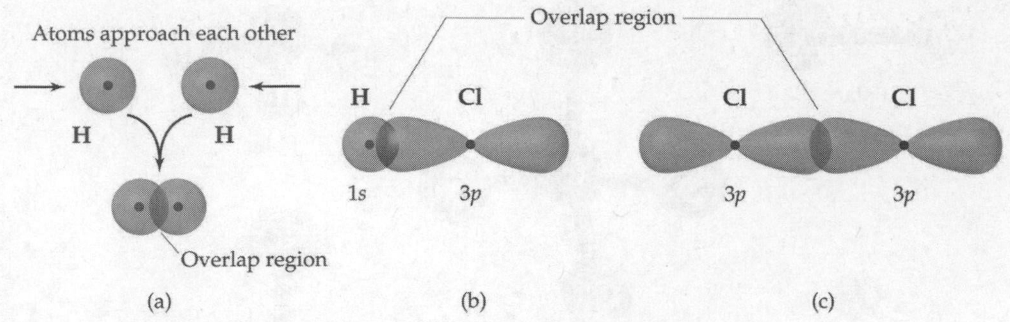

Figure 9.3. Orbitals overlap to form covalent bonds.

Hybrid orbitals result from the mathematical mixing of two or more atomic "unhybridized" atomic orbitals. The shape of any hybrid orbital is different from the shapes of the original unhybridized orbitals and explains the geometries of molecules. For example, the mixing of the valance 2s and 2p orbitals of carbon can result in sp^3, sp^2 or sp hybridized orbitals as shown in Figure 9.4.

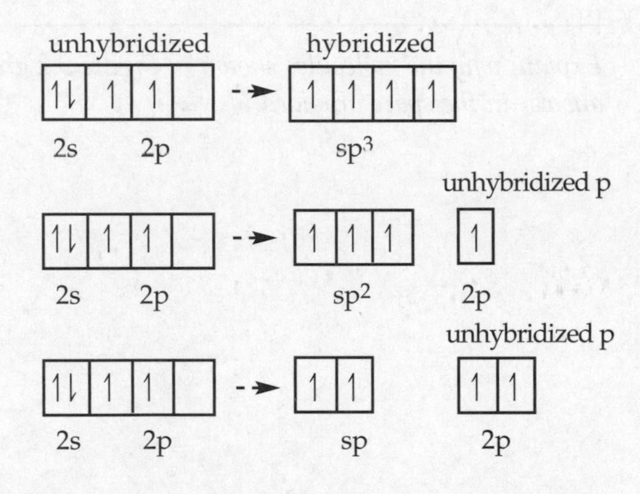

Figure 9.4. Mixing one s and three p orbitals results in an sp^3 hybrid. Mixing one s and two p orbitals results in an sp^2 hybrid. Mixing one s and one p orbital yields an sp hybrid.

Common misconception: Remember that the number of orbitals mixed always equals the number of hybrid orbitals produced. Left over orbitals that are "unhybridized" often take part in bonding, especially with double and triple bonds.

Your Turn 9.3

Use a diagram like that in Figure 9.4 to show the valence unhybridized and sp^3 hybridized orbitals of oxygen. Write your answer in the space provided.

Atomic Orbital Set	Hybrid Orbital Set	Geometry	Examples
s,p	Two *sp*	180° Linear	BeF_2, $HgCl_2$
s,p,p	Three sp^2	120° Trigonal planar	BF_3, SO_3
s,p,p,p	Four sp^3	109.5° Tetrahedral	CH_4, NH_3, H_2O, NH_4^+
s,p,p,p,d	Five sp^3d	90° 120° Trigonal bipyramidal	PF_5, SF_4, BrF_3
s,p,p,p,d,d	Six sp^3d^2	90° 90° Octahedral	SF_6, ClF_5, XeF_4, PF_6^-

Table 9.5 illustrates the various geometrical arrangements associated with hybrid orbitals.

Section 9.6

Multiple Bonds

A **sigma (σ) bond** is formed by the end-to-end overlap of orbitals. All single bonds are sigma bonds.

A **pi (π) bond** results from a side-to-side overlap of orbitals. A double bond consists of one sigma and one pi bond. A triple bond results from one sigma and two pi bonds.

Delocalized electrons are spread out over a number of atoms in a molecule rather than localized between a pair of atoms. Delocalized pi bonding electrons are characteristic of molecules that have resonance structures involving double bonds. For example, benzene, C_6H_6, is a ring of six carbons, each bonded to the next by a sigma bond. Delocalized pi bonds blend the two resonance structures into a single structure as shown in Figure 9.4.

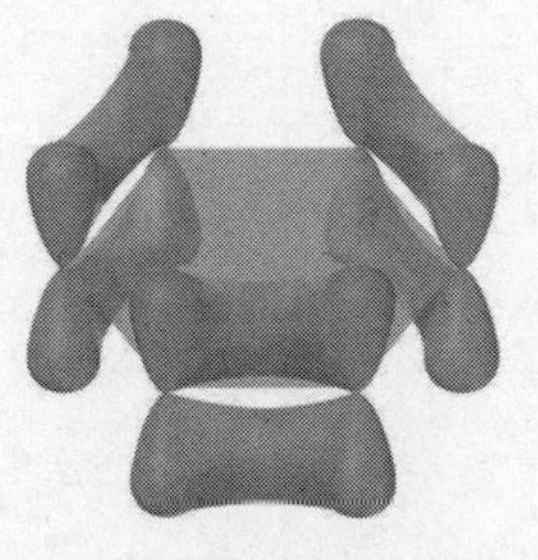

(a) Localized π bonds

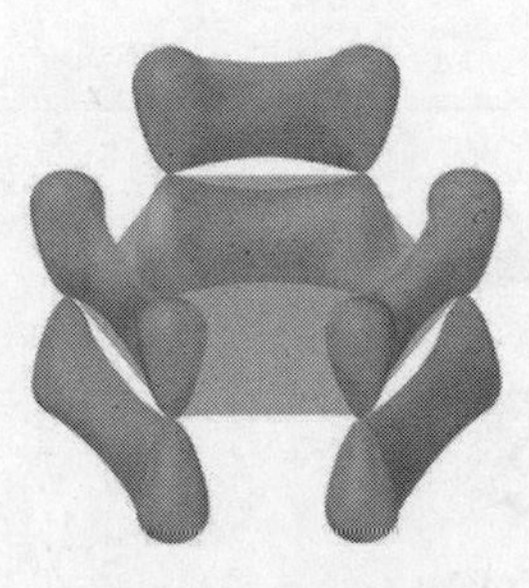

(b) Localized π bonds

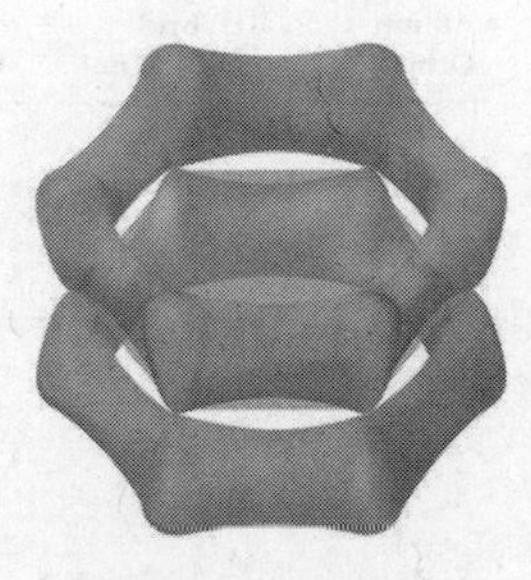

(c) Delocalized π bonds

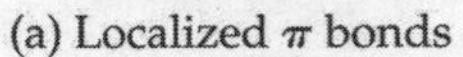

Figure 9.4. Two resonance structures of benzene (a and b) showing localized pi bonds differ from a more accurate representation (c) showing delocalized pi bonds.

Multiple Choice Questions

1. *Which of these molecules is not polar?*

A) H_2O

B) CO_2

C) NO_2

D) SO_2

E) NH_3

2. *Which species contains a central atom with sp^3 hybridization?*

A) C_2H_2

B) SO_3^{2-}

C) O_3

D) BrI_3

E) NH_3

3. *For ClF_3, the electron domain geometry of Cl and the molecular geometry are, respectively,*

A) *trigonal planar and trigonal planar.*

B) *trigonal planar and trigonal bipyramidal.*

C) *trigonal bipyramidal and trigonal planar.*

D) *trigonal bipyramidal and T-shaped.*

E) *trigonal planar and T-shaped.*

4. *The size of the H-N-H bond angles of the following species increases in which order?*

A) $NH_3 < NH_4^+ < NH_2^-$

B) $NH_3 < NH_2^- < NH_4^+$

C) $NH_2^- < NH_3 < NH_4^+$

D) $NH_2^- < NH_4^+ < NH_3$

E) $NH_4^+ < NH_3 < NH_2^-$

5. *What is the molecular geometry and polarity of the BF_3 molecule?*

A) *trigonal pyramidal and polar*

B) *trigonal pyramidal and nonpolar*

C) *trigonal planar and polar*

D) *trigonal planar and nonpolar*

E) *T-shaped and polar*

6. *In which species is the F-X-F bond angle the smallest?*

A) NF_3

B) BF_3

C) CF_4

D) BrF_3

E) OF_2

7. *Which set does not contain a linear species?*

A) CO_2, SO_2, NO_2

B) H_2O, HCN, BeI_2,

C) OCN^-, C_2H_2, OF_2

D) I_3^-, BrF_3, SCN^-

E) H_2S, ClO_2^-, NH_2^-

8. *The hybrid orbitals of nitrogen in* N_2O_4 *are*

A) *sp*

B) sp^2

C) sp^3

D) dsp^3

E) d^2sp^3

9. *How many sigma and how many pi bonds are in* $CH_2{=}CHCH_2C({=}O)CH_3$*?*

A) *5 sigma and 2 pi.*

B) *8 sigma and 4 pi.*

C) *11 sigma and 2 pi.*

D) *11 sigma and 4 pi.*

E) *13 sigma and 2 pi.*

10. *What is the best estimate of the H-O-H bond angle in H_3O^+?*

A) *109.5°*

B) *107°*

C) *104.5°*

D) *116°*

E) *120°*

Free Response Questions

1. *Consider the chemical species IF_5 and IF_4^+.*

a. *Draw the Lewis structure and make a rough three-dimensional sketch of each species.*

b. *Identify the orbital hybridization, the electron domain geometry and the molecular geometry of each.*

c. *Identify the approximate bond angles of each species.*

d. *Predict which, if any is a polar species. Justify your answer.*

e. *Predict the most probable oxidation number of the iodine atom in each species. Give an example of another chemical species having the same oxidation number as IF_4^+.*

f. *Would you expect that the conversion of IF_5 to IF_4^+ to be exothermic or endothermic? Explain.*

2. *Consider each of these molecules: C_3H_4, C_3H_6 and C_3H_8.*

a. *Draw the Lewis structure for each molecule and identify the orbital hybridization of each carbon atom.*

b. *Specify the geometry of each central carbon atom.*

c. *Write a balanced chemical for the complete combustion of each molecule.*

d. *Use the following bond enthalpies to determine the heat of combustion of each molecule. Specify your answer in kJ/mol.*

Bond	kJ/mol	bond	kJ/mol
C-H	413	O-H	463
C-C	348	O_2	495
C=C	614	C=O	799
C≡C	839	C-O	358

Multiple-Choice Answers and Explanations

1. *B. The bond dipoles in linear CO_2 cancel leaving the molecule with a zero dipole moment. H_2O, NO_2, and SO_2 are all nonlinear and have a net dipole moment. NH_3 is trigonal pyramidal and all three bond dipoles point toward nitrogen.*

2. *C. A trigonal planar electron domain geometry exhibited by O_3 is characteristic of the sp^2 hybrid set. C_2H_2 is linear and sp hybridized; SO_3^{2-} is tetrahedral and sp^3, BrI_3 is trigonal bipyramidal and dsp^3 and NH_3 is tetrahedral and sp^3.*

3. *D. ClF_3 has five pairs of electrons surrounding the central chlorine atom, three bonding pairs and two nonbonding pairs giving it a trigonal bipyramidal electron domain geometry. Nonbonding pairs orient themselves in the radial positions because the number of close interactions with other domains is less than in the axial positions. The result is a t-shaped molecular geometry.*

4. *C. All three species have a tetrahedral electron domain geometry. NH_4^+ has no nonbonding pairs and had a tetrahedral molecular geometry with bond angles of 109.5°. NH_3 has one nonbonding pair which repels more than the bonding pairs making the bond angle about 107°. NH_2^- has two nonbonding pairs which repel even more than one nonbonding pair making the bond angle about 105°.*

5. *D. BF_3 has a non-octet Lewis structure and VSEPR analysis shows it is a trigonal planar molecule having bond angles of 120°. The three B-F bond dipoles are oriented toward the corners of an equilateral triangle and cancel each other giving the molecule a zero dipole moment.*

6. *D. VSEPR analysis shows that BrF_3 is a t-shaped molecule having bond angles of about 90°. NF_3 is trigonal pyramidal having 107° bond angles; CF_4 is tetrahedral, 109.5°; OF_2 is non-linear (from tetrahedral), 104.5°.*

7. *E. Lewis structures and VSEPR show that the following are all linear: CO_2, HCN, OCN^-, I_3^-, and SCN^-. (Notice that CO_2, OCN^- and SCN^- are isoelectronic and therefore have the same Lewis structure and geometry.)*

8. *B. Each N is the center of a trigonal plane in the Lewis structure of N_2O_4 The sp^2 hybrid set is characteristic of trigonal planar geometry.*

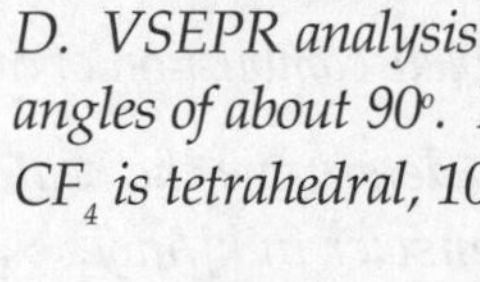

O O
N-N
O O

9. E. *Each single bond is a sigma bond and each double bond consists of one sigma bond and one pi bond. There are eight C-H sigma bonds, four C-C sigma bonds and one C-O sigma bond. There is one C-C pi bond and one C-O pi bond.*

10. B. *The electron domain geometry of the hydronium ion, H_3O^+, is tetrahedral. However, its Lewis structure shows that it one nonbonding pair of electrons making the molecular geometry trigonal pyramidal. The lone pair of electrons exerts a more repulsive force than the bonding pairs forcing a slight decrease in the bond angles. Hydronium ion is isoelectronic with ammonia, NH_3.*

Free Response Answers

1.

a.

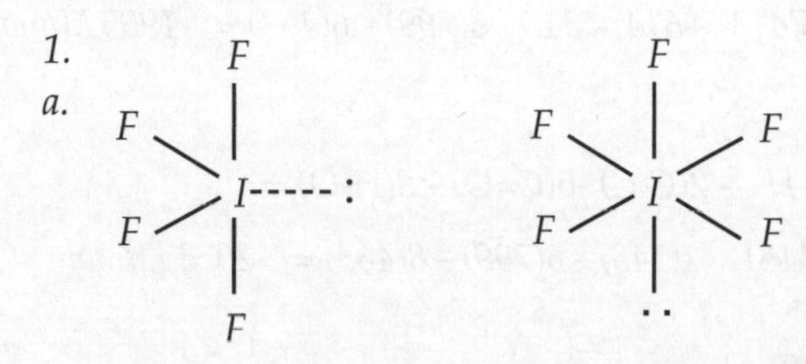

b. *IF_4^+ is dsp^3 hybridized, has a trigonal bipyramidal electron domain geometry and a see-saw molecular geometry.*
IF_5 is d^2sp^3 hybridized, has an octahedral electron domain geometry and a square pyramid molecular geometry.

c. *The radial F-I-F bond angle in IF_4^+ is about 120° or slightly less due to a larger repulsive force exerted by the radial nonbonding pair. The axial F-I-F bond angle is 180°.*

All the bond angles in IF_5 are 90°.

d. *Both species are polar. Each I-F bond is polar due to a large difference in electronegativity of the two atoms. The three dimensional arrangement of bond dipoles in both species contribute to a net molecular dipole.*

e. *The oxidation number of iodine is most probably 5+ in both species. Iodine in the iodate ion, IO_3^- also has an oxidation number of 5+.*

f. *The reaction: $IF_5 \rightarrow IF_4^+ + F^-$ would be endothermic The reaction requires that an I-F bond be broken. Bond breaking is always endothermic.*

2a. $CH_3C{\equiv}CH$ — *sp^3, sp, sp* $CH_3CH{=}CH_2$ — *sp^3, sp^2, sp* $CH_3CH_2CH_3$ — *all sp^3*

b. *All sp hybridized carbons are linear, all sp^2's are trigonal planar, all sp^3's are tetrahedral.*

c. $C_3H_4(g) + 4O_2(g) \rightarrow 3CO_2(g) + 2H_2O(l)$

$C_3H_6(g) + 9/2\ O_2(g) \rightarrow 3CO_2(g) + 3H_2O(l)$

$C_3H_8(g) + 5O_2(g) \rightarrow 3CO_2(g) + 4H_2O(l)$

d. for C_3H_4:

$\Delta H_{comb} = \sum \Delta H_{BDE}$ bonds broken - $\sum \Delta H_{BDE}$ bonds formed

$= 4(O_2) + 4(C\text{-}H) + 1(C{\equiv}C) + 1(C\text{-}C) - 6(C{=}O) - 4(H\text{-}O) =$

$+ 4(495) + 4(413) + 839 + 348 - 6(799) - 4(463) = -1872$ kJ/mol

for C_3H_6

$= 9/2(O_2) + 6(C\text{-}H) + 1(C{=}C) + 1(C\text{-}C) - 6(C{=}O) - 6(H\text{-}O) =$

$+ 4.5(495) + 6(413) + 614 + 348 - 6(799) - 6(463) = -1905$ kJ/mol

for C_3H_8

$= 5(O_2) + 8(C\text{-}H) + 2(C\text{-}C) - 6(C{=}O) - 8(H\text{-}O) =$

$+ 5(495) + 8(413) + 2(348) - 6(799) - 8(463) = -2023$ kJ/mol

Your Turn Answers

9.1. *Linear = 180°; trigonal planar = 120°; trigonal bipyramidal = 90° and 120°, octahedral = 90°.*

9.2. *The nonpolar molecules have bonds geometrically oriented so that their bond dipoles cancel. The bond dipoles of the polar molecules do not cancel.*

9.3. *Unhybridized O* → *sp³hybridized O*

1↓	1↓	1	1

2s 2p

→

1	1	1	1

sp³

TOPIC 10

GASES

The ideal gas equation is a way to calculate moles and is useful in solving stoichiometry problems involving gas densities and molar mass. Mastering partial pressures is essential in understanding the concepts of kinetics and equilibrium later in the course. Pay close attention to the effect of water vapor pressure on stoichiometry calculations. The kinetic molecular theory is a valuable tool in explaining both the qualitative and quantitative macroscopic behaviors of solids, liquids and gases in terms the interactions of atoms and molecules. The gas laws, including Graham's law of effusion are quantitative models that explain the behavior of gases. Pay close attention to the interpretation of Figures 10.23 and 10.24 for a qualitative understanding of the differences between ideal and real gases. Only a qualitative knowledge of the van der Waals equation is important. Your study should focus on the following sections:

10.3 The Gas Laws

10.4 The Ideal-Gas Equation

10.5 Further Applications of the Ideal Gas Equation

10.6 Gas Mixtures and Partial Pressures

10.7 Kinetic Molecular Theory

10.8 Molecular Effusion and Diffusion

10.9 Real Gases: Deviations from Ideal Behavior

Pressure

Section 10.2

Pressure is the force that acts on a given area: Pressure = force/area. Gases exert a pressure on any surface in which they come in contact.

Standard pressure corresponds to a typical pressure that the gases of the atmosphere exert at sea level. The SI value and unit for standard pressure are 1.01×10^5 Pascals.

Pressure units most commonly used in chemistry and their relationships are:
1 atm = 760 mm Hg = 760 torr = 1.01×10^5 Pa = 101 kPa

Standard temperature is 0°C = 273 K.

Common misconception: In all gas law calculations, temperature units must be expressed in Kelvin. K = 273 + °C.

Even though one Celsius degree is the same size at one Kelvin, the two units are 273° apart on their respective temperature scales. All the gas equations are proportional to Kelvin temperature but not Celsius temperature.

Section 10.3

The Gas Laws

Boyle's Law: the volume of a fixed quantity of gas at constant temperature is inversely proportional to the pressure.

$$P_1V_1 = P_2V_2$$

Figure 10.1 graphically illustrates how the volume of a gas depends on the pressure at constant temperature.

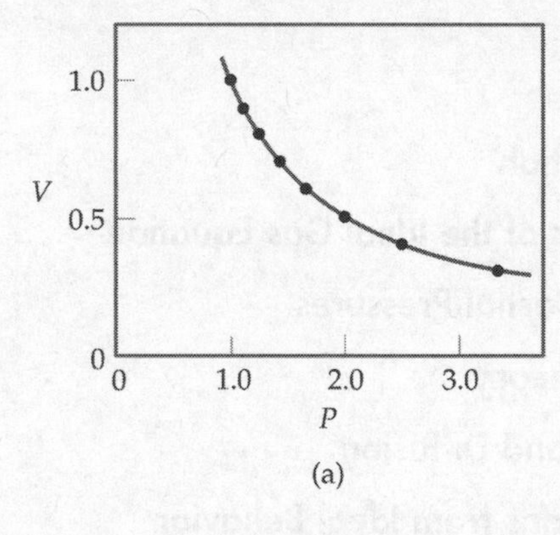

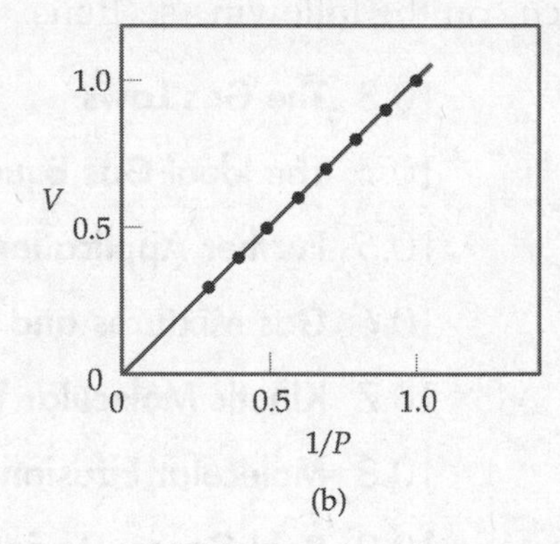

Figure 10.1. Boyle's law states that the volume of a gas decreases with increasing pressure. (a) volume vs. pressure, (b) volume vs. 1/P.

Your Turn 10.1

If the pressure of a fixed amount of gas doubles at constant temperature, what happens to its volume? Write your answer in the space provided.

Charles' Law: The volume of a fixed amount of gas at constant pressure is directly proportional to the absolute temperature.

$$V_1/T_1 = V_2/T_2$$

Figure 10.2 shows how volume decreases with decreasing temperature. Notice how extrapolation of the graph to a theoretical zero volume gives the value of absolute zero temperature (0 K or -273 °C).

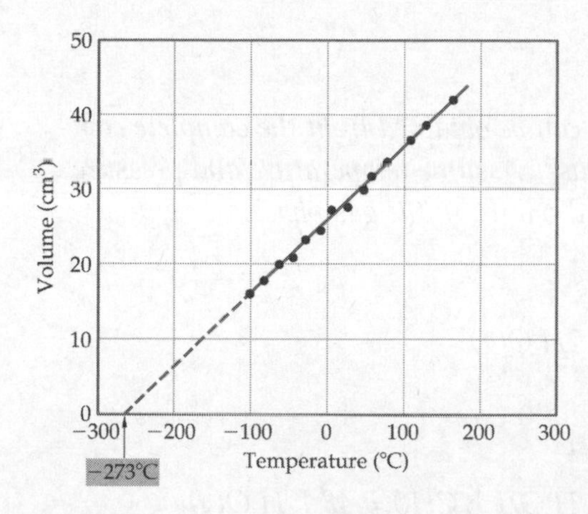

Figure 10.2. Charles' law states that the volume of an enclosed gas increases with increasing temperature. Extrapolation to zero volume yields the value of absolute zero.

Your Turn 10.2

When the temperature of a fixed amount of gas doubles from 20°C to 40°C, what happens to the volume at constant pressure? Write your answer in the space provided.

Avogadro's Hypothesis: Equal volumes of gases under the same conditions of temperature and pressure contain equal number of molecules.

Figure 10.3 shows that volumes of equal moles of gases at the same temperature and pressure is independent on the identity of the gas.

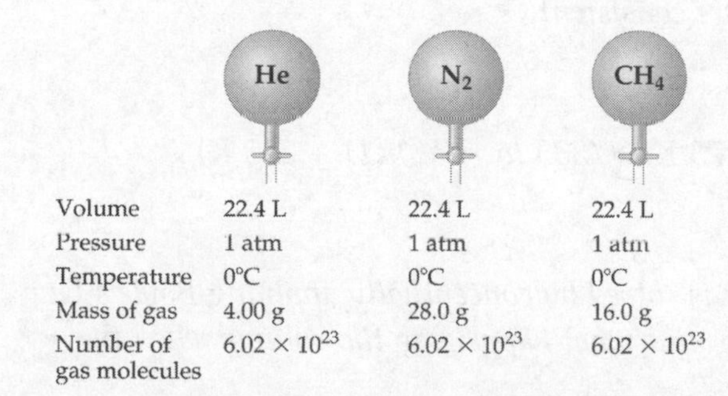

	He	N_2	CH_4
Volume	22.4 L	22.4 L	22.4 L
Pressure	1 atm	1 atm	1 atm
Temperature	0°C	0°C	0°C
Mass of gas	4.00 g	28.0 g	16.0 g
Number of gas molecules	6.02×10^{23}	6.02×10^{23}	6.02×10^{23}

Figure 10.3. Avogadro's hypothesis: Equal number of moles of any gas at the same temperature and pressure occupy equal volumes.

Avogadro's law: The volume of a gas at constant temperature and pressure is directly proportional to the number of moles of the gas.

$$V_1/n_1 = V_2/n_2$$

Since volume is directly proportional to moles, the coefficients that balance chemical equations for gas reactions can be taken as ratios of moles or liters at constant temperature and pressure.

Example:

How many liters of water vapor can be obtained from the complete combustion of 24 liters of methane gas? Assume temperature and pressure are such that all the water formed will be in the gas phase.

Solution:

$$CH_4(g) + 2O_2(g) \dashrightarrow CO_2(g) + 2H_2O(g)$$

$$24\ L \xrightarrow{X\ 2\ L\ /\ 1\ L} ?\ L$$

$$x \text{ liters } H_2O(g) = 24\ L\ CII_4\ (2\ L\ H_2O/1\ L\ CH_4) = 48\ L\ H_2O(g)$$

In what is sometimes called the **combined gas law**, volume, temperature and pressure all change for a fixed amount of gas according to the following equation:

$$P_1V_1/T_1 = P_2V_2/T_2$$

Example:

A weather balloon on the ground contains 25.8 L of He at 29°C and 741 torr. What is the volume of the balloon when it rises to an altitude where the temperature is 11.0°C and the pressure is 535 torr?

Solution:

This problem can be solved by substituting each value into the equation for the combined gas law and solving for V_2. Keep in mind that the temperature must be expressed in Kelvin, but the pressure can be in any suitable unit as long as the units are consistent.

$$P_1V_1/T_1 = P_2V_2/T_2$$

$$(741\ \text{torr})(25.8\ L)/(29 + 273\ K) = (535\ \text{torr})(V_2)/(11 + 273\ K)$$

$$V_2 = 33.6\ L$$

Alternatively, the problem is solved by conceptually applying Boyle's Law and Charles's Law to determine what happens to the volume when the temperature and pressure change. The pressure goes down so the volume goes up and the temperature goes down so the volume goes down. Multiply the original volume by a ratio of pressures greater than one and a ratio of temperatures less than one:

V_2 = *(25.8 L) x (741 torr)/(535 torr) x (11 + 273 K)/(29 + 273 K) = 33.6 L*

The Ideal Gas Equation — Section 10.4

The **ideal gas equation** combines Boyle's, Charles's and Avogadro's laws:

PV = nRT

P= pressure in atmospheres

V = volume in liters

n = amount of substance in moles

T = absolute temperature in Kelvin

R = 0.0821 L-atm/mol-K.

Common misconception: When using the ideal gas equation, pressure, volume, amount of substance, and temperature must always be expressed in atmospheres, liters, moles, and Kelvin, respectively, when the constant, R, has those units. In the other equations, volume and pressure may be expressed in any units as long as they are consistent.

An **ideal gas** is a hypothetical gas whose pressure, volume, and temperature behavior is completely described by the ideal gas equation.

Common misconception: No gas is an ideal gas. All gases are real gases. However, at temperatures at or above 25°C and pressures at or below one atmosphere, generally most real gases behave ideally.

Further Applications of the Ideal Gas Equation — Section 10.5

The ideal gas equation is useful in stoichiometry calculations involving gases.

Example:

Gasoline is a mixture of many hydrocarbon compounds but its chemical formula can be approximated as C_8H_{18}*. How many liters of carbon dioxide gas are formed at 25.0 °C and 712 torr when 1.00 gallon of liquid gasoline is burned in excess air? Liquid gasoline has a density of 0.690 g/mL. One gallon is 3.80 L.*

Solution:

First write and balance an equation. Then calculate the number of moles of CO_2 *using the mole road. Finally use the ideal gas equation to convert moles of* CO_2 *to volume in liters at the given temperature and pressure. Be sure to convert °C to K and torr to atmospheres.*

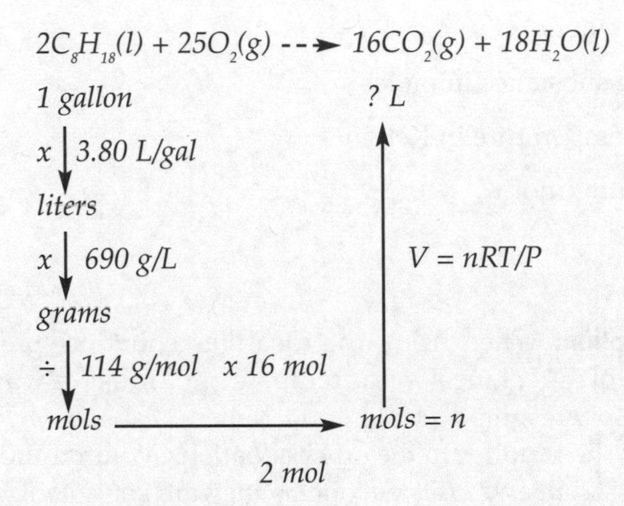

x *mol* CO_2 =

1.00 gal(3.80 L/gal)(690 g/gal) (1 mol/114 g)(16 mol CO_2*/2 mol* C_8H_{18}*) = 184 moles* CO_2*.*

$V = nRT/P =$

(184 mol)(0.0821 L-atm/mol-K)(25 + 273 K)/(712 torr/760 torr/atm) = 4810 L.

The density of a gas at any given temperature and pressure is directly proportional to its molar mass:

$$d = PM/RT$$

where d is density in grams per liter and M is the molar mass in grams per mole.

Example:

What is the density of sulfur dioxide gas at 35°C and 1270 torr?

Solution:

The molar mass of SO_2 is 64.0 g/mol. Use the equation for density. Make sure temperature is in Kelvin and pressure is in atmospheres.

d = PM/RT =

(1270 torr/760 torr/atm)(64.0 g/mol)/(0.0821 L-atm/mol K x (35 + 273K) = 4.23 g/L.

Common misconception: Because the densities of most gases are so low, they are commonly expressed in units of g/L rather than g/mL or g/cm³. At common temperatures and pressures, most gas densities fall between 1 and 10 g/L.

Gas Mixtures and Partial Pressures

Section 10.6

A **partial pressure** of a gas is the pressure exerted by an individual gas in a mixture of gases. Water vapor pressure is a partial pressure.

Dalton's Law of Partial Pressures states that the total pressure of a mixture of gases equals the sum of the pressures that each gas would exert if it were present alone.

The equation is:

$$P_{total} = P_1 + P_2 + P_3 + \ldots$$

where P_{total} is the total pressure of a mixture of gases and P_1, P_2, and P_3 are the **partial pressures** of the individual gases in the mixture.

The **mole fraction** of a gas in a mixture is the ratio of moles of that gas to the total moles of gas in the mixture. $X_1 = n_1/n_t$.

Each gas in a mixture of gases behaves independently of all the other gases in the mixture. Hence, each ideal gas in a mixture obeys the ideal gas law and all other gas laws. Each gas can be treated separately from the other gases in any mixture. Table 10.1 summarizes some equations that are useful in partial pressure calculations.

Table 10.1. Some Useful Equations Involving Partial Pressure

	Equation	**Comments**
Dalton's Law	$P_{total} = P_1 + P_2 + P_3 + \ldots$	P_{total} = the total pressure of the mixture, P_1, P_2 and P_3 = the partial pressures of the individual gases.
Mole fraction, X	$X_1 = n_1/n_t$	n_1 = number of moles of an individual gas. n_t = total moles of gas.
Ideal gas Equation	$P_1 = n_1RT/V$	P_1 = partial pressure of an individual gas. n_1 = the number of moles of that gas.
Partial pressure	$P_1 = (n_1/n_t)(P_t)$	A convenient way to calculate partial pressure from total pressure.

Example:

A mixture of 9.00 g of oxygen, 18.0 g of argon and 25.0 grams of carbon dioxide exert a pressure of 2.54 atm. What is the partial pressure of argon in the mixture?

Solution:

First calculate the number of moles of each gas. Then calculate the mole fraction of argon and multiply it by the total pressure to get the partial pressure of argon.

Moles of oxygen = 9.00g / 32.0 g/mol = 0.281 mol O_2.

Moles of argon = 18.0 g / 39.9 g/mol = 0.451 mol Ar.

Moles of carbon dioxide = 25.0 g / 44.0 g/mol = 0.568 mol CO_2.

Mole fraction of Ar = X_{Ar} *= moles of Ar/total moles of gases =* n_{Ar}/n_t *= 0.451/(0.281 + 0.451 + 0.568) = 0.451/1.30 = 0.347.*

Partial pressure of Ar = $P_{Ar} = (X_{Ar})(P_t)$ *= 0.347 x 2.54 atm = 0.881 atm.*

For convenience in the laboratory, gases are often collected by bubbling them through water. This process causes the collected gas to be "contaminated" with water vapor. That is, the collected gas and the water vapor both exert their own partial pressures. The total pressure inside the collection vessel (usually the atmospheric pressure) is equal to the combined partial pressures of the gas and the water vapor.

$P_{total} = P_{gas} + P_{water}$

Example:

Hydrogen is produced by the action of sulfuric acid on zinc metal and collected over water in a 255 mL container at 24.0°C and 718 torr. The vapor pressure of water at 24°C is 22.38 torr. How many moles of hydrogen gas are produced and how many grams of zinc react?

Solution:

First, write and balance an equation. Then subtract the vapor pressure of water (given in the question or obtained from Appendix B of Chemistry the Central Science) from the atmospheric pressure to obtain the partial pressure of hydrogen gas. Finally, calculate the number of moles of hydrogen using the ideal gas equation. Any time you use the value of R, be sure to convert pressure to atmospheres and volume to liters.

$Zn(s) + 2H^+(aq) \rightarrow Zn^{2+}(aq) + H_2(g)$

$P_{H2} = P_{atm} - P_{water} = 718\ torr - 22.38\ torr = 696\ torr$

$n_{H2} = PV/RT = (696\ torr/760\ torr/atm)(0.255\ L)/(0.0821\ L\text{-}atm/mol\text{-}K)/(24 + 273\ K) = 0.00958\ mol\ H_2$

The balanced equation tells us that the number of moles of hydrogen produced is equal to the number of moles of Zinc reacted.

$x\ g\ Zn = 0.00958\ mol\ Zn \times 65.4\ g/mol = 0.627\ g.$

Kinetic-Molecular Theory

Section 10.7

Kinetic-molecular theory is a model that explains the macroscopic behavior of gases at the atomic and molecular level. The following statements summarize the model for an ideal gas:

1. Gases consist of atoms or molecules in continuous, random motion.
2. The volume of the gas particles is negligible relative to the volume of their container.
3. Attractive and repulsive forces between gas particles are negligible.
4. Collisions between gas particles and between gas particles and their containers are perfectly elastic (they happen without loss of energy).
5. The average kinetic energy of particles is proportional to the absolute temperature.

Pressure of a gas is caused by the collisions of the gas particles with the walls of their container. The magnitude of the pressure is related to how often and how forcefully the particles strike the walls.

The absolute temperature of a gas is a measure of the average kinetic energy of the particles. Molecular motion increases with increasing temperature.

The kinetic-molecular theory explains the gas laws at the molecular level. For example, if the volume of a fixed amount of gas is increased at constant temperature, the container walls have a larger surface area. Thus, the pressure decreases because there are fewer collisions per unit time per unit area on the walls of the container.

Increasing temperature of a fixed amount of gas at constant pressure increases the average kinetic energy of the molecules. Thus the increased frequency and energy of collisions with the container have the effect of pushing back the walls of the container expanding the volume.

Your Turn 10.3

Use kinetic-molecular theory to explain why when the temperature of a fixed amount of gas increases, the pressure increases. Write your answer in the space provided.

Section 10.8

Molecular Effusion and Diffusion

Effusion is the escape of gas molecules through a tiny hole into an evacuated space.

Diffusion is the spread of one substance through space or through another substance.

Graham's law of effusion states that the effusion rate of a gas is inversely proportional to the square root of its molar mass. The equation for Graham's law relates the ratio of effusion rates of two gases to the ratio of the square root of their molar masses:

$$r_1/r_2 = (M_2/M_1)^{1/2}$$

Common misconception: The **rate** at which a particular gas effuses is inversely proportional to the **time** it takes that gas to effuse. Rate is often expressed in moles per unit time.

Example:

At a particular temperature and pressure, neon gas effuses at a rate of 16.0 mol/s.

a. What is the rate at which argon effuses under the same conditions?

b. Under a different set of conditions, 3.0 mol of argon effuse in 49.0 seconds. How long will it take an equal amount of helium to effuse?

Solution:

a. Substitute the given information into the equation for Graham's law.

$r_{Ar}/r_{Ne} = (M_{Ne}/M_{Ar})^{1/2}$

$r_{Ar}/16.0 = (20.2\ g/mol/39.9\ g/mol)^{1/2} = 11.4\ mol/s.$

b. Time is inversely proportional to rate so the equation becomes:

$T_{He}/T_{Ar} = (M_{He}/M_{Ar})^{1/2}$

$T_{He}/49.0\ s = (4.00\ g/mol/39.9\ g/mol)^{1/2} = 15.5\ s.$

Real Gases: Deviations from Ideal Behavior

Section 10.9

Real gases differ from ideal gases largely because the volumes of real gas particles are finite and their attractive forces and repulsive forces are non-zero.

A real gas does not behave ideally at high pressure because the finite volume of its particles is significant compared to the volume of its container. Also attractive forces come into play at short distances when gas particles are crowded together.

A real gas does not behave ideally at low temperature because its non-zero attractive forces are significant at low kinetic energy. Eventually as the temperature (average kinetic energy) drops, attractive forces cause the gas to condense to a liquid.

Deviations from ideal behavior for gases generally increase with increasing molecular complexity and increasing mass. Mass is directly related to molecular volume, and molecular complexity is associated with both volume and attractive forces.

The van der Waals equation predicts the behavior of real gases under conditions of low temperature and high pressure by employing correction factors for finite volume and attractive forces.

$$P = \underbrace{nRT/(V-nb)}_{\text{correction for volume}} - \underbrace{n^2a/V^2}_{\text{correction for attractive forces}}$$

The constants, a and b are correction factors for molecular attraction and volume, respectively, and n is the number of moles. Table 10.2 gives van der Waals correction factors for common gases.

Table 10.2 van der Waals Constants for Gas Molecules.

Substance	*a* (L^2-atm/mol^2)	*b* (L/mol)
He	0.0341	0.02370
Ne	0.211	0.0171
Ar	1.34	0.0322
Kr	2.32	0.0398
Xe	4.19	0.0510
H_2	0.244	0.0266
N_2	1.39	0.0391
O_2	1.36	0.0318
Cl_2	6.49	0.0562
H_2O	5.46	0.0305
CH_4	2.25	0.0428
CO_2	3.59	0.0427
CCl_4	20.4	0.1383

Your Turn 10.4

Which of the noble gases will deviate the most from ideal behavior? Explain your reasoning. Write your answer in the space provided.

Table 10.3. Summary of the Gas Laws and their Equations

Gas Law	Statement	What is constant	Equation
Boyle's Law	The volume of a fixed quantity of gas at constant temperature is inversely proportional to the pressure.	n, T	$P_1V_1 = P_2V_2$
Charles' Law	The volume of a fixed amount of gas at constant pressure is directly proportional to the absolute temperature.	n, P	$V_1/T_1 = V_2/T_2$
Un-named Law	The pressure of a fixed quantity of gas at constant volume is directly proportional to the absolute temperature.	n, V	$P_1/T_1 = P_2/T_2$
Combined Gas Law	Calculates the effect on temperature, pressure or volume when the other two are changed for a fixed amount of gas.	n	$P_1V_1/T_1 = P_2V_2/T_2$
Avogadro's Law	The volume of a gas at constant temperature and pressure is directly proportional to the number of moles.	T, P	$V_1/n_1 = V_2/n_2$
Ideal Gas Law	Combines the above laws into one equation	R=0.0821 L-atm/mol K	$PV = nRT$
Graham's Law of Effusion	The effusion rate of gases is inversely proportional to the square root of their molar masses.		$r_1/r_2 = (M_2/M_1)^{1/2}$
Dalton's Law of Partial Pressures	The total pressure of a mixture of gases equals the sum of the partial pressure of each component.		$Pt = P_1 + P_2 + P_3 \ldots$

Your Turn 10.4

Arrange the following gases in order of increasing deviation from ideality. Justify your answer. H_2O, CH_4, *Ne. Write your answer in the space provided.*

Multiple Choice Questions

1. *Hydrogen peroxide, H_2O_2, in the presence of a catalyst decomposes into water and oxygen gas. How many liters of O_2 at STP are produced from the decomposition of 34.0 g of H_2O_2?*

A) 1.00

B) 5.60

C) 11.2

D) 22.4

E) 44.8

2. *A mixture of 6.02×10^{23} molecules $NH_3(g)$ and 3.01×10^{23} molecules $H_2O(g)$ has a total pressure of 6.00 atm. What is the partial pressure of NH_3?*

A) 1.00 atm

B) 2.00 atm

C) 3.00 atm

D) 4.00 atm

E) 9.00 atm

3. *What will happen to the volume of a bubble of air submerged in water under a lake at 10.0°C and 2.00 atm if it rises to the surface where the temperature is 20.0°C and the pressure is 1.00 atm?*

A) *The volume will increase by a factor of 4.00.*

B) *The volume will increase by a factor of 2.07*

C) *The volume will increase by a factor of 2.00.*

D) *The volume will remain the same.*

E) *The volume will decrease by a factor of 1.93.*

4. *A gas has a density of 3.74 g L^{-1} at 0°C and a pressure of 1.00 atm. Which gas best fits these data?*

A) H_2

B) He

C) N_2

D) Kr

E) Rn

5. *Methanol, CH_3OH, burns in oxygen to form carbon dioxide and water. What volume of oxygen is required to burn 6.00 L of gaseous methanol measured at the same temperature and pressure?*

A) 4.00 L

B) 8.00 L

C) 9.00 L

D) 12.0 L

E) 24.0 L

6. *Which increases as a gas is heated at constant volume?*

I. *Pressure*

II. *Kinetic energy of molecules*

III. *Attractive forces between molecules*

A) *I only*

B) *II only*

C) *III only*

D) *I and II only*

E) *II and III only*

7. *At room temperature and 1 atm pressure the molecules are farthest apart in*

A) *fluorine*

B) *bromine*

C) *iodine*

D) *mercury*

E) *water*

8. *What is the name of the process?* $H_2O(s) \rightarrow H_2O(g)$

A) *condensation*

B) *evaporation*

C) *fusion*

D) *sublimation*

E) *freezing*

9. *Which statement is true of a measured pressure of a sample of hydrogen gas collected over water at constant temperature?*

A) *The measured pressure is greater than the pressure of dry hydrogen.*

B) *The measured pressure is less than the pressure of dry hydrogen.*

C) *The measured pressure is equal to the pressure of dry hydrogen.*

D) *The measured pressure varies inversely with the pressure of dry hydrogen.*

E) *The measured pressure is not related to the pressure of dry hydrogen.*

10. *A gas is most likely to behave as an ideal gas in which instance?*

A) *At low temperature, because the particles have insufficient kinetic energy to overcome intermolecular attractions.*

B) *When the molecules are highly polar, because intermolecular forces are more likely.*

C) *At room temperature and pressure, because intermolecular interactions are minimized and the particles are relatively far apart.*

D) *At high pressures, because the distance between molecules is likely to be small in relation to the size of the molecules.*

E) *At high temperatures, because the molecules are always far apart.*

Free Response Questions

1. *Equal masses (0.500 g each) of hydrogen and oxygen are placed in an evaculated 4.00 L flask at 25.0 °C. The mixture is allowed to react to completion and the flask is returned to 25.0 °C and allowed to come to equilibrium. The equilibrium vapor pressure of water at 25 °C is 23.76 torr.*

a. *Write and balance an equation for the reaction.*

b. *What is the total pressure inside the flask before the reaction begins?*

c. *What is the mass of water vapor in the flask at equilibrium?*

d. *How many grams of which reactant gas remains at equilibrium?*

e. *What is the total pressure inside the flask at equilibrium?*

f. *After the reaction, is there any liquid water present? If so, how many grams? If not, why not?*

2. *A 2.00 L flask at 27°C contains 3.00 grams each of Ar(g), SO_2(g), and He(g). Answer the following questions about the gases and in each case explain your reasoning.*

a. *Which gas has particles with the highest average kinetic energy?*

b. *Which gas has particles with the highest average velocity?*

c. *Which gas has the highest partial pressure?*

d. *Which gas will deviate the most from ideal behavior?*

e. *Which substance will have the highest boiling point?*

f. *What changes in temperature and pressure will increase the deviations of all the gases from ideal behavior?*

Additional Practice in Chemistry the Central Science

For more practice answering questions in preparation for the Advanced Placement examination try these problems in Chapter 10 of Chemistry the Central Science.

Additional Exercises: 10.92, 10.95, 10.96, 10.97, 10.98, 10.102.

Integrative Exercises: 10.105, 10.106, 10.110, 10.112, 10.113.

Multiple Choice Answers and Explanations

1. *C. The balanced equation is:* $H_2O_2 \rightarrow H_2O + 1/2\,O_2$
One mole of hydrogen peroxide produces one half mole of oxygen gas. One mole of H_2O_2 *is 34.0 g because the molar mass of hydrogen peroxide is 34.0 g/mol. If one mole of gas at STP occupies 22.4 liters then one half mole has a volume of 11.4 L.*

2. *D. Because there are twice as many molecules of ammonia than of water, there are twice as many moles of ammonia. The mole fraction of ammonia is 2/3. The partial pressure of ammonia is given by:* $P_{NH3} = X_{NH3} \times P_t = 2/3 \times 6.00\ atm = 4.00\ atm.$

3. *B. Halving the pressure doubles the volume. However, doubling the Celsius temperatures does not double the volume because the volume is proportional to the absolute temperature, not the Celsius temperature. The absolute temperature increase is modest increasing the volume by the factor (20 +273)/(10 + 273) which is a little more than one. The pressure decrease and the temperature increase couple to give a volume increase of just over a factor of 2, which makes 2.07 the most likely correct answer.*

4. *D. Even without a calculator the product of (3.74 g/L)(22.4 L/mol) can be estimated to be about 80 g/mol (4 x 20 = 80). The molar mass*

of krypton from the periodic table is 83.8 g/mol while the closest other noble gases are argon (39.9 g/mol) and xenon (131 g/mol).

5. *C. The balanced equation is: $2CH_3OH(g) + 3O_2(g) \rightarrow 2CO_2(g) + 4H_2O(g)$. The coefficients for gases that balance an equation are ratios of liters as well as moles at constant temperature and pressure. If two liters of methanol requires 3 liters of oxygen, then six liters of methanol requires nine liters of oxygen.*
   $x\ L\ O_2 = 8\ L\ CH_3OH\ (3\ L\ O_2/2\ L\ CH_3OH) = 9.00\ L$

6. *D Heating causes the temperature to increase, and temperature is a measure of kinetic energy. The faster moving molecules strike the walls of the container more frequently and with greater force increasing the pressure. Intermolecular forces are a function of the structures of the molecules present and not of the temperature or pressure.*

7. *A. Kinetic molecular theory says that liquids and solids are in close contact with each other because of the attractive forces between molecules. Gases have little or no attractive forces and the molecules are far apart. This question is really asking, "Which of the following is a gas?" Fluorine, at room temperature and 1 atmosphere, is the only gas on the list.*

8. *D. Sublimation is the process by which a solid changes directly to a gas.*

9. *A. Whenever a gas is collected over water, the gas is contaminated with water vapor. The measured pressure is equal to the pressure of the dry gas plus the partial pressure of water vapor at a given temperature: $P_{total} = P_{H2} + P_{H2O}$. The pressure of the hydrogen sample is greater than the pressure of the dry gas.*

10. *C. Gases behave ideally at low pressures because the molecular volumes are small compared to the volume of the container and at high temperature because the kinetic energy overcomes the intermolecular forces making them insignificant. Most gases behave as ideal gases at room temperature and pressure.*

Free Response Answers

1a $2H_2(g) + O_2(g) \rightarrow 2H_2O$

b. $mol\ H_2 = 0.500\ g / 2.00\ g/mol = 0.250\ mol\ H_2$
$mol\ O_2 = 0.500\ g / 32.0\ g/mol = 0.0156\ mol\ O_2$

$P_{total} = n_{total}\ RT/V =$
(0.250 + 0.0156 mol)(0.0821 L atm/mol K)(25 + 273 K)/(4.00 L) =
1.62 atm

c. *n = PV/RT =*
(23.76 torr / 760 torr/atm)(4.00L)/(0.0821 L atm/mol K)(298 K) =
0.00511 mol H_2O vapor
0.00511 mol x 18.0 g/mol = 0.0920 g H_2O vapor.

d. *O_2 is limiting because 0.250 mol of hydrogen would require 0.125 mol of oxygen to react and there are only 0.0156 mol of oxygen. Twice as many moles of hydrogen react:*
2 x 0.0156 mol = 0.0312 mol H_2 react.
0.250 – 0.0312 mol = 0.219 mol H_2 remain
0.219 mol x 2.00 g/mol = 0.438 g H_2 remain

e. $P_{H2} = (n_{H2})RT/V$
= (0.219 mol)(0.0821 L atm/mol K)(298 K)/4.00L) = 1.34 atm

$Pt = P_{H2O} + P_{H2}$
(23.76 torr/760 torr/atm) + 1.34 atm = 1.37 atm

f. *Yes, liquid water remains in the flask.*
0.0312 moles of hydrogen will produce 0.0312 moles water. There are only 0.00511 moles of water vapor present so the rest is liquid.
0.0312 mol – 0.00511 mol = 0.0261 mol water is liquid
0.0261 mol x 18.0 g/mol = 0.470 g liquid water.

2a. *All three gases have the same average kinetic energy because the temperature is the same for each and temperature is a measure of average kinetic energy.*

b. *The helium atoms will have the highest average velocity because they are the smallest. At the same temperature, the rms velocity of a gas is inversely proportional to the square root of the molar mass.*

c. *Helium has the highest partial pressure because partial pressure is proportional to the number of moles of gas present and the number of moles increases with decreasing molar mass.*

d. *Sulfur dioxide will display the greatest deviation from ideal behavior because SO_2 molecules are the largest and most complex of the three different kinds of gas particles. It is also polar and will have the highest attractive forces.*

e. *Sulfur dioxide will have the highest boiling point because its polarity provides higher attractive forces.*

f. *Decreasing the temperature and increasing the pressure will increase deviations from ideal behavior for all gases.*

Your Turn Answers

10.1. *The volume of a fixed amount of gas halves when its pressure is doubled at constant temperature.*

10.2. *The volume of a fixed amount of gas at constant pressure rises by a factor of (40 + 273)/(20 + 273) when the temperature increases from 20°C to 40°C.*

10.3. *Increasing temperature increases the average kinetic energy and therefore velocity of the molecules. Pressure increases because the faster moving molecules strike the walls of the container more often and with greater force.*

10.4. *Of the noble gases, radon will deviate most from ideal behavior because it is the heaviest and has the largest finite volume for its particles.*

10.5. *Expected deviation from ideal behavior:* $Ne < CH_4 < H_2O$
All have similar molar masses but neon is the simplest in complexity, followed by methane which has perfect tetrahedral geometry, and water which is non-linear. Water is also very polar and forms hydrogen bonds causing its attractive forces to be much greater than the two nonpolar species.

TOPIC 11

INTERMOLECULAR FORCES, LIQUIDS AND SOLIDS

Intermolecular forces, especially hydrogen bonding, explain many macroscopic behaviors of matter in terms of unseen atoms, ions and molecules including vapor pressures of liquids, the structure of ice and the melting and boiling points of substances. Know what kinds of intermolecular forces exist, how they affect the properties of molecules, and how to interpret phase diagrams. Focus especially on these sections:

11.2 **Intermolecular Forces**

11.5 **Vapor Pressure**

11.6 **Phase Diagrams**

Intermolecular Forces

Section 11.2

Intramolecular forces are the attractive forces **within** molecules that we call covalent bonds. Intramolecular forces give rise to many of the chemical properties of molecules.

Intermolecular forces are the forces that exist **between** molecules. They are largely responsible for the physical properties of solids and liquids.

The physical state of a substance depends largely on the balance between the kinetic energies of the particles and the attractive forces between the particles. For example, a gas condenses to a liquid at low temperature because the kinetic energy of the particles decreases to a point where the intermolecular attractive forces become significant. Figure 11.1 illustrates the molecular level differences between solids, liquids, and gases.

In general, the higher the intermolecular forces, the higher the melting points of solids and the higher the boiling points of liquids.

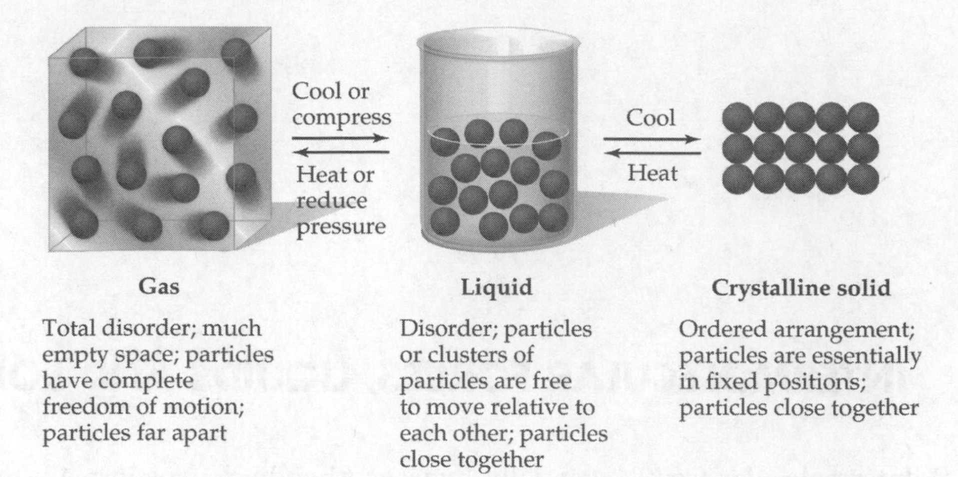

Figure 11.1. Each phase of a substance, solid, liquid and gas, differs in kinetic energy. Higher temperature impart higher kinetic energy which makes attractive forces less significant.

Ion-dipole forces exist between an ion and a polar molecule. For example, when sodium chloride dissolves in water, the partial negative end of the water molecule and the positive sodium ion are held together by an ion-dipole force of attraction. (See Figure 11.2a)

Figure 11.2. (a) Ion-dipole forces between water molecules and sodium and chloride ions. (b) dipole-dipole forces between molecules of water and polar methanol molecules.

Dipole-dipole forces exist between neutral polar molecules in the liquid or solid state. (See Figure 11.2b)

London Dispersion forces are the result of attractions between induced dipoles. Even though no dipole-dipole forces exist between nonpolar atoms or molecules, they do have attractive forces because they do form liquids and solids.

In close proximity, an induced or temporary dipole forms between nonpolar atoms and molecules when the positive nucleus of one atom or molecule attracts the electrons of another atom or molecule. London dispersion forces exist for all atoms and molecules and are the weakest of the intermolecular forces.

Polarizability refers to the degree to which a dipole can be induced in a nonpolar species. Polarizability and thus dispersion forces tend to increase with increasing molar mass.

Your Turn 11.1

Order the halogens according to increasing boiling points. Explain your reasoning. Write your answer in the space provided.

Figure 11.3 illustrates the various kinds of intermolecular forces.

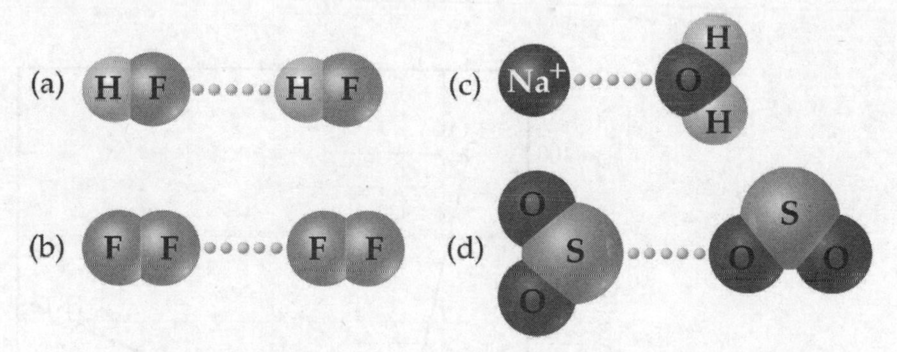

Figure 11.3. Four major intermolecular forces: (a) hydrogen bonding, (b) London Dispersion, (c) ion-dipole, (d) dipole-dipole.

Hydrogen bonding is an especially strong form of a dipole-dipole force. Hydrogen bonding exists only between hydrogen atoms bonded to F, O or N of one molecule and F, O and N of another molecule. The small size of electropositive hydrogen allows it to become very close to and form a strong force of attraction with a nonbonding electron pair on very electronegative F, O or N atoms.

Figure 11.4 shows some examples of hydrogen bonding.

H—Ö:···H—Ö:
| |
H   H

H   H
| |
H—N:···H—N:
| |
H   H

H
|
H—N:···H—Ö:
| |
H   H

H
|
H—Ö:···H—N:
| |
H   H

Figure 11.4. Hydrogen bonding exists between electropositive hydrogen atoms and nonbonding pairs of electrons on electronegative oxygen, nitrogen, or fluorine atoms.

Hydrogen bonding accounts for many unique properties of water. For example, Figure 11.5 illustrates the exceptionally high boiling point of water and Figure 11.6 shows the open, hexagonal arrangement of ice, which causes solid water to float on liquid water.

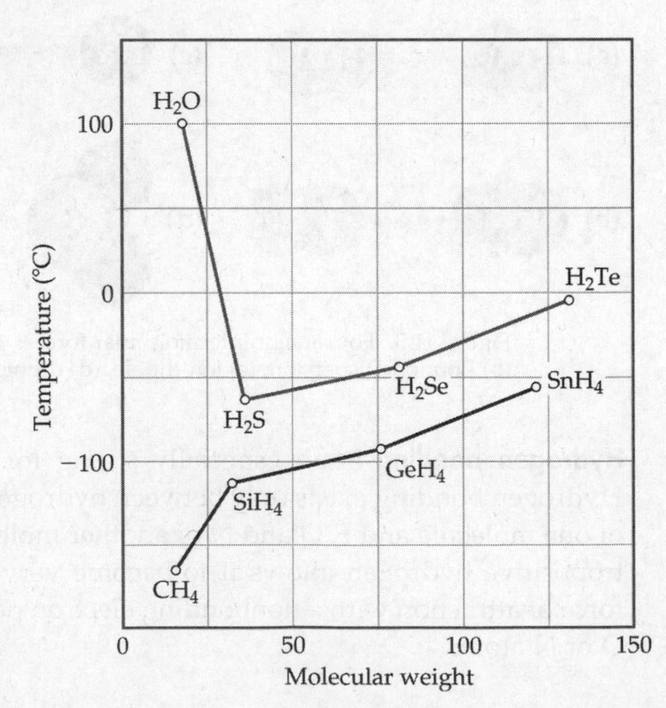

Figure 11.5. Boiling point generally increases with molar mass. Strong hydrogen bonding in water accounts for its unusually high boiling point.

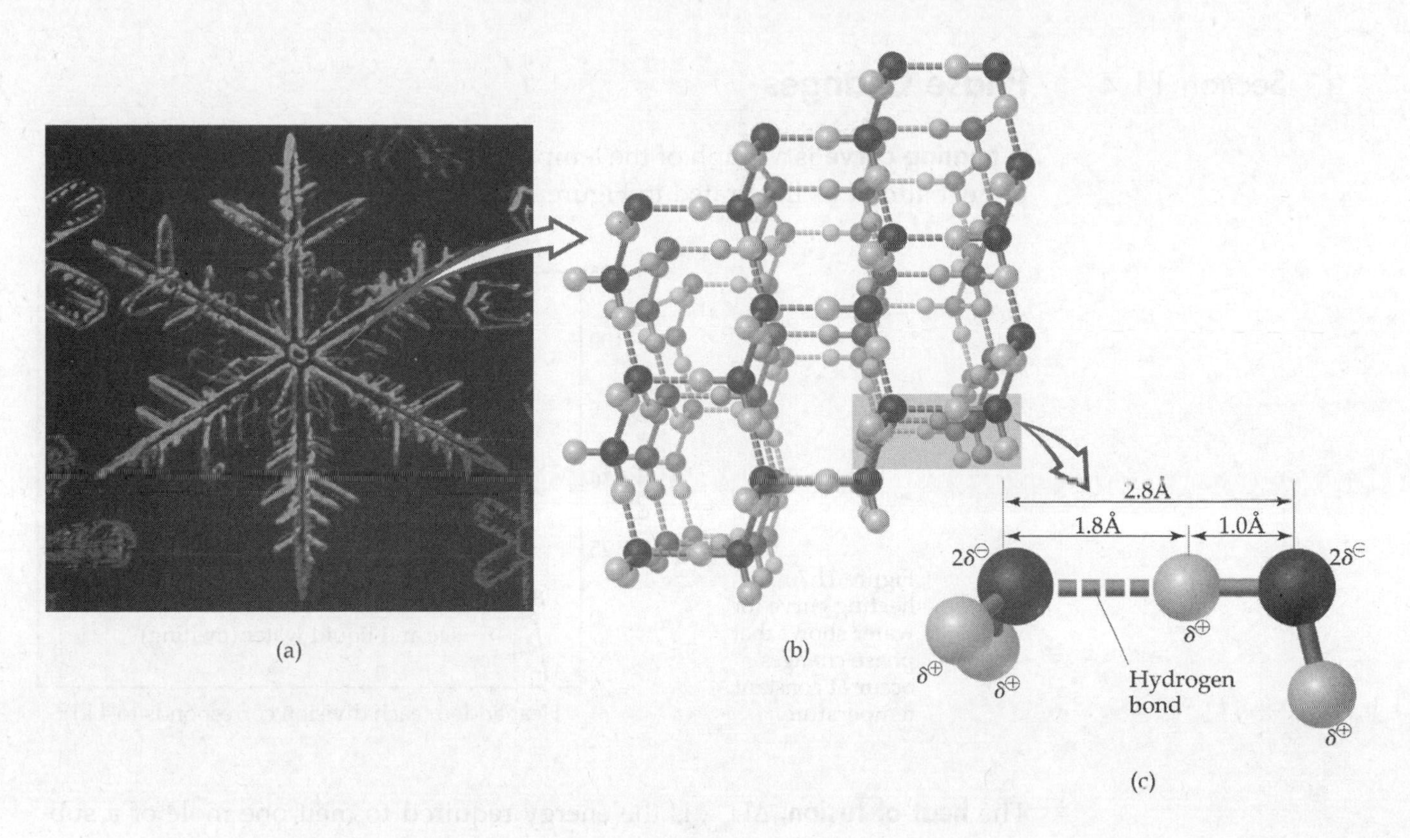

Figure 11.6. The hexagonal geometry of snowflakes (a) and the open, low density structure of ice (b) is due to strong hydrogen bonding of water molecules (c).

Some Properties of Liquids

Section 11.3

Viscosity is the resistance of a liquid to flow. For a series of related compounds, viscosity increases with molar mass.

Surface tension is the energy required to increase the surface area of a liquid by a unit amount. Surface tension is caused by an imbalance of intermolecular forces at the surface of the liquid.

Cohesive forces are intermolecular forces that bind similar molecules to one another, such as the hydrogen bonding in water.

Adhesive forces are intermolecular forces that bind a substance to a surface. For example, water placed in a glass tube adheres to the glass because the adhesive forces between the water and the glass are stronger than the cohesive forces between the water molecules. A curved upper surface or **meniscus** results.

Section 11.4 Phase Changes

A **heating curve** is a graph of the temperature of a system versus the amount of heat added as illustrated in Figure 11.7.

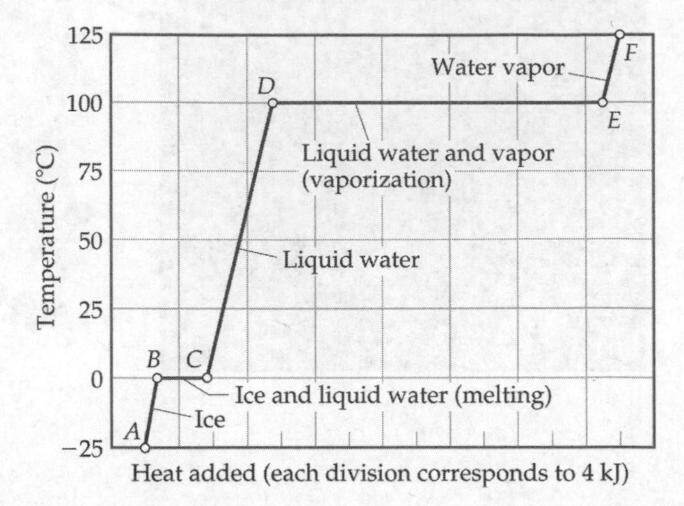

Figure 11.7. The heating curve for water shows that phase changes occur at constant temperature.

The **heat of fusion,** ΔH_{fus}, is the energy required to melt one mole of a substance at constant temperature.

The **heat of vaporization,** ΔH_{vap}, is the energy required to vaporize one mole of a substance at constant temperature.

The horizontal lines of a heating curve represent the heat of fusion, ΔH_{fus}, and heat of vaporization, ΔH_{vap}, of the substance. Notice that the temperature does not change during melting or vaporization. The nearly vertical lines represent the heat required to effect the corresponding temperature change of a single phase.

The **critical temperature** of a substance is the highest temperature at which a liquid can exist. At temperatures higher than the critical temperature, the kinetic energy of the molecules is so great that the substance can only be in the gas phase.

The **critical pressure** is the pressure required to bring about liquefaction at the critical temperature. This is the pressure necessary to bring the molecules sufficiently close together so that the forces of attraction between them can operate at the critical temperature.

Nonpolar substances and those with low molar masses tend to have low intermolecular forces of attraction and correspondingly low critical temperatures and pressures. Polar substances and substances with higher molar masses have higher critical temperatures and pressures because they tend to have higher intermolecular forces of attraction.

Vapor Pressure

Section 11.5

Vapor pressure is the partial pressure exerted by a vapor in a closed system when it is in equilibrium with its liquid or solid phase. For example, water placed in a closed container will evaporate at a constant rate and produce a partial pressure in the gas phase above the liquid. The partial pressure will increase until the rate of evaporation equals the rate of condensation. The system is said to be in a state of dynamic equilibrium at this point and the partial pressure of water vapor remains constant. This partial pressure is the equilibrium vapor pressure of water. See Figure 11.8.

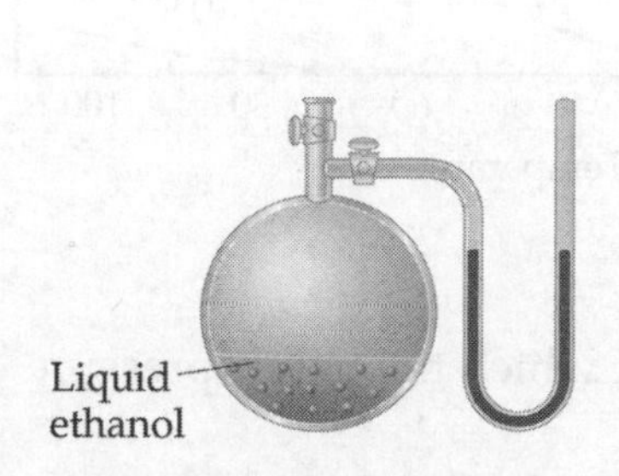

Liquid before any evaporation

(a)

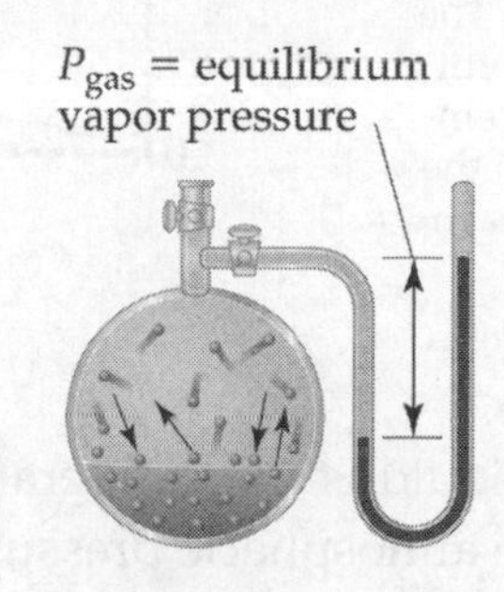

At equilibrium, molecules enter and leave liquid at the same rate.

(b)

Figure 11.8. Equilibrium vapor pressure arises when a liquid in a closed container evaporates producing a partial pressure of vapor above the liquid. At equilibrium the rate of evaporation equals the rate of condensation.

Liquid substances of low molar mass and weak intermolecular forces tend to have high vapor pressures.

Your Turn 11.2

Place the following compounds in order of increasing vapor pressure: CCl_4, CI_4, CBr_4. Explain your answer. Write your answer in the space provided.

Increasing temperature will increase vapor pressure of a liquid or solid as shown in Figure 11.9. Higher temperatures provide greater kinetic energy of molecules giving them higher energies to overcome the attractive forces that hold them together.

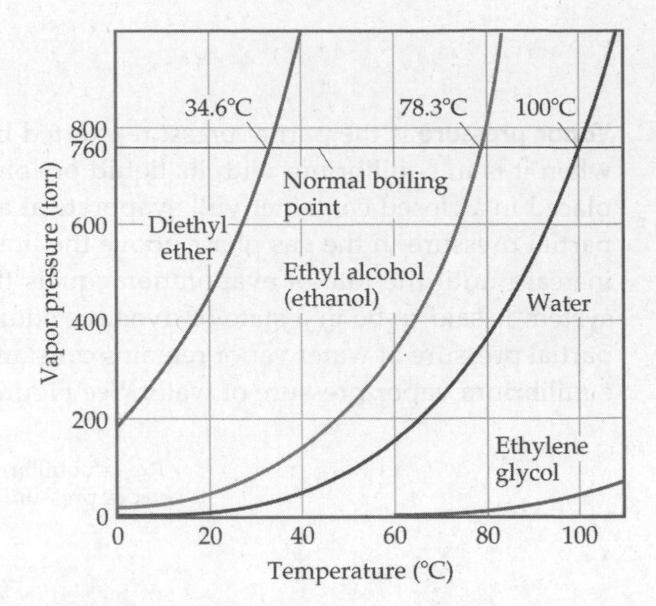

Figure 11.9. Vapor pressure of liquids increase with increasing temperature. The normal boiling point of a liquid is the temperature at which the vapor pressure reaches 760 torr.

The **boiling point** of a liquid is the temperature at which the vapor pressure of the liquid equals the atmospheric pressure.

The **normal boiling point** of a liquid is the temperature at which the vapor pressure equals one atmosphere.

Section 11.6 Phase Diagrams

A phase diagram is a graphical representation of the equilibria among the solid, liquid and gas phases of a substance. Figure 11.10 shows a typical phase diagram. The solid lines represents the temperatures and pressures where the phases of the substance are in equilibrium.

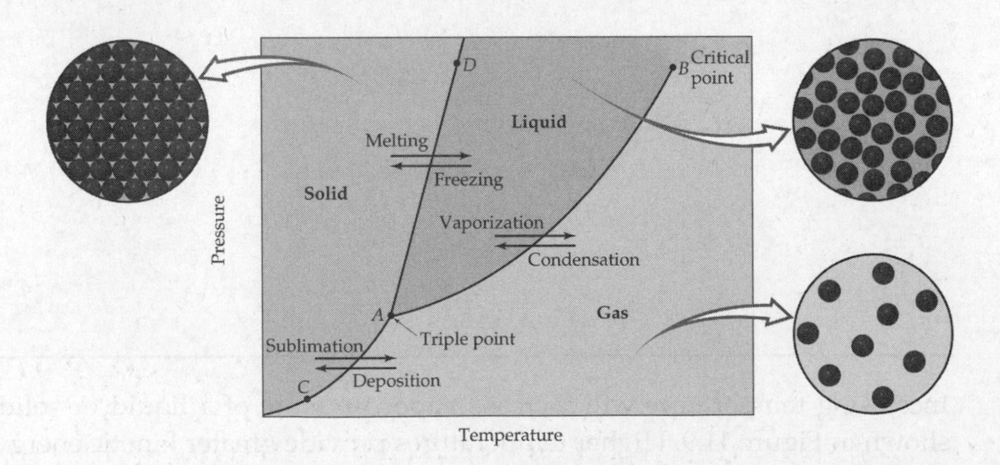

Figure 11.10. A typical phase diagram shows that a substance can exist as a solid, liquid, or gas depending on pressure and temperature.

Your Turn 11.3

What will happen to the system illustrated in Figure 11.10 if: a) The pressure is decreased at point D? b) The temperature is increased at Point C? c) The temperature is decreased at Point B? Write your answers in the space provided.

Figure 11.11 shows the phase diagrams for water and for carbon dioxide. The phase diagram for water is unusual because the slope of the ice-solid equilibrium line is different from most substances and it illustrates that ice melts under pressure. The phase diagram for carbon dioxide shows that liquid CO_2 does not exist at pressures below about five atmospheres.

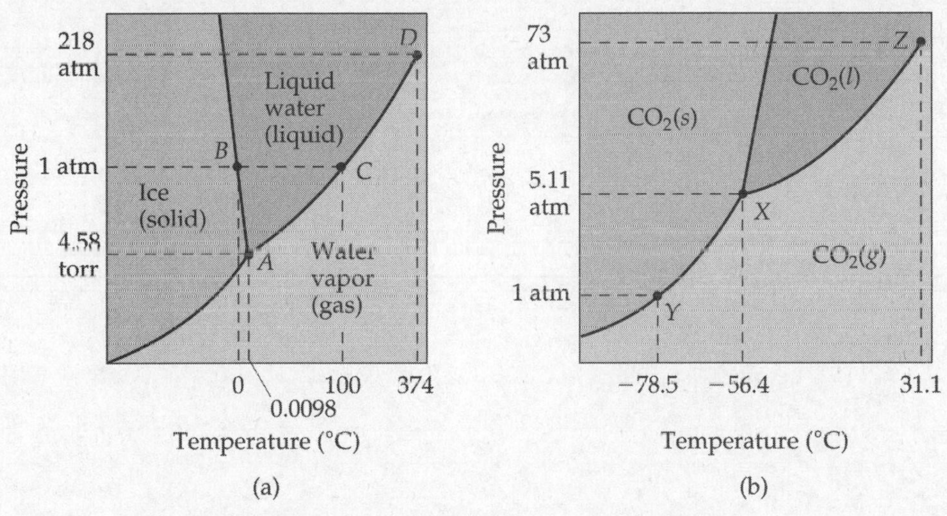

Figure 11.11. Phase diagrams of water and carbon dioxide.

Your Turn 11.4

The phase diagram for water shows that ice melts under pressure. What does this say about the relative densities of ice and liquid water? What other common observation supports your answer? Write your answer in the space provided.

Section 11.8

Bonding in Solids

Molecular solids are atoms or molecules that are held together by relatively weak intermolecular forces. They tend to be soft, low-melting substances.

Covalent-network solids consist of atoms held together by large networks or chains of covalent bonds. They tend to be very hard and have high melting points. The two allotropes of carbon, diamond and graphite, as well as quartz, SiO_2, and silicon carbide, SiC, are examples of covalent-network solids.

Ionic solids are held together by relatively strong ionic bonds. They are characterized by high melting point, brittle structures.

Metallic solids consist entirely of metal atoms. Bonding in metallic solids is due to attractions of metal nuclei to delocalized electrons throughout the solid. The loosely held electrons give metals their characteristic properties of malleability, ductility, and conduction of heat and electricity.

Multiple Choice Questions

1. *Which compound is most likely to form intermolecular hydrogen bonds?*

A) C_4H_{10}

B) NaH

C) C_2H_5OH

D) C_2H_5SH

E) CH_4

2. *Which best explains why bromine is soluble in mineral oil?*

A) *Both substances are liquids.*

B) *Both substances have similar densities.*

C) *Both substances are made up of nonpolar molecules.*

D) *One substance is made up of polar molecules and the other substance is made up of nonpolar molecules.*

E) *Both substances dissolve in water.*

3. *The strongest interaction between hexane and iodine is:*

A) *An instantaneous dipole-dipole interaction.*

B) *A dipole-dipole interaction.*

C) *A hydrogen bond.*

D) *A covalent bond.*

E) *An ionic bond.*

4. *In general, the strongest interaction with water molecules in aqueous solution are for ions that have*

A) *Large charge and large size.*

B) *Large charge and small size.*

C) *Small charge and large size.*

D) *Small charge and small size.*

E) *Zero charge and small size.*

5. *Water and ethanol are completely miscible largely due to which inter molecular forces ?*

A) *Covalent bonds.*

B) *London dispersions.*

C) *Ionic bonds.*

D) *Hydrogen bonds.*

E) *Ion-dipole attractions.*

6. *The energy absorbed when dry ice sublimes is required to overcome which type of interaction?*

A) *Covalent bonds.*

B) *Ion-dipole forces.*

C) *Dipole-dipole forces.*

D) *Dispersion forces.*

E) *Hydrogen bonds.*

7. *A container is half filled with a liquid and sealed at room temperature and atmospheric pressure. What happens inside the container?*

A) *Evaporation stops.*

B) *Evaporation continues for a time then stops.*

C) *The pressure in the container remains constant.*

D) *The pressure inside the container increases for a time and then remains constant.*

E) *The liquid evaporates until it is all in the vapor phase.*

8. *Acetone, $(CH_3)_2C{=}O$, is a volatile, flammable liquid. The central carbon is sp^2 hybridized. The strongest intermolecular forces present in acetone are*

A) *Dipole-dipole forces.*

B) *London dispersions.*

C) *Hydrogen bonds.*

D) *Covalent bonds.*

E) *Ion-dipole forces.*

9. *Which of the factors affect the vapor pressure of a liquid at equilibrium?*

I Intermolecular forces of attraction within the liquid.

II The volume and/or surface area of liquid present

III The temperature of the liquid.

A) I only

B) II only

C) III only

D) I and II only

E) I and III only

10. *The molar masses of a series of similar polar molecules increases in this order: A < B < C < D < E. The boiling points, in degrees Celcius, of molecules A, B, C, D, and E are respectively, 20°, 50°, 150°, 100° and 200°. Which molecule is likely to form hydrogen bonds?*

A) A

B) B

C) C

D) D

E) E

Free Response Questions

1. *Use concepts of chemical bonding and/or intermolecular forces to account for each of the following observations:*

a) The boiling points of water, ammonia, and methane are 100°C, -33°C and -164°C, respectively.

b) At 25°C and 1.0 atm, chlorine is a gas, bromine a liquid, and iodine is a solid.

c) Calcium oxide (2615°C) melts at a much higher temperature than does potassium chloride (770°C).

d) *Propane is a gas and ethanol is a liquid, even though they have similar molar masses.*

2. *Answer the following questions about water using principles of solids, liquids, and gases and intermolecular forces.*

a) *Why does water boil at a lower temperature in Denver, Colorado than in New York City?*

b) *For substances of similar molar mass, why does water have unusually high values for boiling point, heat of vaporization, and surface tension?*

c) *What structural features of ice cause it to float on liquid water?*

d) *Why does calcium chloride dissolve exothermically in water?*

Additional Practice in Chemistry the Central Science

For more practice answering questions in preparation for the Advanced Placement examination try thiese problems in Chapter 11 of Chemistry the Central Science:

Additional Exercises: 11.79, 11.81, 11.83, 11.84, 11.85, 11.88.

Integrative Exercises: 11.100, 11.101, 11.102, 11.105, 11.106.

Multiple Choice Answers and Explanations

1. C. *Hydrogen bonding occurs between molecules that have H-O, H-N or H-F bonds. The hydrogen atom on one molecule attracts a nonbonding electron pair on an O, N or F atom of another molecule.*

2. C. *The adage, "Like dissolves like," means that polar substances will dissolve other polar substances and nonpolar substances will dissolve other nonpolar substances. Polar substances dissolve in each other because of strong dipole-dipole interactions. Nonpolar substances dissolve in each other because of London dispersions. Polar substances generally do not dissolve in nonpolar substances because dipole-dipole interactions among the polar molecules exclude the nonpolar molecules.*

3. A. *Hexane, $CH_3CH_2CH_2CH_2CH_2CH_3$, and iodine, I_2, are both nonpolar molecules. The strongest intermolecular forces acting between*

them are London dispersion forces. London dispersion forces arise when a positive nucleus of one molecule attracts the electrons of another molecule inducing an instantaneous dipole, a momentary shift in electron density.

4. B. *The equation,* $F=KQ_1Q_2/d^2$, *where* Q_1 *and* Q_2 *are ionic charge and d is the distance between ions, describes the force of attraction between two charged particles. It's also useful in estimating the relative strengths of ion-dipole interactions, such as those between a polar water molecules and an aqueous ion. The larger the charge and the smaller the ion, the larger the intermolecular force.*

5. D. *Water and ethanol,* CH_3CH_2OH, *are both small molecules and both have O-H bonds. The hydrogen atoms on one molecule attract nonbonding electron pairs on an oxygen atom of another molecule, forming an especially strong dipole-dipole interaction called a hydrogen bond.*

6. D. *Carbon dioxide molecules are nonpolar. They are held together only by London dispersion forces. Sublimation is the process where a solid changes directly into a gas. Energy is required to overcome the forces of attraction that hold the molecules together in the solid phase.*

7. D. *The liquid will continue to evaporate without stopping. The pressure will rise until the partial pressure of the vapor equals the vapor pressure of the liquid at the given temperature. Then a dynamic equilibrium is established where the rate of evaporation equals the rate of condensation.*

8. A. *The C=O bond in trigonal planar acetone is very polar making the molecule a strong dipole. The absence of H-O bonds rules out hydrogen bonding but strong dipole-dipole interactions exist.*

9. E. *Intermolecular forces and temperature both affect the vapor pressure of a liquid but vapor pressure is independent of volume and surface area.*

10. C. *Because of increased dispersion forces, the boiling points of a series of similar molecules will increase regularly with increasing molar mass. Molecule C is lighter than molecule D, yet it has a higher boiling point. This can be explained by especially strong dipole-dipole forces called hydrogen bonding in molecule C.*

Free Response Answers

1. a) *Water, H_2O, and ammonia, NH_3, both have higher boiling points than methane, CH_4, because water and ammonia form hydrogen bonds among their molecules and methane does not. Hydrogen bonds are relatively strong intermolecular attractive forces, which tend to hold molecules together in liquids requiring more energy to separate them. Water has a much higher boiling point than ammonia because water has two nonbonding pairs of electrons per molecule versus one nonbonding pair for ammonia. Additionally oxygen is more electronegative than is nitrogen. These facts work together to allow for stronger and much greater hydrogen bonding in water than in ammonia.*

   b) *Chlorine, bromine, and iodine are all nonpolar diatomic molecules held together by London dispersion forces. The dispersion forces become greater as the molar masses increase. Iodine has the highest molar mass and the largest dispersion forces, large enough to hold the molecules together in a molecular solid. Bromine has weaker dispersion forces but strong enough to hold the molecules together in a liquid. Chlorine has the weakest dispersion forces, too weak to hold the molecules together so it is a gas.*

   c) *To melt a solid, the temperature must be sufficient to overcome the attractive forces holding the solid together. Calcium oxide, CaO, is composed of small ions of relatively large charge so they are held together with relatively large attractive forces requiring a high temperature to melt. Potassium chloride is composed of larger ions of smaller charge requiring a relatively low temperature to provide sufficient kinetic energy to overcome the relatively low attractive forces.*

   d) *Propane, $CH_3CH_2CH_3$, is a nonpolar molecule held together by relatively weak London dispersion forces. Ethanol, CH_3CH_2OH, is a polar molecule held together by stronger dipole-dipole forces. Additionally, ethanol can form hydrogen bonds which are even stronger forces of attraction.*

2. a) *Boiling point is the temperature at which the vapor pressure of a liquid equals the external pressure. Vapor pressure increases with increasing temperature. Due to its relatively high elevation above sea level, the atmospheric (external) pressure in Denver is lower than that in New York. Consequently, a lower temperature is required to reach the vapor pressure equal to the lower pressure in Denver.*

b) *Water forms very strong hydrogen bonds among its molecules which hold the molecules together in the liquid phase much more than liquids having the relatively weak dipole-dipole or dispersion attractive forces.*

c) *The hydrogen bonding in water causes ice to form a relatively open hexagonal structure that is less dense than the more loosely held liquid.*

d) *The process of dissolving an ionic substance in water breaks the ion-ion bonds (bond breaking is always endothermic) and forms ion-dipole interactions (bond formation is always exothermic) with water. The fact $CaCl_2$ dissolves exothermically means that the ion-dipole interactions formed in the solution are stronger than the ion-ion interactions of the solid ionic lattice.*

Your Turn Aanswers

11.1 *The boiling points of the halogens will increase in this order: $F_2 < Cl_2 < Br_2 < I_2 < At$. As the molar mass increases the polarizability of the molecules increases causing greater induced dipoles and stronger London dispersion forces that hold the molecules together.*

11.2 *The vapor pressures increase in this order: $CI_4 < CBr_4 < CCl_4$. Higher molar mass substances have higher dispersion forces and are held together more strongly than are lower molar mass substances.*

11.3 *a) A pressure decrease at Point D will cause any solid present to melt. b) A temperature increase at Point C will cause any solid present to sublime. c) A temperature decrease at Point B will cause most of the vapor present to condense.*

11.4 *The fact that ice melts under pressure indicates that ice is less dense than liquid water. This conclusion is supported by the fact that ice floats on liquid water.*

TOPIC 12

CHEMISTRY DIAGNOSTIC TEST

Multiple Choice Questions   Section 1

Part A

60 minutes

You **may not** use a calculator for Section I.

Directions: Each set of lettered responses refers to the numbered statements or questions immediately below it. Choose the one lettered response that best fits each statement or question. You may use a response once, more than once or not at all.

Questions 1-3 refer to the following pairs of elements.

A. Be & B
B. P & S
C. Cl & F
D. Cr & Mo
E. Au & Ag

1. *Show a reversal in the trend for first ionization energy because of electron-electron repulsion.*

2. *Show a reversal in the trend for first ionization energy because of screening by full orbitals.*

3. *Exhibit an anomaly in outer electron configuration because full d orbitals are especially stable.*

Questions 4-8 refer to the following elements:

A. Na
B. S
C. Mg
D. N-
E. O

4. *Is the most electronegative.*

5. *Exhibits the most number of oxidation states.*

6. *Has the largest atomic radius.*

7. *Has the smallest ionic radius for its most common ion.*

8. *Has the largest difference between the first and second ionization energies.*

Questions 9-12 refer to equal molar quantities of four gases placed into an evacuated container at 25°C:

A) CH_4 B) H_2O
C) Ne D) Ar
E) all are the same

9. *Which gas has the largest mole fraction?*

10. *Which gas has the slowest average velocity?*

11. *Which gas has the greatest density?*

12. *Which gas has the highest average kinetic energy?*

Part B

Directions: For each of the following questions or incomplete statements select the letter of the best answer or completion directly below it.

13. *Which is not found as a free element in nature?*
A) *Au*
B) *Ag*
C) *Cu*
D) *C*
E) *P*

14. *Which is not true for the transition metal elements?*
A) *Most, but not all, are solids.*
B) *All have oxidation states of 2+ and many have multiple oxidation states.*
C) *Many of their compounds are colored due to partially filled d orbitals.*
D) *Many form complex ions in aqueous solution.*
E) *Many are used industrially.*

15. *Which of the following intermolecular forces are found in all substances?*
A) *dispersions*
B) *dipole-dipole*
C) *hydrogen bonding*
D) *ion-dipole*
E) *covalent bonds*

16. *Which of these molecules can form hydrogen bonds within their pure liquids?*
I. CH_3NH_2 II. CH_3OH
III. $(CH_3)_2C{=}O$ IV. CH_3OCH_3
A) *I, II, III and IV*

B) I, II and III only
C) II only
D) I and II only
E) I, II and IV only

17. *What is the correct order of boiling points, lowest to highest, for these liquids?*
A) $CH_3CH_2CH_3 < CH_3CH_2OH < H_2O$
B) $CH_3CH_2OH < CH_3CH_2CH_3 < H_2O$
C) $H_2O < CH_3CH_2CH_3 < CH_3CH_2OH$
D) $CH_3CH_2CH_3 < H_2O < CH_3CH_2OH$
E) $CH_3CH_2OH < H_2O < CH_3CH_2CH_3$

18. *Hydrogen chloride molecules have associated with them:*
I. *covalent bonds* II. *hydrogen bonds*
III. *dipole-dipole forces* IV *dispersion forces*

A) *I, II, III and IV*
B) *I, II and III only*
C) *I, III and IV only*
D) *I, II and IV only*
E) *I and III only*

19. *What is the maximum number of grams of sodium chloride that can be obtained from the reaction of 40.0 grams of sodium carbonate and 30.0 grams of hydrochloric acid?*
A) (40.0/106)(2)(58.5)
B) (30.0/36.5)(2)(58.5)
C) (30.0/36.5)(58.5)
D) (30.0/36.5)(½)(58.5)
E) (40.0/106)(58.5)

20. *What volume of 0.500 M $AlCl_3$ solution is needed to prepare 300 mL of solution that has a chloride concentration of 0.300 M?*
A) 30.0 mL
B) 50.0 mL
C) 60.0 mL
D) 120 mL
E) 180 mL

21. *How many grams of $CaBr_2$ are needed to prepare 250 mL of a 0.500 M solution?*
A) 12.5
B) 25.0
C) 37.5
D) 50.0
E) 6.25

22. *When the equation describing the complete combustion of propanol is balanced using whole number coefficients, the coefficient of oxygen is:*
A) 3
B) 4
C) 5

D) 9
E) 10

23. *An atom, Z, has a mass that is three times that of a carbon atom. What mass of Z will combine with 3.00 grams of carbon in forming the compound Z_2C?*
A) 6.00 g
B) 12.0 g
C) 18.0 g
D) 24.0 g
E) 36.0 g

24. *2.00 mol glycine, H_2NCH_2COOH, contains*
A) 5.00 mol H
B) 20.0 g H
C) 6.02×10^{23} atoms N
D) 32.0 g 0
E) 64.0 g 0

25. *When potassium metal reacts with an excess of water the products are*
A) $K_2O(s)$ and $H_2(g)$
B) $KH(s)$ and $O_2(g)$
C) $K^+(aq)$, $OH^-(aq)$ and $H_2(g)$
D) $K_2O_2(s)$ and $H_2(g)$
E) $K^+(aq)$, $OH^-(aq)$, and $O_2(g)$

26. *Exactly 50.0 mL of a 0.100 M sulfuric acid solution is combined with exactly 100.0 mL of 0.05 M sodium hydroxide. The mixture is then evaporated to dryness. What solid remains after evaporation?*
A) H_2SO_4
B) NaOH
C) Na_2SO_4
D) $NaHSO_4$
E) $NaSO_4$

27. *Formulas of five compounds of a particular ion X are listed below. Which one of them must be incorrect?*
A) X_2O_3
B) $X_2(CO_3)_3$
C) XPO_4
D) XCl_3
E) X_3N

28. *When butane is combusted in limited air the possible products can include water and*
I. carbon dioxide
II. carbon monoxide
III. carbon
A) I only
B) II only
C) II only
D) I and II only
E) I, II and III

29. *When 20.0 grams of magnesium is added to 1.00 L of 1.00 M hydrochloric acid,*
A) *the limiting reagent is magnesium*
B) *the limiting reagent is hydrochloric acid*
C) *the mixture is stoichiometric*
D) *the limiting reagent is hydrogen gas*
E) *the limiting reagent is magnesium chloride*

30. *Which is the formation reaction for HCl(g)?*
(A) $H_2(g) + Cl_2(aq) \dashrightarrow HCl(aq)$
(B) $H(g) + Cl(g) \dashrightarrow HCl(g)$
(C) $H^+(aq) + Cl^-(aq) \dashrightarrow HCl(aq)$
(D) $\frac{1}{2} H_2(g) + \frac{1}{2} Cl_2(g) \dashrightarrow HCl(g)$
(E) $H_2(g) + Cl_2(g) \dashrightarrow 2HCl(g)$

31. *Which of the following processes is not exothermic?*
A) *rain condensing from water vapor in the atmosphere*
B) *the reaction of sodium with water*
C) *dissolving $CaCl_2(s)$ in water causing an increase in temperature*
D) *melting ice*
E) *snow forming from clouds*

32. *The element that is probably most similar in chemical properties to lead is*
A) C
B) Si
C) Ge
D) Sn
E) Sb

33. *Which flame tests color is not correct?*
A) *sodium, violet*
B) *potassium, violet*
C) *lithium, red*
D) *calcium, red-orange*
E) *copper, blue-green*

34. *Which is the correct ground state electronic configuration for the Cr atom?*
A) $[Ar]\ 3d^4 4s^2$
B) $[Ar]\ 3d^5 4s^1$
C) $[Ar]\ 3d^6 4s^0$
D) $[Kr]\ 3d^4 4s^2$
E) $[Kr]\ 3d^5 5s^1$

35. *How many unpaired electrons are found in the most stable electronic state of a sulfur atom?*
A) 0
B) 1
C) 2
D) 4
E) 6

36. *Which of the following sets of quantum numbers (n, l, m_l, m_s) best describes a valence electron of highest energy in a ground-state gallium atom?*
(A) 4, 0, 0, ½
(B) 4, 0, 1, ½
(C) 4, 1, 1, ½
(D) 4, 1, 2, ½
(E) 4, 2, 0, ½

37. *Which gas will have the highest rate of effusion?*
A) CH_4
B) H_2O
C) *Ne*
D) *Ar*
E) NH_3

38. *Which molecule has the greatest bond dissociation energy?*
A) CO
B) O_2
C) NO
D) F_2
E) Cl_2

39. *Which molecule has the shortest bond length?*
A) N_2
B) O_2
C) Cl_2
D) Br_2
E) F_2

40. *Which molecule has the longest bond length?*
A) F_2
B) Cl_2
C) Br_2
D) I_2
E) N_2

41. *For SF_4, the electron domain geometry of S and the molecular geometry are, respectively,*
A) *tetrahedral and trigonal planar.*
B) *octahedral and trigonal bipyramidal.*
C) *trigonal bipyramidal and see-saw.*
D) *trigonal pyramidal and T-shaped*
E) *tetrahedral and tetrahedral.*

42. *Which is planar?*
I. NO_3^- II. SO_3
III. CO_3^{2-} IV. SO_3^{2-}
A) *I only*
B) *III only*
C) *I, II and III only*
D) *I, III and IV only*
E) *I, II, III, and IV*

43 *A bond angle commonly found in molecules having a central atom that is sp^3 hybridized is*

A) *180°*
B) *120°*
C) *109.5°*
D) *90°*
E) *75°*

44. *Which gas is most soluble in water?*
A) CO_2
B) CH_4
C) O_2
D) NH_3
E) Cl_2

45. *What is the molecular geometry of atoms in the SO_3^{2-} ion?*
A) *tetrahedral*
B) *trigonal planar*
C) *trigonal pyramidal*
D) *trigonal bipyramidal*
E) *t-shape*

46. *Which change represents a reduction process?*
A) ClO_2^- --➤ ClO_3^-
B) CO --➤ CO_2
C) N_2 --➤ NH_3
D) $HCrO_4^-$ --➤ CrO_4^{2-}
E) MnO_2 --➤ MnO_4^-

47. *How many electrons are needed to convert one sulfate ion into one sulfide?*
A) 1
B) 2
C) 4
D) 6
E) 8

48. *A standard solution of 0.154 M HCl is used to determine the concentration of a NaOH solution whose concentration is unknown. If 33.5 mL of the acid solution are required to neutralize 25.0 mL of the base solution, the concentration of the NaOH in moles/liter is*
A) *25.0 / (0.154)(33.5)*
B) *(0.154)(25.0) /33.5*
C) *(33.5 + 25.0) / 0.154*
D) *(0.154)(33.5) / 25.0*
E) *(25.0)(33.5)/ 0.154*

49. *What is the molar concentration of HCl if 125.0 mL of 2.00 M HCl is diluted to 500.0 mL?*
A) *0.250 M*
B) *0.500 M*
C) *1.00 M*

D) 1.50 M
E) 8.00 M

50. Which compound is a strong electrolyte in aqueous solution?
A) HNO_3
B) NH_3
C) CH_3COOH
D) C_2H_5OH
E) H_2O

Section 2

Free Response Questions

Part A

55 minutes

You **may** use a calculator for Part A.

Directions: Answer each of the following questions, clearly showing the methods you use and the steps involved at arriving at the answers. Partial credit will be given for work shown and little or no credit will be given for not showing your work, even if the answers are correct.

Question 1

The percent composition of a solid mixture of calcium carbonate, calcium hydroxide, and calcium chloride is to be determined. When 350.0 mL of 0.250 M hydrochloric acid is added to 3.506 g of the mixture, 385.7 mL of carbon dioxide gas is collected at 35.0°C and 711 torr. It requires 65.4 mL of 0.500 M sodium hydroxide to titrate the remaining acidic mixture to a pink phenolphthalein end point.

a. *Write and balance the net ionic equations for the reactions that take place when hydrochloric acid is added to the mixture. Which component of the mixture does not react with hydrochloric acid?*
b. *How many moles of carbon dioxide are produced by the reaction?*
c. *How many grams of calcium carbonate are contained in the mixture?*
d. *How many moles of excess acid remained after the initial reactions and before the titration?*
e. *How many grams of calcium chloride are contained in the mixture?*
f. *What is the percent composition of the mixture?*

Question 2

a. *The bond dissociation energy for H-H is 436 kJ/mol. Absorption of light will cause molecular hydrogen gas to break into individual atoms in a process called photodissociation.*
   i. Write a formation reaction for H(g).
   ii. Calculate the molar heat of formation of H(g).
b. *The bond dissociation energy for H-O is 463 kJ/mol.*
   In the stratosphere, water vapor photodissociates as follows:
   $H_2O(g) + h\nu \dashrightarrow H(g) + OH(g)$

Calculate the heat of formation of OH(g) given the standard heat of formation of $H_2O(g)$ is -242 kJ/mol.

c. *Hydroxyl radical, OH, can react with ozone in the stratosphere giving the following reactions:*

$OH(g) + O_3(g) \dashrightarrow HO_2(g) + O_2(g)$

$HO_2(g) + O(g) \dashrightarrow OH(g) + O_2(g)$

i. Write the overall reaction that results from these two reactions.

ii. Write the Lewis structures including any resonance forms for ozone, O_3, and for HO_2.

iii. The bond dissociation energy for $O_2(g)$ is 495 kJ/mol and the heat of formation for ozone is 142 kJ/mol. Calculate ΔH for the overall reaction in Part c i.

Question 3

In the laboratory you are to determine the identity of a pure unknown white solid.

a. *A 19.2 gram sample of the volatile compound decomposes upon heating to yield 6.80 grams of ammonia, 8.80 grams of carbon dioxide and the only other product is water. Calculate the simplest formula of the compound in the form: $C_wH_xN_yO_z$*

b. *When a sample of the compound is dissolved in water and made basic with sodium hydroxide, wetted pH paper held above the solution indicates a pH of about 9 and the solution gives off a distinct smell of ammonia. Write and balance a net ionic equation for the reaction of sodium hydroxide with the compound that explains this result.*

c. *When another sample is dissolved in water and made acidic with hydrochloric acid, the solution effervesces. Write and balance a net ionic equation that could explain this result.*

d. *Based on your answers to Parts b and c, rearrange the simplest formula you determined in Part a to identify the compound. Name the compound..*

e. *Use principles of intermolecular forces and polarity to explain why effervescence was observed in Part c, but not in Part b.*

Part B

50 minutes

You **may not** use a calculator for Part B.

Question 4

Write and balance net ionic equations for each of the following laboratory situations. You can assume a reaction happens in each case.

a. *A slightly acidic potassium iodide solution is added to dilute hydrogen peroxide solution.*
What color would the mixture appear in the presence of starch?

b. *Nitric acid is added to solid magnesium oxide.*
What would you observe if the resulting mixture is made strongly (basic with) sodium hydroxide?

c. *Calcium metal is added to water.*
What would you expect the pH of the final solution to be?

Question 5

Oxalic acid, $H_2C_2O_4$, imparts a sour taste to rhubarb. It is a water soluble diprotic acid, one that has two ionizable hydrogens.

a. *Write a net ionic equation for the reaction that occurs when:*
i. equal molar aqueous solutions of oxalic acid and sodium hydroxide are mixed.
ii. Solid oxalic acid is mixed with excess sodium hydroxide solution.

b. *One product of the reaction in a. ii. is the oxalate ion.*
i. Draw a Lewis structure for the oxalate ion.
ii. What is the hybridization of the two carbon atoms?
iii. Identify the molecular geometry around each carbon of the oxalate ion. Specify the bond angles.

c. *Discuss the overall geometry of the oxalate ion in terms of resonance and delocalized electrons. What can you say about the geometrical arrangement of all the atoms relative to each other? What can you conclude about the relative lengths of the various C-O bonds in the oxalate ion?*

Question 6

Answer the following questions about these compounds and elements

a. *Manganese forms five distinct oxides: MnO, MnO_2, MnO_3, Mn_2O_3 and Mn_2O_7.*
i. Identify the oxidation number in manganese in each of these oxides.
ii. A sixth manganese oxide, the mineral hausmanite, Mn_3O_4, contains manganese in different oxidation states. What are the most likely oxidation states of manganese in Mn_3O_4 and how many manganese atoms per formula unit have each of the oxidation states?
iii. Write the electron configuration of the element manganese and of the Mn ion in Mn_2O_3. (Assume the manganese ion is a distinct, isolated ion.)
iv. Would you expect solid MnO to react with nitric acid or an aqueous solution of sodium hydroxide? Explain your answer and write and balance a chemical equation.

b. *Except for scandium and titanium, 2+ is a common oxidation number for all the Period 4 transition elements. Explain why 2+ is a common oxidation number for transition elements.*

c. *Explain why the ionic radius of potassium is smaller than its atomic radius. Why is the ionic radius of chlorine larger than its atomic radius?*

Multiple Choice Answers and Explanations

1. *B. The p^4 configuration of sulfur is less stable than the p^3 configuration of phosphorus because of electron-electron repulsion of the paired electrons.*

2. *A. Boron has a lower ionization energy than beryllium because the two 2s electrons partially screen the 2p electron from the nucleus.*

3. *E. Gold and silver both have ten d electrons and only one s electron in their outer orbitals because of the special stability of full d orbitals.*

4. *E. In general, electronegativity increases from lower left to upper right on the periodic table.*

5. *B. Besides zero, sulfur displays oxidation numbers of 2-, 2+, 4+ and 6+.*

6. *A. Atomic radius increases from right to left along a period and from top to bottom within a group.*

7. *C. Cations are generally small because they have lost their valence electrons and the inner core electrons are poorly screened from the positive nucleus. Within any period, the higher the charge, the smaller the cation.*

8. *A. Atoms from Group 1 have one valence electron and have very low first ionization energies. The second ionization energy is very large for Group 1 cations because the second electron is an inner (noble gas) core electron that is very poorly screened from the nucleus.*

9. *E. Each gas has a mole fraction of 0.25.*

10. *D. Argon has the largest molar mass so its molecules move, on average, with the slowest speed.*

11. *D. All gases occupy the same volume but one mole of argon has the greatest mass of all the gasses.*

12. *E. Temperature is a measure of the average kinetic energy of molecules. All four gases are at the same temperature so they have the same average kinetic energy.*

13. *E. Phosphorus, because of its high reactivity especially with oxygen, is not found in nature.*

14. *B. While most transition metals exhibit oxidation numbers of 2+ because their two "s" electrons are lost first, not all do. Scandium and yttrium each display a 3+ oxidation number but not 2+.*

15. *A. All substances contain dispersion forces.*

16. *D. The criteria for hydrogen bonding is that a hydrogen atom be bonded directly to a nitrogen, oxygen or fluorine atom.*

17. *A. Propane (molar mass = 44) is held together by dispersion forces only and will boil at the lowest temperature. Ethanol (molar mass = 46) has a higher boiling point than propane because it can effectively hydrogen bond. Water (molar mass = 18) forms very strong hydrogen bonds and has a very high boiling point for its molar mass.*

18 *C. Dispersion forces are associated with all matter. HCl bonds by sharing a pair of electrons in a covalent bond. The significant difference in electronegativity between H and Cl make the bond a polar bond. HCl cannot hydrogen bond because, while chlorine is sufficiently electronegative, it is too large.*

19. *A. The balanced equation is:*
$Na_2CO_3 + 2HCl \dashrightarrow 2NaCl + CO_2 + H_2O$.
The limiting reactant is sodium carbonate.
x g NaCl = 40.0 g Na_2CO_3(1 mol/106 g)(2 mol NaCl/1 mol Na_2CO_3)
(58.5 g/mol)

20. C. *The number of moles of chloride ion needed is:* *(0.300 mol/L)(.300 L) = 0.0900 mol* Cl^- *ion. 0.0900 mol of chloride ion requires only one third as much aluminum chloride or 0.0300 mol. To get 0.0300 moles of* $AlCl_3$, *take 0.0600 L of solution:* *.0300 mol/0.500 mol/L = 0.0600 L = 60.0 mL.*

21. B. *x g* $CaBr_2$ *= 250 mL(1 L / 1000 mL)(0.500 mol/L)(200 g/mol) = 25.0 g*

22. D. *Propanol is an alcohol containing three carbons:* $CH_3CH_2CH_2OH$ *or* C_3H_8O. *The balanced equation is:*

$$2C_3H_8O + 9O_2 \longrightarrow 6CO_2 + 8H_2O$$

23. C. *3.00 g of carbon represents ¼ mol of carbon so ½ mol of Z will combine. 0.5 mol Z x 36.0 g/mol = 18.0g Z.*

24. E. *2.00 mol of* H_2NCH_2COOH *contain 2 x 2 = 4 moles of oxygen. 4 mol O x 16.0 g/mol = 64.0 g O*

25. C. *An alkali metal reacts with water to produce hydrogen gas and an aqueous alkali metal hydroxide.*

26. D. *Each initial solution contains 0.005 moles of reactants. The reaction is one between a diprotic acid and a monobasic base leaving only sodium hydrogen sulfate and water as products:*

$$H_2SO_4 + NaOH \longrightarrow NaHSO_4 + H_2O$$

27. E. *Responses A, B, C and D all indicate that X has a 3+ charge, given the ratios it combines with oxide,* O^{2-}; *carbonate,* CO_3^{2-} *phosphate,* PO_4^{3-}; *and chloride,* Cl^-. *Nitride,* N^{3-}, *would combine with* X^{3+} *as XN, not* X_3N.

28. E. *Limiting the amount of air necessary for the combustion of hydrocar bons can yield a wide range of products including all three species listed.*

29. B. *The reaction is:* $Mg(s) + 2H^+(aq) \longrightarrow Mg^{2+}(aq) + H_2(g)$. *Almost one mole of magnesium is present which will require almost two moles of HCl completely react making HCl the limiting reagent.*

30. D. *A formation reaction is one that produces one mole of a product from its elements in their most thermodynamically stable forms at 25°C and one atmosphere. Both hydrogen and chlorine are diatomic gases under these conditions.*

31. D. *A change of phase from gas to liquid to solid is always exothermic. Phase changes from solid to liquid to gas are always endothermic.*

32. D. *Of the elements in the same family as lead only tin is a metal. The others are either metalloids or nonmetals.*

33. A. *Sodium has a bright yellow flame.*

34. B. *Chromium, because half-full d orbitals tend to be more stable, has an anom alous electron configuration different from that predicted by the aufbau process. A half-filled 3d level is more stable than a completely filled 4s level.*

35. C. *The electron configuration of sulfur is* $[Ne]3s^2 3p^4$. *The "s" electrons are*

paired and two of the four "p" electrons are paired and two are unpaired.

36. C. *The ground state electron configuration of gallium is: [Ar]$4s^2 4p^1$. The electron in the 4p orbital is the highest energy valence electron. The quantum numbers 4, 1, 1, ½ describe an electron in a 4 p sublevel.*

37. A. *The rate of effusion of any gas is inversely proportional to the square root of its molar mass. Methane has the smallest molar mass of the gases listed.*

38. A. *Multiple bonds are stronger than single bonds. The Lewis structure for CO shows that it has a triple bond, whereas O_2 and NO have double bonds and F_2 and Cl_2 have single bonds.*

39. A. · *Triple bonds are shorter than double bonds which are shorter than single bonds.vLewis structures show that N_2 has a triple bond, O_2 has a double bond and the rest all have single bonds.*

40. D. *Except for N_2 which has a triple bond, the Lewis structures for all the molecules show that they have single bonds. The bond length is determined by the size of the bonded atoms. The larger atoms have the longer bonds. Atoms become larger from top to bottom wthin any group on the periodic table.*

41. C. *The Lewis structure for SF_4 shows four bonding pairs and one nonbonding pair around sulfur.*

42. C. *Nitrate, sulfur trioxide and carbonate are all isoelectronic having 24 valence electrons each. Their Lewis structures are identical: two single bonds to oxygen and one double bond to oxygen.Sulfite has 26 electrons with three single bonds to oxygen and a nonbonding pair on sulfur.*

43. C. *Orbital hybridizations of sp, sp^2, dsp^3 and d^2sp^3 have associated with them these bond angles respectively: 180°, 120°, 90° and 120°, 90°.*

44. D. *Ammonia is very polar and can hydrogen bond very effectively with water. All the others are non-polar and cannot hydrogen bond.*

45. C. *The Lewis structure of sulfite shows that the central atom has single bonds to each of three oxygen atoms and a nonbonding pair of electrons.*

46. C. *Reduction is the gain of electrons. The oxidation number of an atom is reduced in a reduction process. Nitrogen changes from an oxidation mumer of 0 to -3 when N_2 becomes NH_3.*

47. E. *In sulfate, SO_4^{2-}, sulfur has an oxidation number of 6+. In sulfide, S^{2-}, the oxidation number falls to 2-, a difference of eight electrons.*

48. D. *NaOH + HCl* $\dashrightarrow$ *NaCl + H_2O*

x mol/L NaOH = (0.0335 L HCl / 0.0250 L NaOH)(0.154 mol HCl/ L HCl)(1 mol NaOH/ 1 mol HCl) = (0.0335)(0.154)/(0.0250) = (33.5)(0.154)/25.0

49. B. *M = (2.00 M)(125.0 mL/500.0 mL) = 0.500 M*

50. A. *Nitric acid is a strong acid and therefore a strong electrolyte. It ionizes completely in aqueous solution.*

Free Response Answers

Answers to Question 1

a. $CaCO_3 + 2H^+(aq) \rightarrow CO_2(g) + H_2O(l) + Ca^{2+}(aq)$
$Ca(OH)_2 + 2H^+(aq) \rightarrow Ca^{2+}(aq) + 2H_2O(l)$
Calcium chloride does not react with hydrochloric acid.

b. $n = PV/RT = (711 torr / 760 torr/atm)(0.3857 L)/(0.0821 L atm/mol K)(273 + 35 K) = 0.01427 mol CO_2$.

c. $x g CaCO_3 = 0.01427 mol CO_2(1 mol CaCO_3/1 mol CO_2)(100.0 g/mol) = 1.427 g CaCO_3$

d. $x mol HCl = 0.0654 L NaOH(0.500 mol/L)(1 mol HCl/1 mol NaOH) = 0.0327 mol HCl$

e. *Total moles of HCl = 0.350 L x 0.250 mol/L = 0.0875 mol HCl.*
Moles HCl that reacted with $CaCO_3$ = 2 x moles of $CaCO_3$ =
2 x 0.01427 mol = 0.02854 mol
Moles of HCl that reacted with $Ca(OH)_2$ = 0.0875 – 0.02854 – 0.0327 = 0.0263 mol.
$x g Ca(OH)_2 = (0.0263 mol HCl)(1 mol Ca(OH)_2 / 2 mol HCl)(74.0 g/mol) = 0.973 g Ca(OH)_2$.
$x g CaCl_2 = 3.506 g sample - 0.973 g Ca(OH)_2 - 1.427 g CaCO_3 = 1.106 g CaCl_2$

f. $\% CaCl_2 = 100 x 1.106 g/ 3.506 g = 31.55 \%$
$\% CaCO_3 = 100 x 1.427 g/3.506 g = 40.70 \%$
$\% Ca(OH)_2 = 100 x 0.973 g/3.506 g = 27.75 \%$

Answers to Question 2

a. i. $\frac{1}{2} H_2(g) \rightarrow H(g)$

ii. $H\text{-}H(g) \rightarrow 2H(g)$ $\qquad \Delta H_{BDE} = +436 kJ/mol$
$\frac{1}{2} H_2(g) \rightarrow H(g)$ $\qquad \Delta H = +\frac{1}{2}(436 kJ/mol) = +218 kJ/mol$

b. $\Delta H_{rxn} = \Sigma\Delta H_{products} - \Sigma\Delta H_{reactants}$
$\Delta H_{rxn} = \Delta H_{f(H)} + \Delta H_{f(OH)} - \Delta H_{f(HOH)}$
$463 kJ = 218 kJ + \Delta H_{f(OH)} - (-242 kJ)$
$\Delta H_{f(OH)} = +3 kJ/mol$

c. i. $OH(g) + O_3(g) \rightarrow HO_2(g) + O_2(g)$
$HO_2(g) + O(g) \rightarrow OH(g) + O_2(g)$
$O_3(g) + O(g) \rightarrow 2O_2(g)$

ii. |O=O-Ö| ◄► |Ö-O=O|
H-O-Ö| ◄► H-Ö-O|

iii. For $O_2(g) \rightarrow 2O(g)$ $\qquad \Delta H_{BDE} = +495 kJ$
For $\frac{1}{2} O_2(g) \rightarrow O(g)$ $\qquad \Delta H = \frac{1}{2} (+495 kJ) = +248 kJ$
$O_3(g) + O(g) \rightarrow 2O_2(g)$

$\Delta H_{rxn} = \Sigma\Delta H_{products} - \Sigma\Delta H_{reactants}$
$\Delta H_{rxn} = 2\Delta H_{f(O2)} - \Delta H_{f(O)} - \Delta H_{f(O3)}$
$\Delta H_{rxn} = 2(0) - 248 - 142 = -390 kJ$

Answers to Question 3

a. *mol* H_2O *= (19.2g – 8.80 g – 6.8 g)/(18.0 g/mol) = 0.200 mol* H_2O
x mol C = mol CO_2 *= 8.80 g* CO_2*)(1 mol/44.0 g) = 0.200 mol C*
x mol N = mol NH_3 *= (6.80 g)/(17.0 g/mol) = 0.400 mol N*
x mol O = 2 x mol CO_2 *+ mol* H_2O *= (2 x 0.200) + 0.200*
= 0.600 mol O
x mol H = (3 x mol NH_3*) + (2 x mol* H_2O*) = (3 x 0.400) +*
(2 x 0.200) = 1.60 mol H
C0.200**H**1.60**N**0.400**O**0.600 =
C(0.200/0.200)**H**(1.60/0.200)**N**(0.400/0.200)**O**(0.600/0.200) = $CH_8N_2O_3$

b. $NH_4^+(aq) + OH^-(aq) \dashrightarrow NH_3(g) + H_2O(l)$

c. $2H^+(aq) + CO_3^{2-}(aq) \dashrightarrow CO_2(g) + H_2O(l)$

d. $(NH_3)_2CO_3$ *ammonium carbonate*

e. *Part c produced carbon dioxide, a non-polar gas that is not very soluble in polar water. Polar water molecules can hydrogen bond with each other effectively and exclude the molecules of carbon dioxide, which nucleate and form distinct bubbles.*
Part b produced ammonia, a very polar molecule which can hydrogen bond very effectively as it dissolves in water. Ammonia is a gas, however and escapes the solution without forming bubbles.

Answers to Question 4

a. $2H^+(aq) + 2I^-(aq) + H_2O_2(aq) \dashrightarrow 2H_2O(l) + I_2(aq)$
The mixture would be blue-black in the presence of starch.

b. $2H^+(aq) + MgO(s) \dashrightarrow H_2O(l) + Mg^{2+}(aq)$
A precipitate of $Mg(OH)_2$ *would form when the solution is made strongly basic.*

c. $Ca(s) + H_2O(l) \dashrightarrow Ca^{2+}(aq) + 2OH^-(aq) + H_2(g)$
The pH would higher than 7 due to the formation of the strong base.

Answers to Question 5

a. i. $H_2C_2O_4(aq) + OH^-(aq) \dashrightarrow HC_2O_4^-(aq) + H_2O(l)$
ii. $H_2C_2O_4(s) + 2OH^-(aq) \dashrightarrow C_2O_4^{2-}(aq) + 2H_2O(l)$

b. i.

```
 _          _
|O         |O|
  \\       /
    C  —  C
   /       \\
|O|         O|
 ‾          ‾
```

ii. *Both carbon atoms are* sp^2 *hybridized.*
iii. *The molecular geometry around each carbon atom is trigonal planar with bond angles about 120°.*

c. *All six atoms of the oxalate ion lie in the same plane due to the delocalization of the pi electrons among the four oxygen atoms. Three other resonance structures exist which show this delocalization. All four C-O bonds in oxalate will have the same length due to delocalization of electrons.*

Answers to Question 6

a. i. *The oxidation states of manganese are:* MnO, 2+; MnO_2, 4+; MnO_3, 6+, Mn_2O_3 3+; and Mn_2O_7; 7+.
ii. *Hausmanite is an equal molar mixture of* MnO *and* Mn_2O_3. *Two atoms per formula unit have oxidation number 3+ and one atom has oxidation number 2+.*
iii. Mn: $1s^22s^22p^63s^23p^64s^23d^5$
Mn^{3+}: $1s^22s^22p^63s^23p^64s^03d^4$
iv. *Metal oxides tend to be basic so MnO will react with nitric acid.*
$MnO(s) + 2H^+(aq) \dashrightarrow Mn^{2+}(aq) + H_2O(l)$

b. *Period 4 transition elements tend to lose 4s electrons before 3d electrons. Nearly all the Period 4 transition elements have electron configurations that include* $4s^2$.

c. *Potassium's one valence electron is well screened from the nucleus by 18 inner core electrons making potassium a relatively large atom. Potassium ion has eight valence electrons which are poorly screened and held tight and close to the nucleus making* K^+ *a relatively small cation.*
Cations are smaller than the atoms from which they are derived because the inner core electrons are poorly shielded from the nucleus and thus held closely and tightly.Chlorine is a small atom because its seven valence electrons are poorly screened by only ten inner core electrons. Upon gaining an electron, the resulting chloride anion has an excess of electrons and experiences electron-electron repulsion in its valence shell. Anions are larger than their atoms because of electron-electron repulsion.

TOPIC 13

PROPERTIES OF SOLUTIONS

In preparation for the critical quantitative chapters coming up, it is important to have a fundamental understanding of how solution concentrations are expressed and inter-converted mathematically. The colligative properties of solutions, especially vapor pressure lowering, freezing point depression and boiling point elevation, are often used to calculate the molar mass of compounds. Pay close attention to these sections:

13.4 **Ways of Expressing Concentration**
13.5 **Colligative Properties**

The Solution Process

Section 13.1

A **solution** is formed when one or more substances (the solutes) disperse uniformly throughout the solvent, normally the substance in greatest amount. All of the intermolecular attractive forces discussed in Topic 11 act between solute and solvent particles. For example, ion dipole interactions are the most common forces of attraction when ionic compounds dissolve in water. The positive end of water dipoles surround anions and cations are attracted to the negative end of the water dipole as shown in Figure 13.1

A few ionic compounds dissolve exothermically because the ion dipole forces formed in the solution are greater than the forces required to overcome the ionic bonds of the solid substance. Thermodynamically, exothermic processes tend to be spontaneous.

However, most ionic compounds dissolve endothermically because the energy required to break the ionic bonds is greater than the energy released when the ion dipole interactions are formed in the solution. The driving force of the solution process is the increase in entropy.

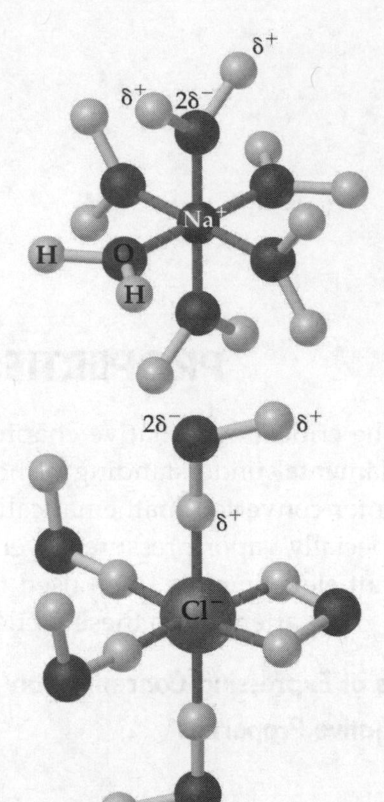

Figure 13.1. Hydrated Na+ and Cl- ions form ion-dipole interactions with water molecules.

Entropy is a state of randomness or disorder of a system. Formation of solutions is favored by the increase in entropy that accompanies mixing. Most processes occur with an increase in entropy.

Your Turn 13.1

Calcium chloride dissolves in water with an increase in temperature of the water. Is this process exothermic or endothermic? Use intermolecular forces to explain. Write your answer in the space provided.

Section 13.2 Saturated Solutions and Solubility

A saturated solution is one in which dissolved solute is in equilibrium with undissolved solute.

Solubility is the amount of solute needed to form a saturated solution in a given amount of solvent. The units of solubility for an aqueous solution are usually grams of solute per 100 milliliters of water, g solute/100 mL water.

An **unsaturated solution** contains less solute than a saturated solution.

A **supersaturated solution** contains more solute than a saturated solution. Supersaturation can be achieved because many substances are more soluble at high temperatures than they are at low temperatures. If a hot, saturated solution is slowly cooled, an unstable supersaturated solution often forms.

Factors Affecting Solubility

Section 13.3

Solubility increases with increasing strength of attractions between solvent and solute particles.

"Like dissolves like."

Substances with similar intermolecular attractive forces tend to be soluble in one another. Polar solutes tend to dissolve in polar solvents. Nonpolar solutes tend to dissolve in nonpolar solvents. Polar solvents, because of their relatively strong attractive forces, tend to exclude nonpolar substances and their relatively weak attractive forces.

Miscible liquids are pairs of liquids that dissolve in all proportions.

Immiscible liquids are those that do not dissolve in each other.

Your Turn 13.2

Explain using intermolecular forces why gasoline and water don't mix. Write your answer in the space provided.

Gases tend to be more soluble in liquids at higher pressure and at lower temperature. The solubility of a gas increases in direct proportion to its partial pressure above the solution.

Your Turn 13.3

Why do bubbles appear in a plastic soda bottle when the cap is removed? Write your answer in the space provided.

Section 13.4

Ways of Expressing Concentration

Concentration is the quantity of solute present in a given quantity of solvent or solution. Table 13.1 defines the various ways to express concentration.

Table 13.1. Ways of Expressing Concentration

Concentration	Abv.	Definition	Example
Mass percent	%	(mass of solute) ÷ (mass of solution) X 100	14.0 % aqueous ethanol = 14.0 g ethanol/100 g solution
Parts per million	ppm	10^6 x (mass of solute) ÷ (mass of solution)	18 ppm Pb^{2+} in water = 18 mg Pb^{2+} / 1000 g solution
Mole fraction	X	(moles of one component) ÷ (total moles of components)	$X_{NH3} = 0.10$ (One tenth of all the moles in solution is ammonia.)
Molarity	M	(moles of solute) ÷ (liters of solution)	0.15 M HCl (aq) = 0.15 moles HCl per liter of solution.
Molality	m	(moles of solute) ÷ (kilograms of solvent)	0.20 m NaCl (aq) = 0.20 moles NaCl per kilogram of water.

Section 13.5

Colligative Properties

Colligative properties are physical properties of solutions that depend only on the quantity of solute particles present in the solution and not on the identity of the solute particles. Colligative properties apply to any solution consisting of a volatile solvent and a nonvolatile solute.

A **volatile** substance is one that has a measurable equilibrium vapor pressure. A nonvolatile substance has no measurable vapor pressure.

The colligative properties are:

1. **Vapor Pressure Lowering**: A solution has a lower vapor pressure than does the pure solvent. The vapor pressure is lower because

the presence of the nonvolatile solute inhibits the escape of solvent molecules. Figure 13.2 compares the phase diagrams for a pure solvent and a solution.

2. **Boiling Point Elevation:** A solution boils at a higher temperature than a pure solvent. Notice that Figure 13.2 shows that a lower vapor pressure increases the boiling point, the temperature at which the equilibrium vapor pressure equals the external pressure.

3. **Freezing Point Lowering:** A solution freezes at a lower temperature than a pure solvent.

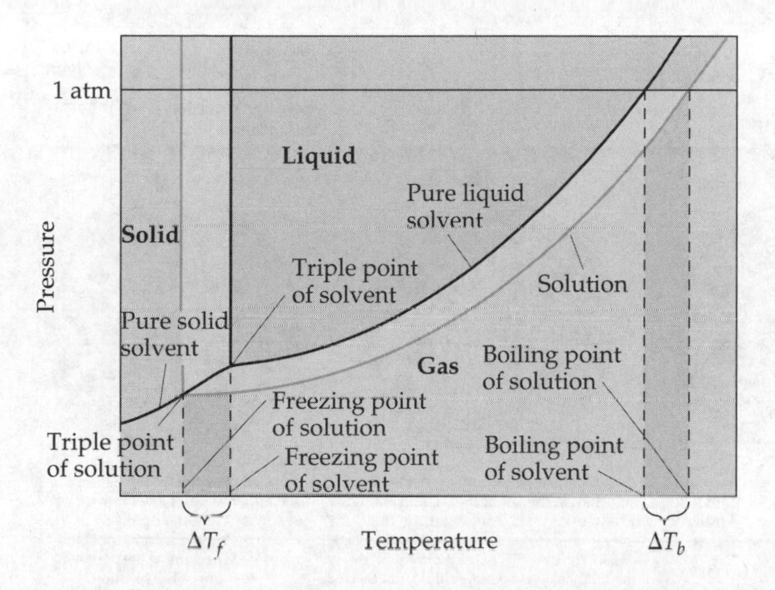

Figure 13.2. Phase diagram for a pure solvent and for a solution of a nonvolatile solute. Notice that the lower vapor pressure of the solution has the effect of increasing the boiling point and decreasing the freezing point.

4. Osmotic Pressure: the pressure required to prevent osmosis. Osmosis is the net movement of solvent through a semipermeable membrane toward a solution of higher concentration. See Figure 13.3.

The **Van't Hoff factor**, i, is a measure of the extent to which electrolytes dissociate. The ideal value for i is the number of ions per formula unit. For example, ideal solutions for the salts, NaCl, $CaCl_2$, $Al(NO_3)_3$ and $Fe_2(SO_4)_3$ have van't Hoff factors of 2, 3, 4 and 5, respectively. Since colligative properties depend on the number of particles dissolved in a solution, the van't Hoff factor, which accounts for the number of ions in solution, must be considered in any colligative property calculations.

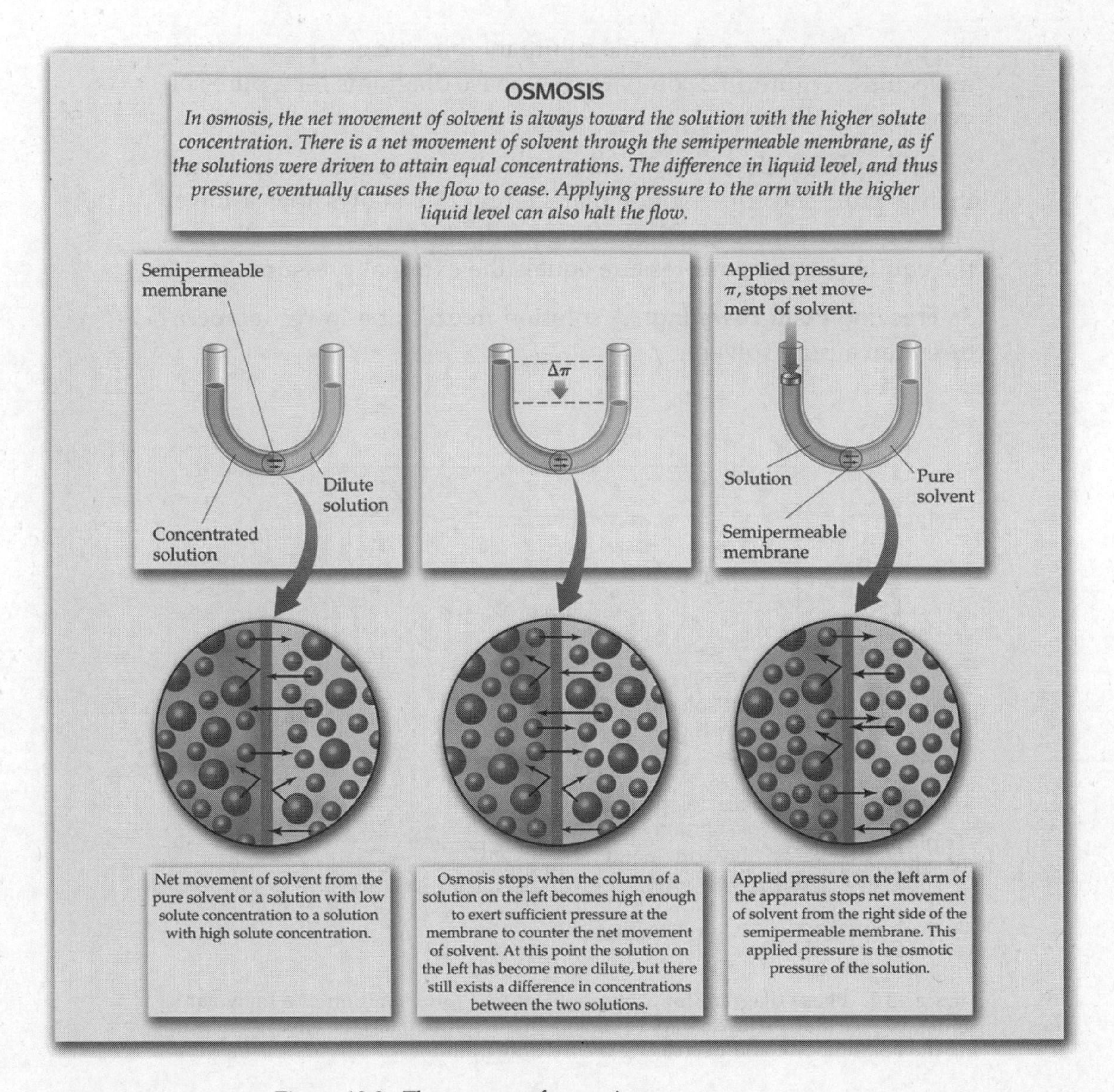

Figure 13.3. The process of osmosis.

Common misconception: The van't Hoff factor is usually less than the predicted factor for ideal solutions of ionic compounds due to ion pairing. Electrostatic forces of attraction between two ions of opposite charge cause them to adhere to one another and behave as a single particle. As a solution concentration decreases, the van't Hoff factor for its electrolytic solute approaches its ideal value.

Table 13.2 shows the quantitative relationships for colligative properties. Notice that each property is directly proportional to the number of moles of solvent or solute particles in the form of mole fraction, molality or molarity.

Table 13.2. Quantitative Relationships for Colligative Properties

Property	Equation	Explanation of terms
Vapor pressure lowering	$P_{A'} = X_A P^o_A$ Raoult's law	$P_{A'}$ = vapor pressure of solution X_A = mole fraction of solvent P^o_A = vapor pressure of pure solvent
Freezing point lowering	$\Delta T_f = K_f m i$	ΔT_f = change in freezing point, °C K_f = freezing point constant, °C/m m = molality of solute, mol solute/kg solvent i = the van't Hoff factor
Boiling point elevation	$\Delta Tb = Kb m i$	ΔT_b = change in boiling point, °C K_b = boiling point constant, °C/m m = molality of solute, mol solute/kg solvent i = the van't Hoff factor
Osmotic pressure	$\Pi = (n/V)RT$ or $\Pi = MRT$	P = osmotic pressure in atm (n/V) = M = molarity of solution in mol/L R = 0.0821 L atm/mol K T = absolute temperature in K

Example:

175 grams of calcium chloride are dissolved in 975 g of water. The density of the resulting solution is 1.10 g per milliliter.

a. What is the vapor pressure of the solution at 25°C? The vapor pressure of pure water at 25°C is 23.76 torr

b. At what temperature will the solution freeze? The freezing point constant for water is 1.86 °C/m.

c. At what temperature will the solution boil? The boiling point constant for water is 0.51 °C/m.

d. What will be the osmotic pressure at 27.0°C?

Solution:

a. moles of $CaCl_2$ *= 175 g/ 111 g/mol = 1.58 mol* $CaCl_2$*.*

However, $CaCl_2$ *dissociates into three moles of ions per mole of* $CaCl_2$*:*

$CaCl_2(s) \rightarrow Ca^{2+}(aq) + 2Cl^{-}(aq)$

The number of moles of ions in solution is three times the number of moles of $CaCl_2$*.*

Moles ions = 3 x 1.58 mol = 4.74 mol.

moles of water = 975 g / 18.0 g/mol = 54.2 mol H_2O*.*

mole fraction of water = moles of water/total moles = 54.2 mol/(54.2 + 4.74) mol = 0.920

This means that the vapor pressure of this solution is only 92% of the vapor pressure of water.

$P_{A'} = X_A P^o_A = 0.920 \times 23.76$ *torr = 21.9 torr*

b. The solvent molality, m = moles of ions/kg water = 4.74 mol/0.975 kg = 4.86 m.

(Notice that in Part a, we already have taken into account the van't Hoff factor of 3 for $CaCl_2$*.)*

$\Delta T_f = K_f\, m\, i = (1.86°C/m)(4.86\ m) = 9.04°C.$

The solution will freeze 9.04°C lower than the freezing point of water: 0.00°C – 9.04°C = -9.04°C.

c. $\Delta T_b = K_b\, m\, i = (0.51°C/m)(4.86\ m) = 2.48°C.$

The solution will boil at 2.48 °C higher than the boiling point of water: 100.00°C + 2.48°C = 102.48°C.

d. Total mass of the solution = 175 g $CaCl_2$ *+ 975 g* H_2O *= 1150 g solution.*

The liters of solution = (1150 g solution/1.10 g/mL solution)(1 L/1000 mL) = 1.05 L

The molarity, M, of the solution = moles of ions/ liters of solution = 4.74/1.05 = 4.51 M.

$\Pi = MRT = (4.51\ mol/L)(0.0821\ L\ atm/mol\ K)(273 + 27\ K) = 111\ atm.$

Note: We have assumed an ideal solution where the van't Hoff factor is three in all cases. The concentration of this solution is such that the true value of the van't Hoff factor is considerably less than three.

Molar Mass Determination

Any of the four colligative properties of solutions can be used to experimentally determine the molar mass of an unknown compound.

Example:

An unknown compound contains only carbon, hydrogen, and oxygen. Combustion analysis of the compound gives mass percents of 31.57% C and 5.30% H. A solution made by dissolving 10.56 g of the compound in 25.0 mL of water freezes at -5.20°C. Determine the empirical formula, molar mass, and molecular formula of the compound. Assume the compound is a nonelectrolyte. The freezing point constant for water is 1.86 °C/m.

Solution:

First calculate the empirical formula by converting the percent in grams of each element to moles. The mass % of oxygen = 100% – 31.57% – 5.30% = 63.13% O.

$C_{(31.57\ g)/(12.0\ g/mol)}\ H_{(5.30\ g)/(1.00\ g/mol)}\ O_{(63.13\ g)/(16.0\ g/mol)} =$

$C_{2.63} H_{5.30} O_{3.95} = C_{(2.63)/(2.63)} H_{(5.30)/(2.63)} O_{(3.95)/(2.53)}$

$= C_1 H_{2.01} O_{1.50} = C_2 H_4 O_3$

Calculate the number of moles by first determining the molality of the solution using the freezing point depression expression.

$\Delta T_f = K_f\, m\, i$

$5.20°C = (1.86\ °C/m)\ m\ (1)$

(The value for i is always 1 for nonelectrolytes.)

$m = (5.20°C)/(1.86°C/m) = 2.80$ *moles unknown/Kg water*

Convert molality to moles by multiplying by the number of kilograms of water.

Moles unknown = (2.80 moles unknown/kg H_2O*)(0.0250 kg* H_2O*) = 0.0700 mol unknown.*

(The density of water is 1.00 g / mL so 25.0 mL = 25 g = 0.0250 kg.)

Molar mass = grams unknown / moles unknown = 10.56 g / 0.0700 mol = 151 g /mol

151 grams per mole is, within significant figures, twice the mass of $C_2H_4O_3$ *so the molecular formula is* $C_4H_8O_6$.

Multiple Choice Questions

1. *Compared to a 1.0 M aqueous solution of glucose, a 1.0 M aqueous solution of calcium chloride will have*

A) the same melting and boiling points.

B) A lower melting point and a lower boiling point.

C) A lower melting point and a higher boiling point.

D) A higher melting point and a higher boiling point.

E) A higher melting point and a lower boiling point.

2. *What is the mass percent of methanol in a solution prepared by adding 32.0 g of methanol to 18.0 grams of water?*

A) 16.0%

B) 32.0%

C) 36.0%

D) 50.0%

E) 64.0%

3. *What is the relative order of freezing points of these three substances?*

I. water

II. 0.1 M aqueous ammonia solution

III. 0.1 M aqueous ammonium chloride

A) III < II < I

B) II < III < I

C) II = III < I

D) II < III = I

E) I < II < III

4. *A one thousand liter sample of water contains one gram of iron (III) ions. What is the concentration of* $Fe^{3+}(aq)$ *in parts per million?*

A) 0.001

B) 0.01

C) 0.1

D) 1

E) 10

5. *Consider a 0.50 M solution of each of the following salts. Which will have the lowest freezing point?*

A) $NaCl$

B) $MgCl_2$

C) K_2SO_4

D) $Cr(NO_3)_3$

E) $CaSO_4$

6. *Enough water is added to 11.5 grams of ethanol to make 2.00 liters of solution. What is the molarity of the ethanol?*

A) 0.125

B) 0.250

C) 0.500

D) 5.75

E) 0.333

7. *Which pairs of substances will dissolve in each other:*

I. CH_3OH

II. C_6H_6

III. CH_3CH_3

A) I and II only

B) II and III only

C) I and II, I and III, II and III.

D) I and III only

E) I and II, II and III only

8. *For a solution containing a non-volatile solute dissolved in a volatile solvent what is true of the vapor pressure, boiling point and freezing point of the solution compared to the pure solvent?*

Vapor pressure	boiling point	freezing point
A) *increases*	*increases*	*increases*
B) *decreases*	*decreases*	*decreases*
C) *increases*	*decreases*	*increases*
D) *decreases*	*decreases*	*increases*
E) *decreases*	*increases*	*decreases*

9. *The freezing point of a 1.0 m aqueous solution of an substance is approximately – 5.6°C. Which of these is the most likely substance dissolved in the solution? The molal freezing point constant for water is -1.86°C m^{-1}.*

A) CH_3CH_2OH

B) $NaCl$

C) $Ca(NO_3)_2$

D) C_6H_6

E) $Al_2(SO_4)_3$

10. *What is the mole fraction of water in a solution that contains 32 grams of methanol in 36 grams of water?*

A) O.33

B) 0.50

C. 0.67

D 1.0

E. 1.1

Free Response Questions

1. *Answer the following questions about these laboratory observations.*

Solid ammonium chloride dissolves in water with a marked decrease in temperature. Calcium chloride solid dissolves in water with a marked increase in temperature. Little or no temperature change is observed when solid sodium chloride dissolves in water.

a. *Write an equation that describes the dissolving process of ammonium chloride.*

b. *Is the dissolving of calcium chloride endothermic or exothermic. Explain.*

c. *Describe the opposing forces of attraction that are at work in the dissolution of calcium chloride. Which are greater? Why?*

d. *What can be said about opposing forces of attraction when sodium chloride dissolves in water. Why?*

e. *Use the observation for ammonium chloride to discuss these seemingly contradictory statements:*

Thermodynamically, exothermic processes tend to be spontaneous.

Most processes occur spontaneously when there is an increase in entropy.

2. *The molecular formula of an unknown compound is determined by combustion analysis and freezing point depression. A solution containing 0.496 g of benzoic acid, C_6H_5COOH, and 25.0 g of camphor, $C_{10}H_{16}O$, freezes at 173.3°C. The freezing point of pure camphor is 179.8°C. An unknown molecular compound is found to contain 80.77% C, 3.846% hydrogen and 15.38 % oxygen. A solution consisting of 0.243 g or unknown compound and 15.1 grams of camphor melts at 176.7°C.*

a. *What is the empirical formula of the unknown compound?*

b. *What is the freezing point constant for camphor? Specify the units.*

c. *What is the concentration of the unknown compound in camphor in units of molality?*

d. *What is the molar mass of the unknown compound?*

e. *What is the molecular formula of the unknown compound?*

Additional Practice in Chemistry the Central Science

For more practice answering questions in preparation for the Advanced Placement examination try these Problems in Chapter 13 of Chemistry the Central Science.

Additional Exercises: 13.89, 13.90, 13.91, 13.92, 13.94, 13.96, 13.97.

Integrative Exercises: 13.99 a-c, 13.101, 13.102, 13.104 a-b, 13.107.

Multiple Choice Answers and Explanations

1. C. *Solutions freeze at lower temperatures and boil at higher temperatures than their pure solutes. Upon dissolving $CaCl_2$ dissociates into three ions giving it effectively a 3.0 M concentration of ions whereas glucose is a molecular substance and dissolves intact. Because there are more particles in the $CaCl_2$ solution, it will freeze at a lower temperature and boil at a higher temperature than the glucose solution.*

2. E. *Percent by mass = grams of one substance divided by total grams of solution = 32.0 g methanol/(32.0 g methanol + 18.0 g water) = 64.0%*

3. A. *The freezing point will be the lowest for the solution with the most dissolved particles. Water has no dissolved particles. Ammonia, a molecular weak electrolyte, remains 99% intact when it dissolves. Ammonium chloride is a strong electrolyte, which dissociates into two ions per mole when it dissolves.*

4. D. *A part per million is one milligram of solute per liter of solution. One gram of Fe^{3+} is 1000 milligrams of Fe^{3+}. 1000 mg Fe^{3+}/1000 L water = 1 mg/L = 1 ppm Fe^{3+}.*

5. D. *The freezing point lowering is proportional to the number of particles dissolved in solution. All of these ionic compounds are strong electrolytes. However, chromium(III) nitrate dissociates into four ions per mole, the most of any of the choices.*

6. A. *Molarity is moles of solute per liter of solution. Ethanol is CH_3CH_2OH and has a molar mass of 46.0 g/mol which is four times 11.5.*

   M = (11.5 g ethanol)/(46.0 g/mol) ÷ 2 L = 0.125 M

7. B. *Substances with similar intermolecular attractive forces will dissolve in each other. CH_3OH is polar and hydrogen bonds. C_6H_6 and CH_3CH_3 are non polar molecules and only form London dispersion forces.*

8. E. *A solution's vapor pressure is lower than the pure solvent which*

increases the temperature at which the vapor pressure will equal the atmospheric pressure (boiling point). The freezing point is also lower.

9. C. *5.6°C is roughly three times the molecular freezing point constant so the compound that dissociates into three ions is the most likely substance that forms the solution.*

10. C. *The mole fraction of water is the number moles of water divided by the total number of moles of substances. Moles of water = 36g/18 g/mol = 2 mol.*

    Moles of CH_3OH = 32 g/32 g/mol == 1 mol.

    Mole fraction of water = 2 moles/ 2 + 1 mol = 0.67.

Free Response Answers

1. a. $NH_4Cl(s) \longrightarrow NH_4^+(aq) + Cl^-(aq)$

   b. *The dissolving of calcium chloride in water is exothermic as evidenced by the increase in temperature of the water. In the dissolving process, heat flows out of the system (ammonium chloride) and into the surroundings (water).*

   c. *Ionic bonds are electrostatic forces of attraction between ions of opposite charge. These ionic bonds are broken when calcium chloride to dissolves in water. When calcium chloride dissolves in water, ion-dipole forces exists between a Ca^{2+} ion and the negative end of a water molecule, and between a Cl^- ion and the positive end of a water molecule. Forming the ion-dipole interactions releases more energy than it takes to break the ionic bonds in calcium chloride. As a result a net amount of heat is released to the system.*

   d. *Because NaCl dissolves with no apparent change in temperature it can be said that the ion-dipole forces between water molecules and sodium and chloride ions are about equal to the ionic bonds in NaCl.*

   e. *The process of ammonium chloride dissolving in water is endothermic based on the observed decrease in temperature. However, the dissolving process occurs with an increase in entropy or randomness of the system. In the case of ammonium chloride and most ionic salts that dissolve in water, the drive toward increasing entropy outweighs the drive toward minimum enthalpy. As a result, the endothermic process is spontaneous.*

2 a. $C_{(80.77\ g)/(12.0\ g/mol)} H_{(3.846\ g)/(1.00\ g/mol)} O_{(15.38\ g/16.0\ g/mol)} =$

$C_{6.73} H_{3.846} {}_{O0.961} = C_{(6.73/.961)} H_{(3.84/0.961)} O_{(0.961/0.961)} = C_7H_4O$

b. $m = (0.496\ g\ benzoic\ acid / 122\ g/mol)/(0.025\ kg\ camphor) = 0.163\ m$

$\Delta T_f = K_f\, m\, i$

$(179.8 - 173.3\,°C) = K_f\,(0.163\ m)(1)$

$Kf = 39.9\,°C/m$

c. $(179.8 - 176.7\,°C) = (39.9\,°C/m)\ m\ (1)$

$m = 0.0777\ m$

d. moles of unknown = (0.0777 mol unknown/kg camphor)(0.0151 kg camphor) = 0.00117 moles unknown

molar mass of unknown = grams unknown / moles of unknown = 0.243 g/0.00117 mol = 208 g/mol.

e. C_7H_4O has a mass of 104 g/mol. 208 is double 104 so the molecular formula is $C_{14}H_8O_2$

Your Turn Answers

13.1. *Calcium chloride dissolving with an increase in temperature is an exothermic process, releasing heat to the water. The ion dipole forces in the solution release more energy than is required to break the ionic bonds of calcium chloride. The excess heat goes to warm the water.*

13.2. *Gasoline is a nonpolar liquid that has only dispersion forces holding its molecules together. Water molecules are tightly held together by hydrogen bonding. Water molecules have greater attractions for each other than they do for molecules in gasoline and so they exclude the gasoline molecules.*

13.3. *Soda is an aqueous solution of carbon dioxide. A sealed soda bottle contains an equilibrium vapor pressure of carbon dioxide above the liquid. When the cap is removed, the CO_2 above the liquid escapes to the environment lowering the partial pressure. Gases are less soluble at lower partial pressures above the solution and bubbles form and begin to escape the solution. Nonpolar carbon dioxide is not very soluble in polar water at lower pressure.*

TOPIC 14

CHEMICAL KINETICS

Chemical kinetics begins a series of topics spanning the next six chapters ending with electrochemistry that comprise the heart and soul of what is expected for students to know for the AP examination. A thorough knowledge of these topics and their interrelationships is essential. Often students are asked to determine a rate law and its units from tabular data, use it to calculate rates and concentrations under specified conditions, and match a suitable mechanism to a rate law. Content in all sections should be mastered.

Reaction Rates

Section 14.1

Reaction rate is a measure of the speed of a chemical reaction. Rate of reaction is expressed as the change in the amount of reactants or products per unit time. Most often the unit for reaction rate is molarity per second (M/s).

Common Misconception: Molarity per second is often expressed as mol $L^{-1} \bullet s^{-1}$.

$M/s = mol/L \bullet s = mol\ L^{-1} \bullet s^{-1}$.

Reaction rate can also be expressed in pressure units per time (atm/s or torr/s) or sometimes in absorbance units per time. (For more on how absorbance can be used to monitor the changing concentration of a reactant or product, see A Closer Look: Using Spectroscopic methods to Measure Reaction Rates in Section 14.2.)

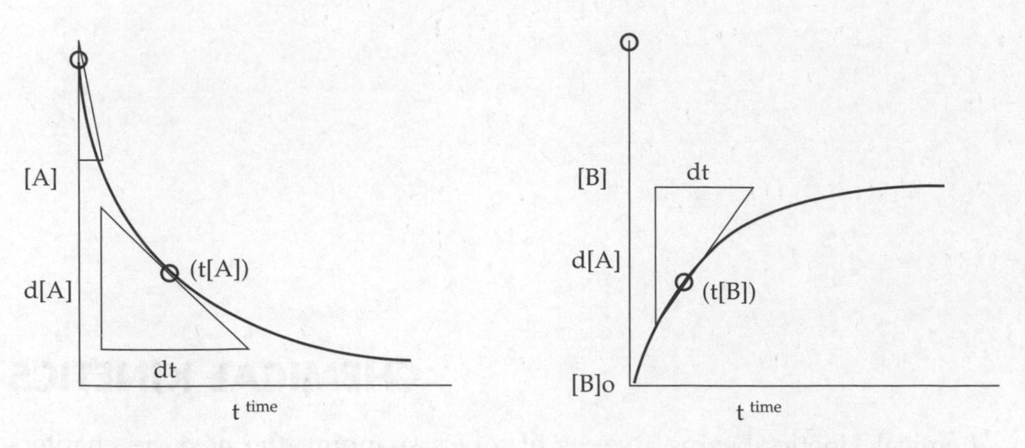

Figure 14.1. Graph of a typical reaction A→B showing how [A] decreases with time and how [B] increases with time.

Concentrations of reactants and products change with time and so does rate. Consider a simple chemical reaction where A → B. Figure 14.1a illustrates how the molar concentration of A changes with time. As time increases the concentration of A decreases. The slope of the tangent line at any point (t, [A]) represents the instantaneous rate of the reaction. The rate is:,

$$\text{Rate} = -\Delta[A] / \Delta t$$

Rates are always expressed as positive quantities. Because the slope in the graph in Figure 14.1a is negative, a negative sign is used to express rate as a positive quantity. A typical reaction rate starts out fast and becomes slower as time goes on. The initial rate at time = 0 is always the largest rate.

Figure 14.1b illustrates how [B] changes with time. As time increases, the concentration of B increases. The slope of the tangent line at any point (t, [B]) represents the instantaneous rate of the reaction. The rate is:,

$$\text{Rate} = +\Delta[B] / \Delta t$$

Notice that the slope of the line is positive so a positive sign in the expression denotes a positive rate.

Section 14.2 Factors that Affect Reaction Rates

On a molecular level, reaction rates depend on the frequency of collisions between molecules and the energy with which the molecules collide. Rate of reaction increases with greater frequency of collisions, and/or greater energy with which the collisions occur.

The major factors that affect reaction rate are:

1. **Concentrations of reactants.** Higher concentrations of reactants produce a faster reaction. As concentration increases, the frequency of collisions increases, thus increasing reaction rate.

2. **Temperature.** Increasing temperature increases reaction rate. Higher temperatures provide increased kinetic energies of molecules so the molecules move more rapidly. Faster moving molecules produce more frequent and higher energy collisions. Additionally, at higher temperatures, there is a greater fraction of collisions resulting in a reaction.

3. **Physical state of the reactants.** In general, homogeneous mixtures of either liquids or gases react faster than heterogeneous mixtures. A solid, for example, tends to react more slowly with either a liquid or a gas because molecular collisions are limited to the surface area of the solid. Increasing the surface area will increase the frequency of collisions, and thus increase the rate of reaction.

4. **The presence of a catalyst.** A catalyst increases the rate of reaction by affecting the kinds of collisions that lead to a reaction. A catalyst acts by changing the mechanism of the reaction. A reaction mechanism, to be discussed in more detail later, is a step-by-step process by which reactants become products.

Your Turn 14.1

Mine explosions from the ignition of powdered coal dust are relatively common, yet lumps of coal burn without exploding. Explain.

Write your answer in the space provided.

Concentration and Rate

Section 14.3

A rate law is a mathematical relationship that shows how rate of reaction depends on the concentrations of reactants. For any general reaction where:

a A + b B $\rightarrow$ products

the rate law takes the form:

$\text{Rate} = k[A]^m[B]^n$

k is the rate constant, [A] is the molar concentration of reactant A, [B] is the molar concentration of reactant B, and the exponents m and n are usually small, whole numbers that relate to the number of molecules of A and B that collide in the step-by-step mechanism.

Common misconception: The coefficients that balance the equation, a and b, are not the same as the exponents, m and n, in the rate law.

Reaction order is the sum of the exponents m and n in a rate law. For example, consider the rate law:

$\text{Rate} = k[A]^1[B]^2$

The reaction is said to be "first order in A" and "second order in B." The sum of m + n is the overall order of the reaction. In this example, 1 + 2 = 3 so the reaction is "third order" overall.

The rate law for any chemical reaction must be determined experimentally, often by observing the effect of changing the initial concentrations of the reactants on the initial rate of the reaction.

Consider the reaction

$A + 2B \rightarrow 2C + D$

The rate law always takes the form:

$\text{Rate} = k[A]^m[B]^n$

A series of experiments measuring initial rate at various concentrations of reactants might give the data in Table 14.1.

Table 14.1. The effect of changing concentrations on the initial rate of a reaction.

Experiment	**[A] (M)**	**[B] (M)**	**rate = -d[A]/dt (M/s)**
1	0.10	0.10	0.04
2	0.10	0.20	0.08
3	0.20	0.20	0.32

Experiments 1 and 2 show that the rate is doubled when [B] is doubled. This means that the exponent of [B] is 1. ($2^1 = 2$ so n=1.)

Experiments 2 and 3 show that the rate is quadrupled when [A] is doubled. This means that the exponent of [A] is 2. ($2^2 = 4$ so m=2.) The data in Table 14.1 show that the rate law is:

Rate = $k[A]^2[B]^1$

The reaction is second order in A, first order in B, and third order overall.

The data in Table 14.1 can be used to calculate the value of the rate constant.

Example:

What is the numerical value of the rate constant for the reaction described in Table 14.1. Specify its units.

Solution:

Rate = $k[A]^2[B]^1$

$0.04\ M/s = k[0.1M]^2[0.1M]^1$

$k = (0.04\ M/s)/(0.01\ M^2 \times 0.1M)$

$= 40\ \ 1/M^2s$

Your Turn 14.2

What are the units for each rate constant for the following rate laws? Assume each rate is expressed in M/s. a. rate = k[A]; b. rate = $k[A]^2$; c. rate = $k[A]^3$. Write your answers in the space provided.

The coefficients that balance a chemical equation are proportional to the rates of appearance or disappearance or reactants and products. Reaction rate relates directly to stoichiometry.

Example:

Consider the reaction between gaseous hydrogen and gaseous nitrogen to produce ammonia gas.

$3H_2(g) + N_2(g) \rightarrow 2NH_3(g)$

At a particluar time during the reaction $H_2(g)$ disappears at the rate of 3.0 M/s.

a. What is the rate of disappearance of $N_2(g)$?

b. What is the rate of appearance of $NH_3(g)$?

Solution:

b. In the balanced equation, $N_2(g)$ has a coefficient of 1 whereas $H_2(g)$ has a coefficient of 3. $N_2(g)$ disappears at one third the rate of $H_2(g)$.

1/3(3.0 M/s) = 1.0 M/s.

b. $NH_3(g)$ appears at two thirds the rate of $H_2(g)$.

2/3(3.0 M/s) = 2.0 M/s.

Your Turn 14.3

If ammonia appears at 2.6 M/s, how fast does hydrogen disappear?

Place your answer in the space provided.

In terms of mathematics, the following equation applies to the reaction:

$$-\Delta H_2(g)/\ \Delta t = -3\Delta N_2(g)/\ \Delta t = +\ 3/2\ \Delta NH_3(g)/\ \Delta t$$

Notice the use and placement of signs and the coefficients that balance the chemical equation. If they seem counter-intuitive, translate the mathematical expression into words: "The disappearance of hydrogen gas is three times the rate as the disappearance of nitrogen gas and three-halves the rate of appearance of ammonia gas."

The application of stoichiometry to the data in Table 14.1 can be used to calculate various rates.

Example:

From the data for Experiment 1 in Table 14.1, calculate the rate of appearance of C.

Solution:

The initial rate of disappearance of A in Experiment 1 is 0.04 M/s. The balanced equation shows that C has a coefficient twice that of A so the rate of appearance of C is twice that of A. $+ \Delta C/\Delta t = -2A/\Delta t = 2(0.04\ M/s) = 0.08\ M/s$.

The Change of Concentration with Time

Section 14.4

A first order reaction is a reaction whose rate depends on the concentration of a single reactant raised to the first power. The rate is expressed in terms of the rate law and the slope of the tangent line at any point t,[A] on the graph in Figure 14.1a.

$\text{Rate} = k[A] = -\Delta A/\Delta t$ differential rate law

Using calculus, this equation is transformed into the equation of the curved line in Figure 14.2a. The equation of the line is

$\ln[A]_t = -kt + \ln [A]_0$ integrated rate law

Notice that the equation is in the form of the simple linear equation, $y = mx + b$ where $y = \ln[A]_t$, $b = \ln [A]_0$, and the slope of the line = -k, the rate constant.

For a **first order reaction** a plot of ln[A] vs. time will yield a straight line with a slope of –k as shown in Figure 14.2 b. This is a useful graphical method for determining rate constants for first order reactions.

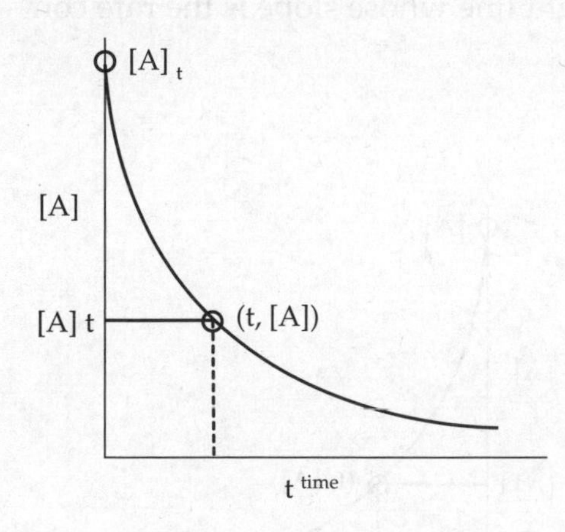

Figure 14.2 a. Plot of [A] vs. time for a first order reaction.

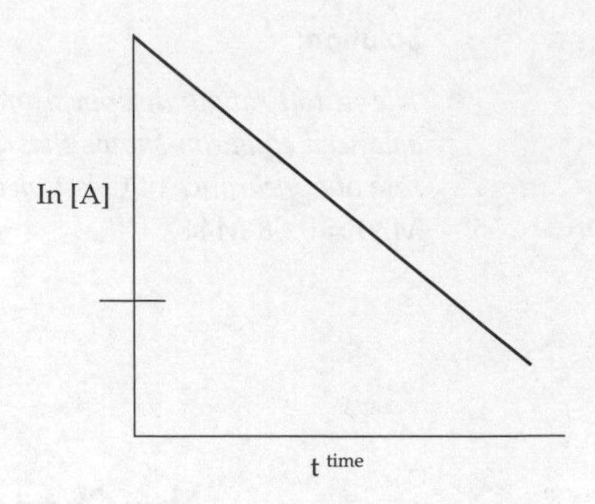

Figure 14.2 b. Plot of ln[A] vs. time for a first order reaction. The negative slope of the line equals the rate constant.

Often the equation of the line for a first order reaction is written in the more familiar expression

$\ln([A]_0/[A]_t) = +kt$

A simple **second-order reaction** is one whose rate depends on the concentration of the reactant raised to the second power.

The characteristic equations are

Rate = $k[A]^2 = -\Delta A/\Delta t$ differential rate law

and

$1/[A]_t = kt + 1/[A]_0$ integrated rate law

Note that the integrated rate law is a linear equation of the form

$y = mx + b$

A plot of 1/[A] vs. time will yield a straight line whose slope is the rate constant k as shown in Figure 14.3b.

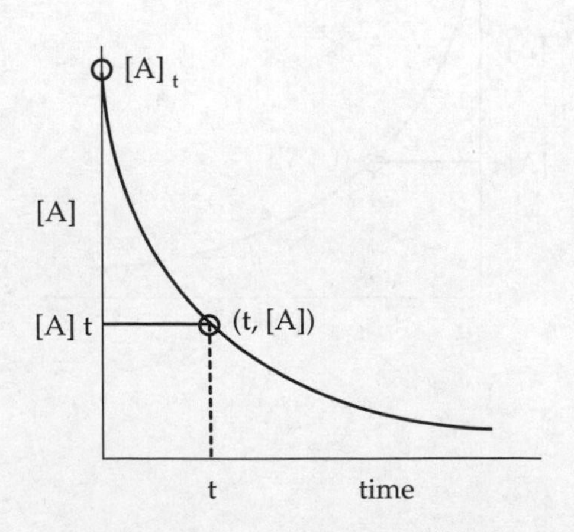

Figure 14.3 a. Plot of [A] vs. time for a second-order reaction.

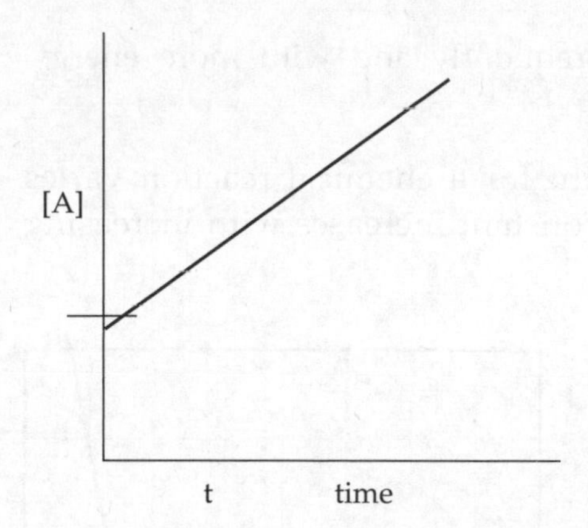

Figure 14.3 b. Plot of 1/[A] vs. time for a second-order reaction. The slope of the line = the rate constant.

The **half life** of a reaction, $t_{1/2}$ is the time required for the initial concentration of a reactant to fall to half its value.

For a first order reaction: $t_{1/2} = 0.693/k$

For a second order reaction: $t_{1/2} = 1/k[A]_0$

Table 14.2 summarizes the mathematical relationships of first and second order reactions.

Table 14.2. Mathematical relationships of simple first and second order kinetics.

Order or Reaction	**First**	**Second**
Differential rate law	Rate = $k[A] = -\Delta A/\Delta t$	Rate = $k[A]^2 = -\Delta A/\Delta t$
Integrated rate law	$\ln[A]_t = -kt + \ln [A]_0$ or $\ln([A]_0/[A]_t) = +kt$	$1/[A]_t = kt + 1/[A]_0$
Half life	$t_{1/2} = 0.693/k$	$t_{1/2} = 1/k[A]_0$
Straight line plot	ln[A] vs. time	1/[A] vs. time
Slope =	-k	k

Section 14.5

Temperature and Rate

Generally, increasing temperature increases reaction rate. The collision model, based on kinetic molecular theory says that molecules must collide in order to react. Temperature increases the speed of molecules and as molecules move faster they collide more frequently and with more energy, increasing reaction rates.

Figure 14.4 shows how the rate constant for a chemical reaction varies with temperature. Generally the rate constant increases with increasing temperature.

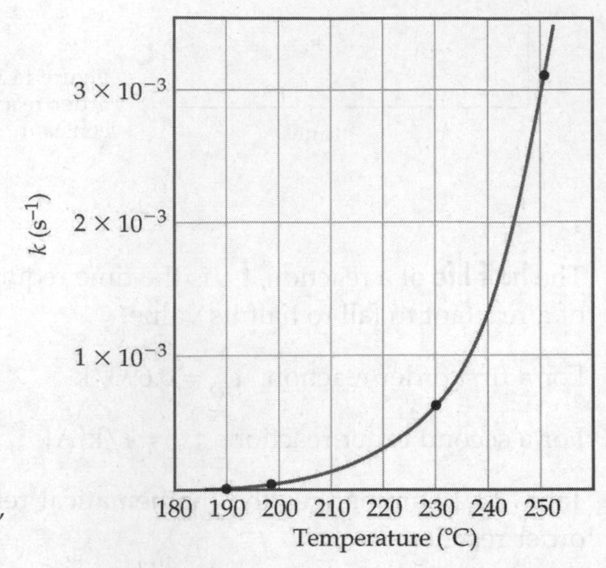

Figure 14.4. The rate constant, k, varies with temperature, T.

The algebraic equation that describes the line in Figure 14.3 relates activation energy, Ea, to the rate constant, k, at various temperatures, T.

$\ln(k_1/k_2) = (Ea/R)(1/T_2 - 1/T_1)$

k_1 = the rate constant at temperature, T_1. k_2 = the rate constant at temperature T_2 and Ea is the activation energy. (All temperatures must be expressed in Kelvin.) R = 8.314 J/K mol.

Activation energy, Ea, is the minimum amount of energy required to initiate a chemical reaction. Figure 14.5 shows the activation energy on the energy profile of a typical exothermic reaction. The activation energy can be considered to be an energy barrier that molecules must get over in order to react. On an energy profile, the activation energy is the energy difference between the reactants and the highest point of the profile.

The **activated complex** is the highest energy arrangement of molecules as they change from reactants to products. The very top of the energy profile represents the energy of the activated complex, also called the transition state.

A **catalyst** acts to lower the activation energy of a chemical reaction and thus increase the rate of reaction.

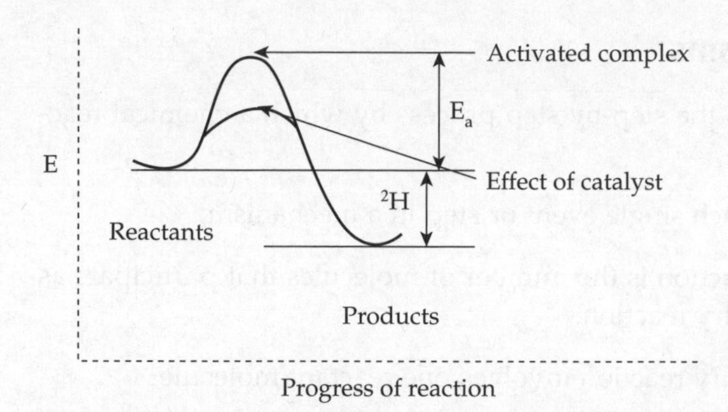

Figure 14.5. Energy profile for a typical exothermic reaction showing activation energy, Ea, change in enthalpy, ΔH, and the effect of a catalyst.

For any reaction, the higher the activation energy, the slower the rate. Very fast reactions have low activation energies and slow reactions have high activation energies. Figure 14.6 illustrates the relationship of activation energy to reaction rate. Profile 1 describes a reaction with the faster rate and the reaction for Profile 2 has the slowest rate.

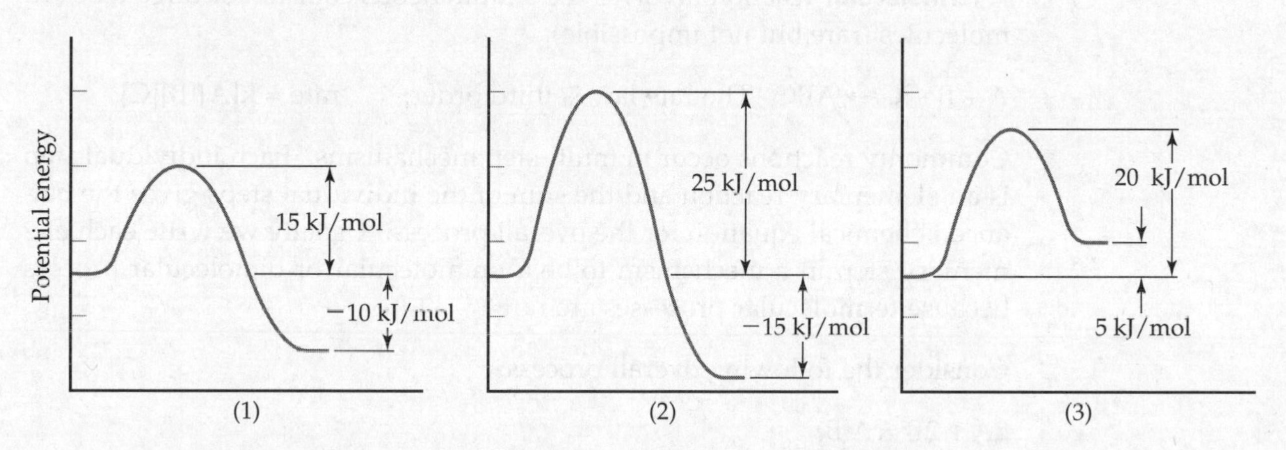

Figure 14.6. Energy profiles of two exothermic reactions and one endothermic reaction showing the activation energies. The higher the activation energy the higher the rate of reaction.

Your Turn 14.4

Explain in terms of collision theory, why temperature affects rate of reaction. Write your answer in the space provided.

Section 14.6

Reaction Mechanisms

A **reaction mechanism** is the step-by-step process by which a chemical reaction occurs.

An **elementary step** is each single event or step in a mechanism.

The **molecularity** of a reaction is the number of molecules that participate as reactants in an elementary reaction.

A **unimolecular** elementary reaction involves one reactant molecule.

$AB \rightarrow A + B$ The rate law is first order: rate = k[AB]

A **bimolecular** elementary reaction involves the collision of two reactant molecules.

$A + B \rightarrow AB$ The rate law is second order: rate = k[A][B]

A **termolecular** reaction involves the simultaneous collision of three reactant molecules (rare but not impossible).

$A + B + C \rightarrow ABC$ The rate law is third order: rate = k[A][B][C]

Commonly reactions occur in multi-step mechanisms. Each individual step is an elementary reaction and the sum of the individual steps gives the balanced chemical equation for the overall process. Usually we write each elementary step in a mechanism to be a unimolecular or bimolecular process because termolecular processes are rare.

Consider the following overall process:

$2A + 2B = A_2B_2$

A possible mechanism that can explain how reactants A and B become the product A_2B_2 is:

Elementary step		rate law for each elementary step
1.	$A + A = A_2$	rate = $k[A]^2$
2.	$A_2 + B = A_2B$	rate = $k[A_2][B]$
3.	$A_2B + B = A_2B_2$	rate = $k[A_2B][B]$

$2A + 2B = A_2B_2$

Notice that the sum of the elementary steps of a mechanism gives the chemical equation for the overall process.

An **intermediate** is a chemical species that is formed in one elementary step of a multi-step mechanism and consumed in another. In the above mechanism, A_2 and A_2B are intermediates. An intermediate is never a reactant or product of the overall reaction. Intermediates never appear in the rate law for the overall reaction.

The **rate determining step** is the slowest step of a multi-step mechanism and governs the rate of the overall reaction.

The slowest step in a multi-step mechanism is reflected in the rate law. The rate law includes only those reactant molecules which react during and before the rate determining step. Therefore, the rate law must be determined experimentally and cannot be determined from the overall balanced equation.

For example, if Step 1 in the above mechanism is the slowest step, the rate law is rate = $k[A]^2$ because two molecules of A have taken part in the mechanism up to this step.

If Step 2 is slowest, the rate law will be rate = $k[A]^2[B]$ because two molecules of A and one molecule of B have taken part taken part through Step 2 of the mechanism.

If Step 3 is slowest, the rate law will include two molecules of A and two molecules of B, because they all have taken part through Step 3 of the mechanism: rate = $k[A]^2[B]^2$.

Your Turn 14.5

Propose an alternate mechanism for the overall reaction: $2A + 2B = A_2B_2$. Write your answer in the space provided.

A **catalyst** is a substance that increases the rate of a chemical reaction without undergoing a permanent change in the process. Catalysts act by changing the mechanism of a reaction so that the slowest step in the un-catalyzed reaction does not exist in the catalyzed process. The effect of a catalyst is to lower the activation energy of the overall process by replacing the slowest step with one or more faster steps having an activation energy lower than that of the un-catalyzed rate-determining step.

To illustrate the action of a catalyst consider the ozone cycle, the process that cycles diatomic oxygen to ozone and back to diatomic oxygen in the upper atmosphere. In its simplest form, the mechanism might look something like this:

Step 1 $O_2(g) + h\surd \rightarrow O(g) + O(g)$

Step 2 $O(g) + O_2(g) \rightarrow O_3(g)$

Step 3 $O_3(g) + h\surd \rightarrow O^2(g) + O(g)$

Step 4 $O(g) + O(g) \rightarrow O_2(g)$

Overall: $O_3(g) + O_2(g) \rightarrow O_2(g) + O_3(g)$

(Steps 1 through 4 continually repeat, producing and destroying ozone at the same rate while absorbing harmful ultraviolet radiation (h√) from the sun.

It has been shown that chlorine atoms coming from chlorofluorocarbons released to the atmosphere catalyze the $O_3 \rightarrow O_2$ reaction. The net result is that ozone is depleted faster than it is generated by the natural cycle. Thus, chlorine atoms from chlorofluorocarbons catalytically deplete ozone in the stratosphere. In its simplest form the mechanism that catalyzes O_3 to O_2 is:

Step 1 $2Cl(g) + 2O_3(g) \rightarrow 2ClO(g) + 2O_2(g)$

Step 2 $ClO(g) + ClO(g) \rightarrow O_2(g) + 2Cl(g)$

Overall: $2O_3(g) \rightarrow 3O_2(g)$

Notice that ClO(g) is an intermediate. ClO(g) is generated as a product in one elementary step and is consumed as a reactant in another later elementary step.

By contrast, Cl is a catalyst. It is consumed as a reactant in one elementary step, and is re-generated as a product in a later step. A catalyst increases the rate of a chemical reaction without undergoing a permanent change. A catalyst acts by providing a different mechanism for the reaction, one that has a lower activation energy.

(For more detailed information about the natural ozone cycle and its catalytic depletion by chlorofluorocarbons, see Section 18.3 of Chemistry the Central Science.)

Multiple Choice Questions

1. *Which of these change with time for a first-order reaction?*

I. rate of reaction II. rate constant III. half-life IV. concentration of reactant

A) I only B) III only C) I and II only D) II and III only
E) I and IV only

2. *Under certain conditions, the average rate of appearance of oxygen gas in the reaction:*

$2O_3(g) \rightarrow 3O_2(g)$

is 6.0 torr s^{-1}. What is the average rate expressed in units of torr s^{-1} for the disappearance of O_3?

A) 9.0 B) 6.0 C) 4.0 D) 3.0 E) 1.2

3. *For irreversible chemical reactions, the rate will be affected by changes in all of these factors except:*

A) temperature. B) concentration of reactants.

C) presence of a catalyst. D) concentration of products.

E) surface area of solid reactant

4. *The rate expression for a third-order reaction could be:*

A) rate = $k[X]$ B) rate = $k[X]^2[Y]$ C) rate = $k[X][Y]$
D) rate = $k[X]^2[Y]^2$ E) rate = $k[X]^2$

5. *The slowest step of a reaction mechanism is called the:*

A) elementary step.

B) inhibitor.

C) rate law.

D) rate-determining step.

E) catalyst

6. *The dissociation of XY molecules, as shown below, occurs at a temperature of 800 K. The rate constant, $k = 6.0 \times 10^{-3}\ s^{-1}$.*

   $2XY(g) \rightarrow X_2(g) + Y_2(g)$

What is the reaction order?

A) 0 B) 1 C) 2 D) 3 E) 4

7. *The rate law of a certain reaction is rate = k [X][Y]. The units of k, with time measured in seconds, is*

A) s^{-1} B) $M^{-1}s^{-2}$ C) $M^{-2}s^{-1}$ D) M^{-1} E) $M^{-1}s^{-1}$

8. *For a first-order reaction of half-life 75 min, what is the rate constant in min^{-1}?*

A) (0.693)/75 B) (0.693)/1.25 C) (0.693)(75) D) 75/(0.693) E) 0.693

9. *The half-life of ^{14}C is 5730 years. Approximately how many years will it take for 94% of a sample to decay?*

A) 5,730 B) 2 X 5730 C) 3 X 5730 D) 4 X 5730 E) 5 X 5730

10. *A reaction between X_2 and Y was found to be described by the rate equation rate = $k[X_2][Y]^2$. What can be said about the process?*

A) the balanced chemical equation for the reaction is

   $X_2 + 2Y \rightarrow X_2Y_2$

B) the rate-determining step must be a three-atom collision.

C) the rate-determining step must be the first step of a multi-step mechanism.

D) the mechanism is most likely to be multi-step.

E) the mechanism must consist of just one elementary step.

Free Response Questions

1. *The overall chemical equation for the reaction of nitrogen oxide, NO, with chlorine, Cl_2, is:*

   $2\ NO + Cl_2 \rightarrow 2\ NOCl$

The initial rates of reaction for various concentrations of the reactants were measured and recorded at constant temperature as follows:

Experiment	*[NO] (M)*	*[Cl_2] (M)*	*$-\Delta[Cl_2]/\Delta t$ (M/hr)*
1	0.25	0.50	0.75
2	0.25	1.00	3.02
3	0.50	2.00	24.10

a. Determine the rate law for this reaction.

b. Calculate the numerical value for the rate constant and specify the units.

c. What is the order of this reaction with respect to each reactant and what is the overall order of the reaction?

d. What is the rate of disappearance of Cl_2 when the initial concentrations of the reactants are: [NO] = 0.50 M and [Cl_2] = 0.10 M?

e. When Cl_2 is disappearing at 4.5 M/hr, what is the rate of appearance of NOCl?

f. What is the rate of appearance of NOCl when the initial concentrations of the reactants are: [NO] = 0.20 M and [Cl_2] = 0.30 M?

No calculators are permitted for Question 2.

2. *Consider the proposed mechanism for the reaction between nitrogen monoxide and hydrogen gas. Assume the mechanism is correct.*

Step 1: $2NO \rightarrow N_2O_2$

Step 2: $N_2O_2 + H_2 \rightarrow N_2O + H_2O$

Step 3: $N_2O + H_2 \rightarrow N2 + H_2O$

a. Use the steps in the mechanism to determine the overall balanced equation for the reaction. Clearly show your method.

b. If Step 2 is the rate-determining step, write the rate law for the reaction. Explain your answer.

c. If the observed rate law is rate = k $[NO]^2[H_2]^2$, which step is rate determining? Explain your reasoning.

d. Identify all the intermediates in the mechanism.

e. If the first step is the rate-determining step, what is the order of the reaction with respect to each reactant?

Additional Practice in Chemistry the Central Science

For more practice working kinetics problems in preparation for the Advanced Placement examination, try these problems in Chapter 14 of Chemistry the Central Science:

Additional Exercises: 14.81, 14.82, 14.84, 14.87, 14.89, 14.91, 14.94.

Integrated Exercises: 14.98, 14.100, 14.102, 14.104, 14.105.

Answers and explanations to multiple choice questions:

1. *E. Rate decreases with time because the concentrations of reactants decrease. The half life of a first order reaction remains constant as does the rate constant at constant temperature.*

2. *C. Relative rates of disappearance of reactants and appearance of products are proportional to the stoichiometry of the reactants. Using the coefficients of the balanced equation,* O_2 *appears 3/2 as fast as* O_3 *disappears. Thus, the rate at which* O_3 *disappears = 2/3 X 6.0 torr* s^{-1} *= 4.0 torr* s^{-1}*.*

3 *D. The frequency of collisions of reactant molecules largely determines reaction rate. Any factor that changes the frequency of collisions will affect the rate. Concentrations of products of an irreversible reaction do not affect the rate of a reaction because collisions of product molecules do not aid the forward reaction. Changing temperature, reactant concentrations, and the surface area of a solid reactant all change the frequency of collisions. A catalyst increases the rate of reaction by changing the reaction mechanism, eliminating the slowest step, and lowering the activation energy of the overall process.*

4. *B. The overall order of a reaction is the sum of the exponents of the rate law.*

5. *D. An elementary step is each individual step in a mechanism. The rate law is a mathematical equation that relates rate to concentrations of reactants. A catalyst is a substance that increases the speed of a chemical reaction by changing the mechanism. An inhibitor is a substace that decreases the rate of a chemical reaction.*

6. *B. The units of the rate constant conveniently reveal the order of the reaction. The rate constant for a first order reaction contain one unit, reciprocal time,* s^{-1}*. Second order rate constants contain two units:* $M^{-1}s^{-1}$*. Third order contain three:* $M^{-2}s^{-1}$ *and so fourth. Notice that if units of a third order reaction are given as* $L^2\ mol^{-2}\ s^{-1}$ *this relationship is often obscured. Be sure to change* $L^2 \bullet mol^{-2} \bullet s^{-1}$ *to the more familiar* $M^{-2}s^{-1}$*.*

7. *E. The sum of the exponents of the rate law is 2, so this is a second order reaction. Second order reactions are typified by one concentration unit and one time unit for a total of two units. (See the explanation to Question 6.)*

8. *A. The equation for the half life of a first order reaction is* $t_{1/2} = 0.693/k$*. Substituting, 75 min = 0.693/k or k = (0.693)/(75)* min^{-1}*.*

9. *D. Half life is the time it takes for the amount of a reactant to lose half its value. One half life would leave 50% of the original amount, two half lives, 25%, 3 would leave 12.5%, and 4 would leave 6.25%. 6.25 % left means that about 94% has been consumed in four half lives. The total time elapsed is approximately 4 X 5730 years.*

10. *D. The rate equation does not necessarily relate to the balanced chemical equation in a multi-step mechanism, nor does it suggest the nature of the products. Three-atom collisions are highly unlikely so a single elementary step is not highly probable, so while the rate determining step can be a three-atom collision, it is not required. The best that can be said is that the mechanism takes place in two or more steps. and the first step is not the rate-determining step.*

Answers to Free Response Questions

1a. *The rate law will always take the form: Rate = k[Reactant 1]*^{n} *[Reactant 2]*^{m} *where, in this case, Reactant 1 is NO, and Reactant 2 is* Cl_2*. The object is to use the data in the given table to determine the exponents, n and m.*

Experiments 1 and 2 show that the rate quadruples (apparently within experimental error) when $[Cl_2]$ *doubles while [NO] remains constant. Replacing* Cl_2 *with 2, and the rate with 4, while ignoring the remainder of the equation because nothing else changes, we get* $4 = 2^n$*. So n = 2.*

Experiments 2 and 3 show that the rate increases by a factor of eight when both [NO] and [Cl_2] double. Replacing Cl_2 with 2, NO with 2, and the rate with 8, while ignoring the rest because nothing else changes, we get $8 = 2^m \times 2^2$. So m = 1.

The answer is: Rate = $k[NO]^1[Cl_2]^2$

b. *Substitute the data from Experiment 1 (You can use any of the three experiments!) into the rate law and solve for k.*

Rate = $k[NO]^1[Cl_2]^2$

$0.75\ M/hr = k(0.25M)(0.50M)^2$

$k = 0.75M/hr/(0.25M)(0.25M^2)$

$k = 12\ /M^2hr = 12\ L^2\ mol^{-2}\ hr^{-1}$

c. *The order with respect to each reactant is the exponent of that reactant, and the overall order is the sum of the exponents.*

The reaction is: first order in NO, second order in Cl_2, and third order overall. (Notice that the units for the rate constant, k, are consistent with a third order overall reaction. That is, they contain two units of M, and one unit of time, or a total of three units for a third order reaction.)

d. *Substitute the given values and the calculated rate constant (with all units!) into the rate law and solve for rate.*

Rate = $k[NO]^1[Cl_2]^2$

Rate = $12\ 1/M^2hr(0.50M)(0.10M)^2$

Rate = $12/M^2hr(0.50M)(0.01M^2)$

Rate = 0.060 M/hr

e. *The rate of one reactant or product compared to the rate of another reactant or product is always proportional to the balanced overall equation for the reaction. The balanced equation shows that two moles of NOCl appear when one mole of Cl_2 disappears. So if the rate of disappearance of Cl_2 is 4.5 M/hr then the rate of appearance of NOCl is twice that or 2 X 4.5 M/hr = 9.0 M/hr.*

f. *The table and the rate law are good only for the disappearance of Cl_2. Neither is valid for the appearance of NOCl. The best you can do is substitute the given data and the calculated rate constant into the rate law and solve for the rate of disappearance of Cl_2. Then multiply your answer by two to take into account that the rate of appearance of NOCl is twice the rate of disappearance of NOCl. (Notice how the solution to Part e gives you a clue to the solution to Part f! This is often the case in multi-step quantitative problems on the AP Chemistry examination.)*

Rate = $k[NO]^1[Cl_2]^2$

Rate = 12 1/M^2hr(0.20M)$(0.30M)^2$

Rate = 12/M^2hr(0.20M)(0.09M^2)

Rate = 0.22 M/hr

This calculated rate is for the disappearance of Cl_2*, because that's what the rate law calculates. To find the rate of appearance of NOCl, multiply by 2 to take into account the stoichiometry of the reaction!*

0.22 M/hr X 2 = 0.44 M/hr.

2a. *To determine the overall reaction, add the individual steps and cancel the alike terms that appear on both sides of the equations:*

Step 1: $2NO \rightarrow N_2O_2$

Step 2: $N_2O_2 + H_2 \rightarrow N_2O + H_2O$

Step 3: $N_2O + H_2 \rightarrow N_2 + H_2O$

$$2NO + 2H_2 \rightarrow N_2 + 2H_2O$$

b. *Through Step 2, two molecules of NO, and one molecule of* H_2*, have reacted. so if Step 2 is limiting, the rate law would be:*

Rate = $k[NO]^2[H_2]$

c. *The given rate law tells us that through the rate-determining step, two molecules of NO and two molecules of* H_2 *have reacted. This condition applies only if Step three is rate-determining.*

d. *An intermediate is a substance that is produced in one elementary step and consumed in a later step. Notice that each strike-through in the solution to Step a identifies an intermediate:* N_2O_2 *and* N_2O.

e. *The order of the reaction with respect to an individual reactant is the exponent of that reactant in the rate law. If Step 1 is rate-determining, the rate law is : Rate* = $k[NO]^2$. *The reaction is second order in NO, and zero order in* H_2.

Your Turn Answers

14.1. *The surface area of powdered coal is much greater than the surface area of lumps of coal, causing powdered coal to burn at a much more rapid rate than coal lumps. The small particle size of coal dust allows for more efficient mixing with gaseous oxygen from the air.*

14.2. *a.* rate = $k[A]$, $k = s^{-1}$. *b.* rate = $k[A]^2$; $k = M^{-1}\ s^{-1}$. *c.* rate = $k[A]^3$; $k = M^{-2}\ s^{-1}$.

14.3. *Hydrogen disappears 3/2 as fast as ammonia appears.* $3/2(2.6) = 3.9\ M/s$.

14.4. *Increasing temperature increases the kinetic energy of atoms and molecules. The particles move faster and collide more often with greater energy resulting in more frequent collisions with sufficient energy for reaction.*

14.5.

1. $B + B = B_2$	*or*	1. $A + B = AB$	*or*	1. $A + B = AB$
2. $B_2 + A = AB_2$		2. $AB + AB = A_2B_2$		2. $AB + A = A_2B$
3. $AB_2 + A = A_2B^2$		3. $A_2B + B = A_2B_2$		3. $A_2B + B = A_2B_2$

Others are possible.

TOPIC 15

CHEMICAL EQUILIBRIUM

Quantitative chemical equilibrium for gaseous reactions as described in Chapter 15 of Chemistry the Central Science is often the topic of Question 1 in the free response section of the AP exam in chemistry. Qualitative concepts of chemical equilibrium, especially those involving Le Chatelier's principle, are often the topic of one of the required essay questions of the free response section of the exam. Both quantitative and qualitative questions appear in the multiple choice section. Students should be able to write equilibrium expressions, interconvert Kp and Kc equilibrium constants, calculate equilibrium constants and concentrations from given data, and apply Le Chatelier's principle to equilibrium systems.

The Concept of Equilibrium

Section 15.1

Chemical equilibrium occurs when two opposite reactions proceed at the same rate. A chemical equilibrium is dynamic. That is, even though no macroscopic changes are observable, on the sub-microscopic level, atoms, ions, and molecules continue to change as both a forward and a reverse reaction occur at the same rate. Consider the interaction of dinitrogen tetroxide and nitrogen dioxide:

$N_2O_4(g) \leftrightarrows 2NO_2(g)$

colorless brown

The " $\leftrightarrows$ " denotes that the system is at equilibrium. That is, the forward and reverse reactions occur at the same rate. Even though, at equilibrium, there is no observable change in temperature, pressure, color or any other property, the molecules continue to interconvert.

The Equilibrium Constant

Section 15.2 and 15.4

The **equilibrium constant expression**, also called the law of mass action, for any reaction takes the form of a ratio of the molar concentrations of reactants divided by those of products.

For example, for the reaction:

$N_2O_4(g) \leftrightarrows 2NO_2(g)$

The equilibrium constant expression is

$Kc = [NO_2]^2 / [N_2O_4]$

Each molar concentration is raised to the power of the respective coefficient in the balanced chemical equation.

The "c" in Kc denotes that the amounts of reactants and products are expressed in the molar concentration unit, M, moles per liter.

If, in a gaseous reaction, the reactants and products are expressed in partial pressures, the equilibrium constant expression is written:

$Kp = P^2NO_2 / PN_2O_4$

The "p" in Kp denotes that all the reactants and products are expressed as partial pressures, in atmospheres. The exponents are the coefficients in the balanced equation.

The relationship between Kc and Kp is given:

$Kp = Kc(RT)^{\Delta n}$

R, the universal gas constant = 0.0821 L atm/mol K.

T is the absolute temperature in K.

Δ n is the change in number of moles of gas as reactants become products.

In the example, $N_2O_4(g) \leftrightarrows 2NO_2(g)$, Δ n = +1 because one mole of gaseous reactants becomes two moles of gaseous products, which is a net gain of one mole.

Whenever a pure solid and/or a pure liquid appears in the equilibrium reaction, it's concentration is not included in the equilibrium expression.

For example, the equilibrium expression for the reaction:

$AgCl(s) \leftrightarrows Ag+(aq) + Cl-(aq)$

does not include the solid AgCl(s).

$Kc = [Ag^+][Cl^-]$

Similarly for $2H_2O(l) = H_3O^+(aq) + OH^-(aq)$, the equilibrium expression is:

$Kc = [H_3O^+][OH^-]$

Your Turn 15.1

Write the equilibrium constant expressions, Kc and Kp for the following reaction:

$CaCO_3(s) \leftrightarrows CaO(s) + CO_2(g)$

Write your answer in the space provided.

Section 15.3

Interpreting and Working with Equilibrium Constants

The magnitude of the equilibrium constant for a given reaction roughly reflects the ratio of products to reactants and gives an indication whether products or reactants predominate at equilibrium.

A very large value of an equilibrium constant means that products predominate, and the equilibrium reaction is said to "lie to the right" with products predominating. For example:

$2NO(g) + O_2(g) \leftrightarrows 2NO_2(g)$, Kc = 5.0 X 10^{12}.

A very small value of an equilibrium constant means that reactants predominate, and the reaction "lies to the left." For example:

$2HBr(g) = H_2(g) + Br_2(g)$, Kc = 5.8 X 10^{-18}.

When a reaction is written in reverse, the equilibrium expression for the reverse reaction is the reciprocal of that of the forward reaction.

Example:

Kc for $AgCl(s) \leftrightarrows Ag^+(aq) + Cl^-(aq)$ *is 1.8 X* 10^{-10}.

What is Kc for $Ag^+(aq) + Cl^-(aq) \leftrightarrows AgCl(s)$*?*

Solution:

For the given reaction, Kc = 1.8 X $10^{-10} = [Ag^+][Cl^-]$.

For the reaction in question Kc = $1/[Ag^+][Cl^-] = 1/1.8 \times 10^{-10} = 5.6 \times 10^{+9}$

Your Turn 15.2

For which reaction given in the above example do products predominate? Which reaction lies to the left? Explain. Write your answer in the space provided.

When a reaction is balanced by doubling the coefficients, then the equilibrium constant for the reaction balanced with the doubled coefficients is the square of the equilibrium constant of the original reaction.

Example:

At 1000 K for $SO_2(g) + 1/2\ O_2(g) \leftrightarrows SO_3(g)$, $Kp = 1.85$.

What is Kp at 1000 K for $2SO_2(g) + O_2(g) \leftrightarrows 2SO_3(g)$?

Solution:

For $SO_2(g) + 1/2\ O_2(g) \leftrightarrows SO_3(g)$, $Kp = Po_2{}^{1/2}\ Pso_3 = 1.85$

For $2SO_2(g) + O_2(g) \leftrightarrows 2SO_3(g)$, $Kp = Po_2\ Pso_3{}^2 = (1.85)^2 = 3.42$.

Section 15.5

Calculating Equilibrium Constants

Calculations of equilibrium constants from initial and equilibrium concentrations and/or pressures are perhaps the most important calculations required for mastery of the Advanced Placement exam in chemistry. Often an "ICE" table is used to analyze and manipulate the data. "ICE" stands for "Initial", "Change", and "Equilibrium".

Example:

Initially 0.40 mol of nitrogen, and 0.96 mol of hydrogen are placed in a 2.0 L container at constant temperature. The mixture is allowed to react and at equilibrium, the molar concentration of ammonia is found to be 0.14 M. Calculate the equilibrium constant, Kc, for the reaction.

$N_2(g) + 3H_2(g) \leftrightarrows 2NH_3(g)$

Solution:

We first need to calculate the initial concentrations of the reactants from the initial amounts and the volume of the container.

$[N_2]$ = 0.40 mol/2.0 L = 0.20 M = "I" (Initial concentration) for N_2.

$[H_2]$ = 0.96 mol/2.0 L = 0.48 M= "I" (Initial concentration) for H_2.

Assume the molar concentration of nitrogen that reacts is "x". Then the molar concentration of hydrogen that reacts is 3x because the balanced chemical equation tells us that for each one mole of nitrogen that reacts, three moles of hydrogen react. By similar reasoning, the molar concentration of ammonia formed is 2x. Therefore,

"C" (Change) for nitrogen = -x

"C" (Change) for Hydrogen = -3x.

"C" (Change) for ammonia is +2x.

Next, set up an "ICE" table as follows, and add "I" + "C" for each quantity to obtain "E", the corresponding Equilibrium quantity:

	$N_2(g)$	$+ 3H_2(g) \leftrightarrows$	$2NH_3(g)$
I	0.20M	0.48M	0M
C	-x	-3x	+2x
E	0.20-x	0.48-3x	2x

An equilibrium concentration of ammonia of 0.14 M is given in the problem.

So 2x = 0.14 M and x = 0.070 M.

At equilibrium $[N_2]$ = 0.20 – x = 0.20 – 0.070 = 0.13 M.

At equilibrium $[H_2]$ = 0.48 -3x = 0.48 – 3(0.070) = 0.27 M.

Substituting into the equilibrium constant expression for the reaction:

$Kc = [NH_3]^2/[N_2][H_2]^3 = /(0.140\ M)^2/(0.13\ M)(0.27\ M)^3 = 7.66$

(Note: The units for equilibrium constants are not included.)

Applications of Equilibrium Constants

Section 15.6

The **reaction quotient expression**, Q, for any chemical reaction is defined in the same way as the equilibrium constant expression. However, unlike the equilibrium constant expression, non-equilibrium values for concentrations or partial pressures may be substituted into the reaction quotient expression.

The reaction quotient, Q, is useful in determining the direction (forward or reverse) a chemical reaction will proceed to achieve equilibrium.

To determine the direction the reaction will proceed toward equilibrium, compare the value of Q to the value of Kc (or Kp).

If Kc = Q, the system is already at equilibrium.

If Kc < Q, the system will proceed to the left to achieve equilbrium.

If Kc > Q, the system will proceed to the right to achieve equilbrium.

(Notice that if Kc and Q are placed in alphabetical order, the =, < and > signs point the direction the reaction will proceed toward equilibrium.)

Example:

Consider the interaction of dinitrogen tetroxide and nitrogen dioxide:

$N_2O_4(g) \leftrightarrows 2NO_2(g)$ *at* $100^\circ C$, $Kc = 0.211$

Which direction will the reaction proceed if $[N_2O_4] = 1.0\ M$ *and* $[NO_2] = 0.5M$?

Solution:

Substitute the given initial concentrations into the expression for Q, which is the same as the expression for Kc.

$Q = [NO_2]^2 / [N_2O_4] = (0.50)^2/(1.0) = 0.25$

$Kc = 0.211 < 0.25 = Q$

The reaction proceeds from right to left toward reactants to establish equilibrium. Notice that at the given concentrations, there is too much $NO_2(g)$*, and not enough* $N_2O_4(g)$*. The reaction proceeds toward* $N_2O_4(g)$ *to achieve equilibrium.*

Section 15.7 Le Chatelier's Principle

Le Chatelier's principle summarizes the behavior of a chemical reaction at equilibrium when a stress is imposed on the reaction. It states that if a change is applied to a system at equilibrium, the system will move in a direction that minimizes the change.

To understand Le Chatelier's principle and why changes affect equilibria, it is important to remember that the equilibrium condition is when the rates of the forward and reverse reactions of a chemical system are equal. By a change to a system at equilibrium we mean a change that alters the rate of either the forward or reverse reaction so the system is no longer at equilibrium. The system responds to move in a direction that re-establishes equilibrium. The system is said to "shift right" (toward products) or "shift left" (toward reactants) depending on the change that disturbs the equilibrium.

Changes that affect equilibria are:

1. Change in concentrations of reactants or products.

Adding a reactant shifts the equilibrium toward products. The increased concentration of the reactant makes the forward reaction faster than the reverse reaction, causing the reaction to shift toward products. (Recall that increasing the concentration of a reactant increases the rate of the forward reaction because, at higher concentrations, collisions are more frequent.)

Removing a reactant shifts the equilibrium toward reactants. A decreased concentration of a reactant slows the forward reaction, making the reverse reaction faster than the forward reaction, causing the reaction to shift toward reactants.

Adding a product shifts the equilibrium toward reactants. An increased concentration of product speeds the reverse reaction shifting the reaction toward reactants.

Removing a product shifts the equilibrium toward products. Decreased product concentration slows the reverse reaction causing the reaction to favor products.

Figure 15.1 summarizes the effect of changing the concentrations of reactants or products on a system at equilibrium.

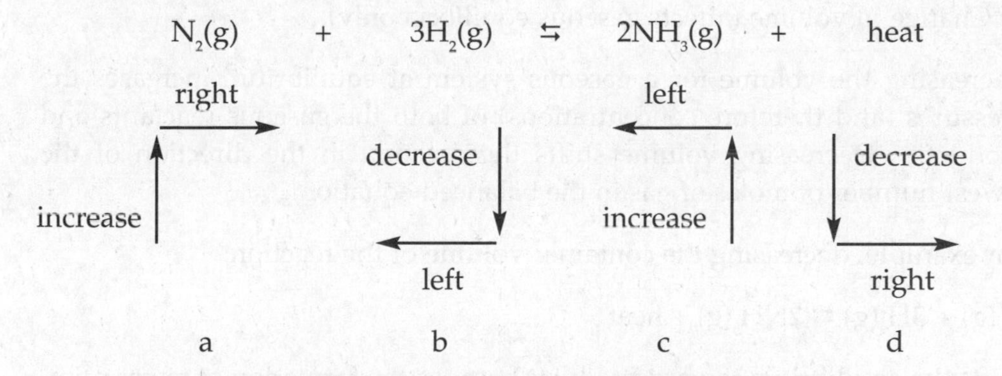

a. Adding reactant shifts equilibrium toward products.
b. Removing reactants shifts equilibrium toward reactants.
c. Adding products shifts equilibrium toward reactants.
d. Removing products shifts equilibrium toward products.

Figure 15.1. The effect of adding or removing reactants or products on a system at equilibrium.

Common misconception: Figure 15.1, tells what direction a system at equilibrium will shift under a given set of conditions. However, questions on the Advanced Placement exam often ask why a system behaves as it does. Be sure to study the discussion of why the rate of the forward or reverse reaction changes and how that change affects the equilibrium position.

Your Turn 15.3

How will decreasing the concentration of hydrogen gas affect the amount of hydrogen iodide present at equilibrium? Explain.

$H_2(g) + I_2(g) \leftrightarrows 2HI(g) + heat$

Write your answer in the space provided.

---

2. Change in volume (affects gaseous equilibria only).

Decreasing the volume for a gaseous system at equilibrium increases the pressures (and therefore concentrations) of both the gaseous reactants and products. Decreasing volume shifts the reaction in the direction of the fewest number of moles of gas in the balanced equation.

For example, decreasing the container volume of the reaction:

$N_2(g) + 3H_2(g) \leftrightarrows 2NH_3(g) + heat$

shifts the equilibrium toward products because the formation of fewer number of moles will decrease the pressure.

Decreasing the volume increases the total pressure. The system responds by moving in a direction that will reduce the total pressure. In this case, the reaction shifts toward NH_3 because the number of moles in the system will decrease and the pressure will decrease. Another way to say this is that when the total pressure increases, the equilibrium shifts in a direction that reduces the number of moles of gas, in this case, to the right.

For example, if the volume of the container is halved, the pressures of each gas will double. However, the equilibrium constant, Kp is defined in terms of the stoichiometry of the reaction and doubling all pressures increases the numerator (a squared term) less than the denominator (the combination of terms raised to the third and first powers). Thus, the equilibrium is out of balance and adjusts to the right to rebalance the mixture keeping Kp a constant.

$Kp = P^2NH_3 / PN_2P^3H_2$

On a molecular level, consider that an increase in pressure increases the rates of both the forward and reverse reactions because the number of collisions for both reactants and products are more frequent. However, the rates of the forward and reverse reactions are not increased equally. The rate of the forward reaction, because it has more moles of gas, increases more than the rate of the reverse reaction. The reaction shifts right consuming enough reactants to again make the rates equal.

Your Turn 15.1

What is the affect on the equilibrium between ozone and oxygen when the volume of the container is increased? Explain your answer on the basis of rates of reaction.

$2O_3(g) \leftrightarrows 3O_2(g)$

Write your answer in the space provided

---

3. Change in temperature

Increasing temperature favors the endothermic reaction. The rate of both the forward and the reverse reactions are increased due to faster moving particles, more frequent collisions and more effective collisions. However, the rate of the endothermic reaction increases more than does the rate of the exothermic reaction. This is because the activation energy of the endothermic reaction is always greater than that of the exothermic reaction. At low temperatures, a sufficient number of exothermic reactants already have enough energy to overcome the relatively low activation barrier. At low temperature, very few endothermic reactants have the requisite energy to go over the relatively high activation barrier. Increasing temperature increases the energy of both reactants and products it aids the lower energy endothermic reactants more than the higher energy exothermic reactants. Thus, increasing temperature increases the rate of the endothermic reaction more than it increases the rate of the exothermic reaction.

Decreasing temperature favors the exothermic reaction. A lower temperature slows the rate of both forward and reverse reactions. However, at low temperatures, a sufficient number of exothermic reactants still have enough energy to surmount the energy barrier and react.

Changing temperature changes the equilibrium constant. In contrast, changes in concentration, volume or pressure, changes the position of an equilibrium without changing the equilibrium constant. Figure 15.1 shows that you can deduce the effect of temperature on the direction of change of an equilibrium system if you treat temperature as a reactant (in an endothermic reaction) or a product (in an exothermic reaction).

The Effect of a Catalyst

A catalyst does not affect the position of the equilibrium but it does increase the rate at which equilibrium is established. A catalyst increases the rate of a reaction by lowering the activation energy. However, a catalyst lowers by equal amounts, the activation energies of both the forward and reverse reactions. (See Figure 15.2.)

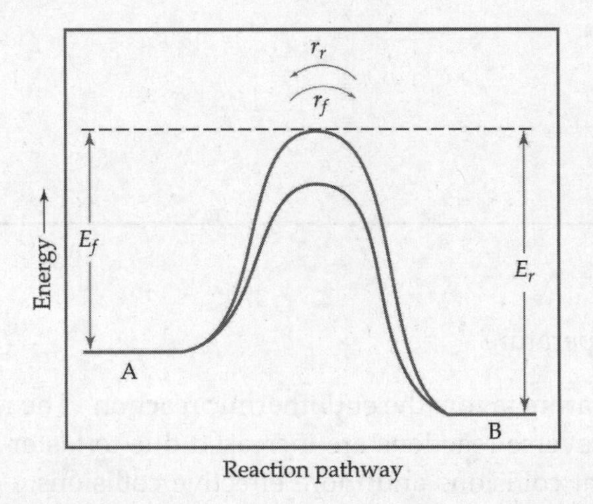

Figure 15.2. A catalyst works by providing a different pathway from reactants to products. The activation energy of both the forward and reverse reactions is lower. A catalyst does not affect the position of the equilibrium.

Multiple Choice Questions

1. $3H_2(g) + N_2(g) = 2NH_3(g)$ $\Delta\Delta H = - 92.2\ kJ$

The number of moles of $H_2(g)$ are decreased by

A) decreasing container size.

B) adding NH_3.

C) increasing temperature.

D) removing N_2.

E) adding a catalyst

2. *Which factor will affect both the value of the equilibrium constant and the position of equilibrium for the formation of calcium carbonate?*

$CaO(s) + CO_2(g) \leftrightarrows CaCO_3(s) + heat$

A) increasing the volume of the container

B) adding CO_2

C) removing CaO(s)

D) raising the temperature

E) adding a catalyst

3. *Consider a reaction $3A(g) + B(s) \leftrightarrows 2C(g)$. If 2.0 mol A, 3.0 mol B and 2.0 mol C are present in a 1.0 L flask at equilibrium, what is the value of K_c?*

A) 4.0

B) 1.0

C) 2.0

D) 0.25

E) 0.50

4. *Consider the following reaction:*

$2SO_2(g) + O_2(g) \leftrightarrows 2SO_3(g)$ *Kp = 9.0 at a certain temperature.*

At the same temperature, what is Kp for $SO_2(g) + 1/2O_2(g) \leftrightarrows SO_3(g)$?

A) 9.0

B) 4.5

C) 3.0

D) 18

E) 2.3

5. *For the chemical reaction, $PCl_3(g) + Cl_2(g) \leftrightarrows PCl_5(g)$, $\Delta Ho_{rxn} = -92.6$ kJ. Which conditions favor maximum conversion of the reactants to product?*

A) high pressure and high temperature

B) high pressure and low temperature

C) low pressure and low temperature

D) low pressure and high temperature

E) adding a catalyst

6. *Which of the following equilibrium constants indicates that its corresponding reaction goes nearly to completion?*

A) $Kc = 1.0 X 10^{-2}$

B) $Kc = 1.0 X 10^{-8}$

C) $Kc = 1.0$

D) $Kc = 1.0 X 10^{+2}$

E) $Kc = 1.0 X 10^{+8}$

7. *What is the equilibrium expression for the decomposition of ammonium carbonate, $(NH_4)_2CO_3$, according to the following equation.*

$(NH_4)_2CO_3(s) \leftrightarrows 2NH_3(g) + CO_2(g) + H_2O(g)$

A) $Kc = [NH_3][CO_2][H_2O]$

B) $Kc = [NH_3]^2[CO_2][H_2O]$

C) $Kc = [NH_3]^2[CO_2][H_2O]/[(NH_4)_2CO_3]$

D) $Kc = (NH_4)_2CO_3]/[NH_3]^2[CO_2][H_2O]$

E) $Kc = [NH_3][CO_2]$

8. *In the presence of a catalyst, sulfur dioxide reacts with oxygen to form sulfur trioxide.*

$2SO_2(g) + O_2(g) \leftrightarrows 2SO_3(g)$

When 2.00 mol of O_2 and 2.00 mol of SO_2 are placed in a one liter container, and allowed to come to equilibrium at a certain temperature, the mixture is found to contain 1.00 mol of SO_3. What is the amount of O_2 at equilibrium?

A) *0.00 mol*

B) *1.00 mol*

C) *1.50 mol*

D) *0.50 mo*

E) *0.75 mol*

9. *Ammonia is placed in a flask and allowed to come to equilibrium at a specified temperature according to the equation.*

$3H_2(g) + N_2(g) \leftrightarrows 2NH_3(g)$

Analysis of the equilibrium mixture shows that it contains 3.00 atm NH_3 and 1.00 atm N_2. What is the value of the equilibrium constant, Kp?

A) *0.333*

B) *27*

C) *3.00*

D) *0.25*

E) *0.50*

10. *Consider the chemical equilibrium: $Na_2CO_3\ (s) \leftrightarrows CO_2(g) + Na_2O(s)$ in a closed container at a certain temperature. The most convenient way to measure the equilibrium constant for the system is to measure:*

A) *the temperature of the reaction.*

B) *the pressure of the CO_2 gas.*

C) *the molar concentrations of all the reactants.*

D) *the forward and reverse rate constants.*

E) *the mass of the solid present*

Free Response Questions

1. *A 40.0 g sample of solid ammonium carbonate is placed in a closed, evacuated 3.00 L flask and heated to 400°C. It decomposes to produce ammonia, water, and carbon dioxide according to the equation:*

   $(NH_4)_2CO_3(s) \leftrightarrows 2NH_3(g) + H_2O(g) + CO_2(g)$

   The equilibrium constant, Kp, for the reaction is 0.295 at 400°C.

   a. *Write the Kp equilibrium constant expression for the reaction.*

   b. *Calculate Kc at 400°C.*

   c. *Calculate the partial pressure of $NH_3(g)$ at equilibrium at 400°C.*

   d. *Calculate the total pressure inside the flask at equilibrium.*

   e. *Calculate the number of grams of solid ammonium carbonate in the flask at equilibrium.*

   f. *What is the minimum amount in grams of solid $(NH_4)_2CO_3$ that is necessary to be placed in the flask in order for the system to come to equilibrium?*

2. *Hydrogen gas reacts with solid sulfur to produce hydrogen sulfide gas.*

   $H_2(g) + S(s) \leftrightarrows H_2S(g) \qquad \Delta\Delta H_{rxn} = -20.17$ kJ/mol

   An amount of solid S and an amount of gaseous H_2 are placed in an evacuated container at 25°C. At equilibrium, some solid S remains in the container. Predict and explain each of the following.

   a. *The effect on the equilibrium partial pressure of H_2S gas when additional solid sulfur is introduced into the container.*

   b. *The effect on the equilibrium partial pressure of H_2 gas when additional H_2S gas is introduced into the container.*

   c. *The effect on the mass of solid sulfur present when the volume of the container is increased*

   d. *The effect on the mass of solid sulfur present when the temperature is decreased.*

   e. *The effect of adding a catalyst to initial amounts of reactants.*

ractice in Chemistry
the Central Science

ems in preparation for the
ese problems in Chapter 15

2, 15.64, 15.65, 15.66,

2, 15.83, 15.84, 15.85.

hoice questions

1. *A. According to Boyles law, decreasing the container size will increase the partial pressures of all three gases and hence the total pressure. Because there are more moles of gaseous reactants (4), than gaseous products (2), reactant molecules will tend to collide more frequently than product molecules. The forward reaction rate will increase more than the reverse reaction rate. The system will shift toward products decreasing the number of moles of $H_2(g)$. A catalyst does not change the position of the equilibrium. All the other choices increase the number of moles of $H_2(g)$.*

2. *D. Temperature is the only variable that will change the equilibrium constant. Increasing the temperature will favor the endothermic (in this case, the reverse) reaction causing the equilibrium to shift to the left increasing the partial pressure of CO_2 and decreasing Kp.*

3. *E. $Kc = [C]^2/[A]^3 = (2.0M)^2/(2.0M)^3 = 0.50$*

   (Note that calculators are not allowed on the multiple choice section so complex numeric problems such as this one require relatively simple arithmetic.)

4. *C. The reaction in question is the same as the given reaction except that it is balanced with coefficients half the value of the given reaction. Since coefficients are exponents in equilibrium constant expressions, kp for the second reaction is the square root of Kp for the given reaction: $(9.0)^{1/2} = 3.0$.*

5. *B. The reaction is exothermic so low temperature will favor products. Additionally, high pressure will shift the reaction toward prod-*

ucts because compressing the gases will cause the forward reaction to become faster than the reverse reaction. This is because at higher pressures more molecular collisions are likely to take place. The two moles of gas on the left will become one mole of gas on the right, lowering the pressure.

6. *E. "Nearly to completion" means that by the time the system reaches equilibrium, most of the reactants have become products. Kc is the ratio of products to reactants, so the larger the Kc, the larger the quantity of products at equilibrium.*

7 *B. The equilibrium constant, Kc, equals the ratio of the concentrations of products to reactants, each raised to the power of the coefficient that balances the equation. Equilibrium constants do not include reactants or products that are pure solids or liquids.*

8. *C. An "ICE" table would look like this:*

	$2SO_2(g)$	$+ O_2(g) =$	$2SO_3(g)$
I	2.00	2.00	0
C	-2x	-x	+2x
E	2.00-2x	2.00-x	2x

$2x = 1.00$ mol/L, $x = 0.50$ mol/L, $2.00 - x = 1.50$ mol/L = 1.50 mol.

9. *A. An "ICE" table would look like this:*

	$3H_2(g) +$	$N_2(g) =$	$2NH_3(g)$
I	0	0	
C	+3x	+x	-2x
E	3x	x	3.00

$x = 1.00$, $3x = 3.00$

$Kp = (P_{NH_3})^2/(P_{H_2})^3(P_{N_2}) = (3.00)^2/(3.00)^3(1.00) = 0.333$

10. *B.* $Kp = P_{CO_2}$. *The partial pressure of the CO_2 at equilibrium is equal to the equilibrium constant, Kp. Equilibrium constants do not include reactants or products that are pure solids or liquids.*

Free Response Answers:

a. $Kp = P^2NH_3 PH_2O\, PCO_2$. *The Kp expression is the product of the partial pressures of products raised to the power of the coefficients that balance the equation. Pure solids and liquids are excluded from any equilibrium expression.*

b. *$Kc = Kp/(RT)^{\Delta n}$ where $R = 0.0821$ Latm/mol K, T is the absolute temperature in Kelvin, and Δn is the change in number of moles of gas as the reaction proceeds left to right. In this case $\Delta n = +4$ because four moles of gas are produced from zero moles of gas.*

$Kc = Kp/(RT)^{\Delta n} = 0.295/[(0.0821)(400+273)]^4 = 3.17 \times 10^{8}$.

c. *If x equals the partial pressure of H_2O formed in the reaction, an ICE table will look like this:*

$(NH_4)_2CO_3(s)$ =	$2NH_3(g)+$	$H_2O(g)+$	$CO_2(g)$
I	0	0	0
C	$+2x$	$+x$	$+x$
E	$2x$	x	x

(Note: It is not important to know how much solid is present initially or at equilibrium.)

$Kp = P^2NH_3\, PH_2O\, PCO_2 = 0.295 = (2x)^2(x)(x) = 4x^4$

$x = 0.521$

$PNH_3 = 2x = 2\,(0.521) = 1.04$ atm

d. $x = PCO_2 = 0.521$ atm $= PH_2O$ and $2x = PNH_3 = 2 \times 0.521 = 1.04$ atm

$Ptotal = PCO_2 + PH_2O + PNH_3 = 0.521 + 0.521 + 1.04 = 2.08$ atm.

e. *To calculate the number of grams of ammonium carbonate remaining at equilibrium, we must calculate the number of grams that reacted and subtract from the original 40.0 g. From the stoichiometry of the reaction, the moles of ammonium carbonate that reacted is equal to the number of moles of water found in the flask at equilibrium. The moles of water can be calculated from the partial pressure of water using the idea gas equation.*

Moles of water at equilibrium $= PV/RT = (0.521\text{atm})(3.00\text{L})/(0.0821$ L atm/mol K$)(400 + 273) = 0.0283$ mol H_2O.
0.283 mol $H_2O = 0.0283$ mol of $(NH_4)2CO_3(s)$ that reacted.

The molar mass of $(NH_4)_2CO_3(s)$ is 96.0 g/mol.

The number of grams of$(NH_4)_2CO_3(s)$ that reacted is 0.0283 mol x 96.0 g/mol = 2.72g.

The number of grams that remain is 40.0 g – 2.72 g = 37.3 g.

f. *If 2.72 g react to establish equilibrium, then just slightly more than 2.72 grams of $(NH_4)_2CO_3(s)$ must be present initially for equilibrium to be established because at equilibrium, some solid must remain.*

2a. *Increasing or decreasing the amount of solid in an equilibrium mixture has no effect on the position of the equilibrium. $S(s)$ does not appear in the equilibrium expression.*

b. *The partial pressure of $H_2(g)$ will increase. Additional $H_2S(g)$ added to the container increases the rate of the reverse reaction causing more reactants to form as the equilibrium system re-balances.*

c. *The mass of the solid sulfur remains the same. Increasing the volume of the container decreases the partial pressures of both gasses, but because there is one mole of each gas on each side of the equilibrium equation, the equilibrium system does not change. If more moles of gas were on one side than the other, the equilibrium would shift toward the side with the greater number of moles of gas.*

d. *The mass of the solid sulfur will decrease. Heating a reaction increases the rate of both forward and reverse reactions because increasing temperature increases the speed with which the molecules move and the kinetic energy they have. Thus, at a higher temperature the collisions between molecules become more frequent. However, increasing temperature always increases the rate of the endothermic reaction more than the rate of the exothermic reaction because, the higher temperature gives more molecules on the endothermic side the sufficient energy required to surmount the energy barrier.*

Notice that this question can be tricky. Because pure solids are not included in the equilibrium expression, their relative amounts do not affect the position of the equilibrium. But their amounts are affected by changes in the equilibrium. Be sure to read each question carefully and answer the question directly.

e. *The final equilibrium position will not be affected even though the system will reach equilibrium faster. But, catalysts increase the rates of both the forward and reverse reactions equally by lowering the activation energy.*

Your Turn Answers

15.1. *Kc = [CO_2]. Kp = P_{CO_2}*

15.2. *In the reaction, $Ag^+(aq) + Cl^-(aq) \leftrightarrows AgCl(s)$, products predominate because the equilibrium constant, Kc = $5.6 \times 10^{+9}$, is very large.*

The reaction, $AgCl(s) \leftrightarrows Ag^+(aq) + Cl^-(aq)$, lies to the left where reactants predominate because the equilibrium constant, Kc = 1.8×10^{-10}, is very small.

15.3. *Decreasing [H_2] will decrease the amount of HI present. The equilibrium will shift left consuming HI because the rate of the forward reaction will decrease due to more infrequent collisions.*

15.4. *Increasing the volume of the container decreases the pressure of both of the gases present. The equilibrium will shift in a direction to increase the pressure (to the right, toward oxygen.) Decreasing the pressure of both of the gases, decreases their concentrations, and decreases the rates of both the forward and reverse reactions. The rate of the forward reaction will decrease less than the rate of the reverse reaction causing the equilibrium to shift toward products.*

TOPIC 16

ACID-BASE EQUILIBRIA

A large portion of the AP exam is taken from the topic of acid base equilibria treated in Chapters 16 and 17 of *Chemistry the Central Science*. It is important to be able to calculate quantitative parameters, especially pH, of solutions of strong and weak acids and bases. Writing and understanding chemical equations illustrating the Brønsted-Lowry definition of acids and bases and their relationships to the K_a and K_b expressions is important. It's also imperative to understand acid-base hydrolysis of salt solutions and how chemical structure affects acid-base behavior. Don't forget to know how to recognize strong and weak acids and bases from their formulas and understand Lewis acids and bases.

Acids and Bases: A Brief Review

Section 16.1

The Arrhenius concept of acids and bases states that acids ionize in water solution to produce hydrogen ions. Acids are substances that increase the hydrogen ion, H^+, concentration when dissolved in water. For example:

$HCl(g) \rightarrow H^+(aq) + Cl^-(aq)$

$HNO_3(aq) \rightarrow H^+(aq) + NO_3^-(aq)$

Bases are substances that increase the concentration of hydroxide ion, OH^-, when dissolved in water. A base dissociates to produce hydroxide ions in water. Notice in the examples below that sodium hydroxide is "monobasic" and barium hydroxide is "dibasic". The prefixes, mono- and di- refer to the number of available hydroxides in each formula.

Sodium hydroxide is monobasic: $NaOH(s) \rightarrow Na^+(aq) + OH^-(aq)$

Barium hydroxide is dibasic: $Ba(OH)_2(s) \rightarrow Ba^{2+}(aq) + 2OH^-(aq)$

Section 16.5 Strong Acids and Bases

Strong acids and strong bases are strong electrolytes. Strong acids and strong bases ionize completely in dilute aqueous solution. There are seven common strong acids, and eight common strong bases. It is useful to memorize their names and formulas given in Table 16.1. Notice that the strong bases are the hydroxides of the Group 1 alkali metals, and the hydroxides of the heavier Group 2 alkaline earth metals. Strong acids and bases are generally, but not always, associated with the Arrhenius model for acids and bases.

Table 16.1. The names and formulas of strong acids and strong bases.

Strong acids		**Strong bases**	
*sulfuric acid	H_2SO_4	lithium hydroxide	LiOH
nitric acid	HNO_3	sodium hydroxide	NaOH
perchloric acid	$HClO_4$	potassium hydroxide	KOH
chloric acid	$HClO_3$	rubidium hydroxide	RbOH
hydrochloric acid	HCl	cesium hydroxide	CsOH
hydrobromic acid	HBr	**calcium hydroxide	$Ca(OH)_2$
hydroiodic acid	HI	**strontium hydroxide	$Sr(OH)_2$
		**barium hydroxide	$Ba(OH)_2$

*Sulfuric acid is a diprotic acid and only the first proton ionizes completely.
** $Ca(OH)_2$, $Sr(OH)_2$ and $Ba(OH)_2$, are dibasic and in each case, only the first hydroxide group dissociates completely.

Common misconception: The strong diprotic acid, sulfuric acid, H_2SO_4, does not dissociate completely. Therefore in an aqueous solution of sulfuric acid the concentration of the H^+ ions are not double the concentration of the acid. The same is true of strong dibasic bases such as calcium hydroxide. In a solution of calcium hydroxide, $Ca(OH)_2$, the concentration of the OH^- ion is not double that of the calcium hydroxide.

Neutral ions, in general, are the aqueous anions of strong acids and the aqueous cations of strong bases. Table 16.2 lists the common cations and anions that are neutral in aqueous solution. In general, most all other anions are slightly basic and most all other cations are slightly acidic.

Table 16.2. Neutral ions are the aqueous anions of strong acids and the aqueous cations of strong bases.

Neutral aqueous anions		**Neutral aqueous cations**	
nitrate	NO_3^-	lithium ion	Li^+
perchlorate	ClO_4^-	sodium ion	Na^+
chlorate	ClO_3^-	potassium ion	K^+
chloride	Cl^-	rubidium ion	Rb^+
bromide	Br^-	cesium ion	Cs^+
iodide	I^-	calcium ion	Ca^{2+}
		strontium ion	Sr^{2+}
		barium ion	Ba^{2+}

*Sulfate ion, SO_4^{2-}, is weakly basic.

Your Turn 16.1

Explain why aqueous solutions of all of the following salts are neutral: NaCl, KNO_3, $LiClO_4$, $BaBr_2$, CsI. Write your answer in the space provided.

Brønsted-Lowry Acids and Bases

Section 16.2

The Brønsted-Lowry theory states:

Acids are proton (H^+) donors .

Bases are proton acceptors.

Consider the interaction of hydrogen chloride gas with water:

$HCl(g) + H_2O(l) \rightarrow H_3O^+(aq) + Cl^-(aq)$

HCl is an acid because it donates a proton to water. H_2O is a base because it accepts a proton.

Hydronium ion, $H_3O^+(aq)$, is a hydrated proton. When water accepts a proton from an acid, the product is a hydronium ion.

Notice that when water accepts a proton from HCl it becomes H_3O^+. $H_3O^+(aq)$ is a hydrated proton called the hydronium ion.

Common misconception: Chemists use $H^+(aq)$ and $H_3O^+(aq)$ interchangeably to represent a hydrated proton, the ion responsible for the acidic properties of an aqueous solution. Both of the following equations are chemically equivalent. Notice how one emphasizes the Brønsted-Lowry model and the other represents the Arrhenius model.

$HCl(g) + H_2O(l) \rightarrow H_3O^+(aq) + Cl^-(aq)$ Brønsted-Lowry model

acid base

$HCl(g) \rightarrow H^+(aq) + Cl^-(aq)$ Arrhenius model

Sodium carbonate, Na_2CO_3, is a Brønsted-Lowry base because the carbonate ion accepts a proton from water. Water is an acid because it donates a proton to carbonate ion. Since the sodium ion is a neutral ion, we ignore it in the equation:

Example: $CO_3^{2-}(aq) + H_2O(l) \leftrightarrows OH^-(aq) + HCO_3^-(aq)$

base acid

Conjugate acid-base pairs are two substances in aqueous solution whose formulas differ by an H^+. The acid is the more positive species having the extra H. (See Table 16.3.)

Table 16.3. Examples of acid-base conjugate pairs

Acid	Base	Equations involving acid-base conjugate pairs
NH_4^+ H_2O	NH_3 OH^-	$NH_3(g) + H_2O(l) \leftrightarrows NH_4^+(aq) + OH^-(aq)$ base acid acid base
H_2SO_3 HSO_3^-	H_3O^- H_2O	$H_2SO_3(aq) + H_2O(l) \leftrightarrows HSO_3^-(aq) + H_3O^+(aq)$ acid base base acid

Acid-base reactions are reversible reactions. The reversible equations in Table 16.3 illustrate the focus of the Brønsted-Lowry model on the transfer of protons. In each reaction there are two sets of acid-base conjugate pairs.

Common misconception: HF is a weak acid so F^- is the conjugate base of HF. The reaction of HF with water is expressed as a reversible equilibrium:

$HF(aq) + H_2O(l) \leftrightarrows F^-(aq) + H_3O^+(aq)$

However, HCl is a strong acid so Cl^- is not a conjugate base of HCl. The reaction of HCl with water is expressed as an irreversible reaction:

$HCl(aq) + H_2O(l) \rightarrow Cl^-(aq) + H_3O^+(aq)$

Because the HCl reaction is not reversible, Cl^- does not accept a proton, and is not a base. The same is true of all anions of strong monoprotic acids.

An **amphoteric** substance is one that can act either as an acid or a base. Notice in Table 16.3 that water acts as an acid when it transfers a proton to ammonia. It acts as a base when accepting a proton from sulfurous acid. Water is amphoteric because it can act as an acid or a base.

Weak Acids and Weak Bases

Section 16.2

Weak acids are acidic substances that only partially ionize in aqueous solution. Weak acids are weak electrolytes. For example, hydrofluoric acid, acetic acid and nitrous acid are all weak acids. (Each ionizes only partially to establish an equilibrium between the acid and its conjugate base).

Table 16.4 shows the ionization of some weak acids in water. A more complete listing of the ionization constants of weak acids can be found in Appendix D, Table D-1 of Chemistry the Central Science.

Table 16.4. Ionization of some weak acids in water.

weak acid conjugate base	equilibrium expression, Ka	value of Ka
$HF(aq) + H_2O(l) \leftrightarrows H_3O^+(aq) + F^-(aq)$	$Ka = [H_3O^+][F^-]/[HF]$	6.8×10^{-4}
$HClO(aq) + H_2O(l) \leftrightarrows H_3O^+(aq) + ClO^-(aq)$	$Ka = [H_3O^+][ClO^-]/[HClO]$	3.0×10^{-8}
$HIO(aq) + H_2O(l) \leftrightarrows H_3O^+(aq) + IO^-(aq)$	$Ka = [H_3O^+][IO^-]/[HIO]$	2.3×10^{-11}
$H_2CO_3(aq) + H_2O(l) \leftrightarrows H_3O^+(aq) + HCO_3^-(aq)$	$Ka_1 = [H_3O^+][HCO_3^-]/[H_2CO_3]$	4.3×10^{-7}
$HCO_3^-(aq) + H_2O(l) \leftrightarrows H_3O^+(aq) + CO_3^{2-}(aq)$	$Ka_2 = [H_3O^+][CO_3^{2-}]/[HCO_3^-]$	5.6×10^{-11}

Weak bases are also weak electrolytes. Weak bases only partially ionize in solution.
Table 16.5 shows the ionization of some weak bases in water. A more complete listing of the ionization constants of weak bases can be found in Appendix D, Table D-2 of Chemistry the Central Science.

Table 16.5. Ionization of some weak bases in water

weak base conjugate acid	equilibrium expression, Kb	value of Kb
$NH_3(aq) + H_2O(l) \leftrightarrows OH^-(aq) + NH_4^+(aq)$	$Kb = [OH^-][NH_4^+]/[NH_3]$	1.8×10^{-5}
$C_5H_5N(aq) + H_2O(l) \leftrightarrows OH^-(aq) + C_5H_5NH^+(aq)$	$Kb = [OH^-][C_5H_5NH^+]/[C_5H_5N]$	1.7×10^{-9}
$HCO_3^-(aq) + H_2O(l) \leftrightarrows OH^-(aq) + H_2CO_3(aq)$	$Kb = [OH^-][H_2CO_3]/[HCO_3^-]$	2.3×10^{-8}
$F^-(aq) + H_2O(l) \leftrightarrows OH^-(aq) + HF(aq)$	$Kb = [OH^-][HF]/[F^-]$	1.5×10^{-11}

Weak bases tend to be neutral nitrogen compounds called amines, and anions other than the anions of strong acids. Amines have a pair of electrons that can attract protons. Anions have negative charges that can also attract protons.

The acid ionization constant, Ka, is the equilibrium constant for the ionization of a weak acid in water.

The base ionization constant, Kb is the ionization constant for a weak base.

Common misconception: Ka and Kb are not new ideas. The respective subscripts "a" and "b" denote that Ka is a special case of Kc used to specify the ionization of a weak acid in water and that Kb is a special case for a weak base. Also, Ka and Kb are sometimes called the acid dissociation constant and the base dissociation constant, respectively.

The value of Ka or Kb indicates the relative extent to which a weak acid or weak base ionizes. For example, the larger the Ka, the greater the extent to which the acid ionizes. When comparing two weak acids, the one with the larger Ka is said to be the "stronger" weak acid. It ionizes to a larger extent. Of the acids listed in Table 16.4, hydrofluoric acid, HF, has the largest Ka, so it ionizes to the greatest extent. HF is said to be the strongest of the weak acids listed. Similarly, of the bases listed in Table 16.5, ammonia, NH_3, has the largest Kb so it is the strongest weak base of those listed.

Your Turn 16.2

Which acid listed in Table 16.4 is the weakest? Which base listed in Table 16.5 is the weakest base? On what do you base your answers? Write your answers in the space provided.

Carbonic acid, H_2CO_3, is an example of a weak diprotic acid. Table 16.4 shows two ionization constants for carbonic acid. Ka_1 is the equilibrium constant for the ionization of the first proton. Ka_2 represents the equilibrium constant for the second proton ionization. For all polyprotic acids, the first proton to ionize is always the most readily ionized. The Ka value becomes successively smaller as successive protons are removed so $Ka_1 > Ka_2 > Ka_3$.

The Autoionization of Water

Section 16.3

The **autoionization of water** is a reversible equilibrium where a water molecule transfers a proton to another water molecule.

$H_2O(l) + H_2O(l) \leftrightarrows OH^-(aq) + H_3O^+(aq)$

acid base base acid

Water is both a weak acid and a weak base. It is amphoteric. It has the ability to act as a proton donor (an acid) or a proton acceptor (a base).

The **ion product constant** for water, Kw, is the equilibrium constant for the autoionization of water. Kw is a special case of Kc.

$Kc = Kw = [OH^-][H_3O^+] = 1.0 \times 10^{-14}$ at 25°C.

The Relationship Between Ka and Kb

Section 16.8

For any conjugate acid-base pair, Kw = Ka X Kb.

Consider, for example, the reaction of carbonic acid with water:

$H_2CO_3(aq) + H_2O(l) \leftrightarrows H_3O^+(aq) + HCO_3^-(aq)$

$Ka = [H_3O^+][HCO_3^-]/[H_2CO_3] = 4.3 \times 10^{-7}$

Hydrogen carbonate ion, HCO_3^-, is the conjugate base of the weak acid, H_2CO_3. Hydrogen carbonate reacts with water according to the following equation:

$HCO_3^-(aq) + H_2O(l) \leftrightarrows OH^-(aq) + H_2CO_3(aq)$

$Kb = [OH^-][H_2CO_3]/[HCO_3^-] =$

Kb for any conjugate base is the ratio of Kw to Ka of the acid:

$$Kb = Kw/Ka = ([OH^-][H_3O^+])/([H_3O^+][HCO_3^-]/[H_2CO_3]) = [OH^-][H_2CO_3]/[HCO_3^-]$$

$Kb = Kw/Ka = 1.0 \times 10^{-14} / 4.3 \times 10^{-7} = 2.3 \times 10^{-8}$

Common misconception: It is customary to tabulate only Ka or Kb for a conjugate pair because, using the relationship, one can be conveniently converted to the other. Typically only the equilibrium constants for non-ionic species appear in a table. For example, the Kb for ammonia, NH_3, is reported in Appendix D of Chemistry the Central Science, but not the Ka for the conjugate acid ammonium ion, NH_4^+. Similarly the Ka for acetic acid, $HC_2H_3O_2$, appears in Appendix D, but not the Kb for the conjugate base acetate ion, $C_2H_3O_2^-$. It is important to be able to recognize the conjugate of any given acid or base and to know how to calculate its corresponding Ka or Kb.

Recall that water and protonated anions (HCO_3^-, HSO_3^-, $H_2PO_4^-$, etc.) tend to be amphoteric. That is, they can act as either acids or bases. To tell whether a protonated anion is acidic or basic in water we can do one of three things. Measure the pH of an aqueous solution containing the anion, use an acid-base indicator, or calculate and compare the Ka and Kb of the anion. If the Ka of the anion is larger than its corresponding Kb, the anion forms an aqueous solution that is acidic. If its Kb is larger, the solution is basic.

Consider the triprotic acid, phosphoric acid, H_3PO_4. The sequential ionization of its three protons and the corresponding Ka's are illustrated by the equations in Table 16.6. Notice that two species involved in the equations are amphoteric protonated anions, $H_2PO_4^-$ and HPO_4^{2-}, and their reaction as bases are also illustrated in Table 16.6.

Table 16.6. The ionization of phosphoric acid

	Acid ionization	Acid ionization constant
1	$H_3PO_4(aq) + H_2O(l) \leftrightharpoons H_3O^+(aq) + H_2PO_4^-(aq)$	$Ka_1 = 7.5 \times 10^{-3}$
2	$H_2PO_4^-(aq) + H_2O(l) \leftrightharpoons H_3O^+(aq) + HPO_4^{2-}(aq)$	$Ka_2 = 6.2 \times 10^{-8}$
3	$HPO_4^{2-}(aq) + H_2O(l) \leftrightharpoons H_3O^+(aq) + PO_4^{3-}(aq)$	$Ka_3 = 4.2 \times 10^{-13}$
	Base ionization	**Base ionization constant**
4	$H_2PO_4^-(aq) + H_2O(l) \leftrightharpoons OH^-(aq) + H_3PO_4(aq)$	$Kb = Kw/Ka_1$
5	$HPO_4^{2-}(aq) + H_2O(l) \leftrightharpoons OH^-(aq) + H_2PO_4^-(aq)$	$Kb = Kw/Ka_2$
6	$PO_4^{3-}(aq) + H_2O(l) \leftrightharpoons OH^-(aq) + HPO_4^{2-}(aq)$	$Kb = Kw/Ka_3$

Example:

Is an aqueous solution of Na_2HPO_4 acidic or basic?

Solution:

Calculate the value of Kb for HPO_4^{2-} and compare it to the value of Ka_3 for HPO_4^{2-}. (Keep in mind that the aqueous sodium ion, $Na^+(aq)$, is neutral.) The larger ionization constant will predict the acid-base characteristic of its aqueous solution.

For Equation 5, $Kb = Kw/Ka_2 = 1.0 \times 10^{-14}/6.2 \times 10^{-8} = 1.6 \times 10^{-7}$

(We use Ka_2 to calculate the Kb for Equation 5 because Equation 5 includes the same conjugate pair as Equation 2.)

$Ka = 4.2 \times 10^{13} < Kb = 1.6 \times 10^{-7}$

Because its Kb is larger than its Ka, HPO_4^{2-} forms a basic solution.

Lewis Acids and Bases Section 16.11

A **Lewis acid** is an electron-pair acceptor.

A **Lewis base** is an electron-pair donor.

Example:

In the following equation, which species acts as a Lewis acid and which acts as a Lewis base? Explain:

$Al^{3+}(aq) + H_2\ddot{O}{:}\ (l) \leftrightharpoons Al{:}\ddot{O}H_2^{3+}(aq)$

Solution:

In the example equation, the aluminum ion acts as a Lewis acid, because it accepts a pair of electrons forming a covalent bond to oxygen. The water molecule acts as a Lewis base because it donates a pair of electrons.

Lewis acids commonly are highly charged metal cations in aqueous solutions. Their high charge allows them to attract a non-bonding pair of electrons belonging to one or more water molecules.

Common misconception: Although aqueous calcium ion, Ca2+, is the cation of a strong base and is generally considered to be neutral, it does show slightly acidic properties consistent with Lewis acid character and ion pairing.

Your Turn 16.3

In the following equation, which species acts as a Lewis acid and which acts as a Lewis base? Explain:

$BF_3(g) + NH_3(g) \leftrightarrows BF_3NH_3$

Write your answer in the space provided.

All three theories, Arrhenius, Brønsted-Lowry, and Lewis, are different ways of visualizing the same concept. The different classifications lead to different insights in understanding acid-base reactions.

The Arrhenius model focuses on what ions (H^+ for acids and OH^- for bases) are produced in solution. The Brønsted-Lowry model demonstrates acids and bases as proton transfer agents in chemical reactions. The Lewis concept emphasizes the pair of electrons that constitute bond-breaking and bond-making in an acid-base reaction.

Consider the reaction ammonia with water:

$:NH_3(g) + H_2O \rightarrow H:NH_3^+(aq) + OH^-(aq)$

Ammonia is clearly an Arrhenius base because it increases the hydroxide ions in solution. It is also a Brønsted-Lowry base because it accepts a proton.

Additionally it is a Lewis base because it donates a pair of electrons to make a new bond between nitrogen and hydrogen. Table 16.7 summarizes the major points of the three acid-base theories.

Table 16.7. Comparing acid-base models with definitions and examples.

Model	**Definition**	**Example equation**
Arrhenius acid Arrhenius base	Increases H^+ in solution Increases OH^- in solution	$HCl(g) \rightarrow H^+(aq) + Cl^-(aq)$ $NaOH(s) \rightarrow Na^+(aq) + OH^-(aq)$
Brønsted acid Brønsted base	Proton donor Proton acceptor	$HNO_2(aq) + H_2O(l) \leftrightarrows H_3O^+(aq) + NO_2^-(aq)$ $NH_3(aq) + H_2O(l) \leftrightarrows NH_4^+(aq) + OH^-(aq)$
Lewis acid Lewis base	Electron pair acceptor Electron pair donor	$Al^{3+}(aq) + H_2O: \leftrightarrows Al:OH_2^{3+}(aq)$ $Ag^+(aq) + :NH_3(aq) \leftrightarrows Ag:NH_3^+(aq)$

The pH Scale — Section 16.4

Molar concentrations of $H^+(aq)$ are often expressed as pH, approximated for most solutions as the negative logarithm (base 10) of $[H^+]$.

$pH = -\log [H^+]$ or $pH = -\log [H_3O^+]$

This mathematical model is accurate for pH's ranging from 2-12 and is often used as an approximation for pH's ranging from 0 to 14. Table 16.8 shows the relationships among $[H^+]$, $[OH^-]$, pH and pOH for various solutions.

Table 16.8. The pH Scale: Relationships of [H+], [OH-], pH and pOH.

$[H^+]$	pH	pOH	$[OH^-]$
1×10^{-14}	14	0	1×10^{-0}
1×10^{-13}	13	1	1×10^{-1}
1×10^{-12}	12	2	1×10^{-2}
1×10^{-11}	11	3	1×10^{-3}
1×10^{-10}	10	4	1×10^{-4}
1×10^{-9}	9	5	1×10^{-5}
1×10^{-8}	8	6	1×10^{-6}
1×10^{-7}	7	7	1×10^{-7}
1×10^{-6}	6	8	1×10^{-8}
1×10^{-5}	5	9	1×10^{-9}
1×10^{-4}	4	10	1×10^{-10}
1×10^{-3}	3	11	1×10^{-11}
1×10^{-2}	2	12	1×10^{-12}
1×10^{-1}	1	13	1×10^{-13}
1×10^{-0}	0	14	1×10^{-14}

Common misconception: It's common to see pH scales ranging from 0 to 14 as illustrated in Table 16.8. However, the assumption that pH = -log[H^+] is valid only for pH's ranging from about 2 through 12. At higher and lower pH's, which represent higher concentrations of acid or base, ion-ion pairing is common and pH = -log[H^+] is invalid.

Table 16.9 shows the mathematical relationships involving pH. Notice that the equations in the right hand column of Table 16.9 are the logarithmic forms of the equations in the left hand column.

Table 16.9. Mathematical relationships for interconverting [H^+], [OH^-], pH and pOH.

pH = - log[H^+]	[H^+] = 10^{-pH}
pOH = -log[OH^-]	[OH^-] = 10^{-pOH}
pH + pOH = 14	[H^+][OH^-] = 1×10^{-14}

Table 16.10. Mathematical relationships for calculating [H^+] and [OH^-] in strong and weak acid and base solutions.

	acid	base
strong	x = I	y = I (monobasic)
		y = 2 x I (dibasic)
weak	Ka = x^2 / I if I > 100 Ka	Kb = y^2 / I if I > 100 Kb
	Ka x Kb = Kw = 1×10^{-14}	

I = initial molar concentration of acid or base

x = the moles per liter of acid that ionizes.

y = the moles per liter of base than ionizes.

Calculations Involving Strong Acids and Strong Bases

Section 16.6

Strong acids ionize completely. Therefore no equilibrium is established because all of the initial concentration of the reactant acid is converted to products. For all strong acids the ICE tables are the same. For example, the ICE table for the ionization of 0.20 M nitric acid is:

	$HNO_3(aq) \rightarrow$	$H^+(aq) +$	$NO_3^-(aq)$
I	0.20	0	0
C	-0.20	+0.20	+0.20
E	0	0.20	0.20

All of the strong acid ionizes so the initial concentration of the acid is same as the final concentration of H^+ ion.

Example:

Calculate the $[H^+]$ and pH in a solution of 0.015M nitric acid, HNO_3.

Solution:

$[HNO_3] = [H^+] = 0.015\ M$

$pH = -\log(0.015) = 1.82.$

By similar reasoning, the $[OH^-]$ of a strong base solution is the same as the concentration of a mono-basic strong base and double the concentration of a di-basic strong base.

I = y for monobasic and

I = 2y for dibasic where y = the moles per liter of base that ionizes.

Example:

Calculate the $[OH^-]$ and the pH in a 0.025 M KOH solution and the $[OH^-]$ and the pH in a 0.025 M $Ba(OH)_2$ solution.

Solution:

For the KOH solution,

$[KOH] = [OH^-] = 0.025M$

$pOH = -\log(0.025) = 1.60.$

$pH = 14 - pOH = 14 - 1.60 = 12.40.$

For the solution of $Ba(OH)_2$,

$[OH^-] = 2 \times [Ba(OH)_2] = 2 \times 0.025M = 0.050M$

$pOH = -\log(0.050) = 1.30.$

$pH = 14 - pOH = 14 - 1.30 = 12.70$

Section 16.6 Calculations Involving Weak Acids

The ionization of a weak acid in water is a reversible equilibrium, and each weak acid ionization produces the same "ICE" table.

Example:

What is the pH of a 0.15 M solution of acetic acid, $HC_2H_3O_2$?

Solution:

First set up an ICE table for the ionization of an initial concentration of acetic acid, $HC_2H_3O_2$. *Let x equal the number of moles per liter of weak acid that ionizes.*

$HC_2H_3O_2(aq) + H_2O(l) \leftrightarrows H_3O^+(aq) + C_2H_3O_2^-(aq) \qquad Ka = 1.8 \times 10^{-5}$

I	I	0	0
C	$-x$	$+x$	$+x$
E	$I - x$	x	x

$Ka = [H_3O^+][C_2H_3O_2^-]/[HC_2H_3O_2] = (x)(x)/(I - x)$

In general, if $I > 100\ Ka$, *then the assumption* $I - x = I$ *is a good approximation. Most weak acids do have a Ka that is less than* 10^{-3} *so I is usually* $> 100\ Ka$.

So the equation used for calculations involving the ionization of weak acids simplifies to:

$Ka = x^2/I$

where Ka is the ionization constant for the weak acid, x is the number of moles per liter of weak acid that ionize and I is the initial molar concentration of weak acid.

$Ka = x^2/I$

$1.8 \times 10^{-5} = x^2/0.15$

$x = [H^+] = 1.6 \times 10^{-3}\ M$

$pH = -\log[H^+] = -\log(1.6 \times 10^{-3}) = 2.78$

Calculations for Weak Bases — Section 16.7

Like weak acids, weak bases all have the same ICE table. Even better, the ICE table for weak bases is the same as that for weak acids, except that base ionization produces hydroxide ion rather than hydrogen ion.

Example:

What is the pH of a 0.75 M solution of aqueous ammonia?

Solution:

Set up an ICE table for the reaction of ammonia with water.

$NH_3(aq) + H_2O(l) \leftrightarrows NH_4^+(aq) + OH^-(aq) \quad Kb = 1.8 \times 10^{-5}$

I	I	0	0
C	-y	+y	+y
E	I - y	y	y

For clarity, "y" is used instead of "x" to represent the number of moles per liter of base that ionizes.

$Kb = [OH^-][NH_4^+]/[NH_3] = (y)(y)/(I - y)$

If I > 100 Kb then y is very small compared to I, so the generic equation for calculations involving weak base ionizations is:

$Kb = y^2/I$

where Kb is the ionization constant for the weak base, y is the number of moles per liter of weak base that ionize and I is the initial molar concentration of weak base.

$Kb = y^2/I$

$1.8 \times 10\text{-}5 = y^2/0.75$

$y = [OH\text{-}] = 3.7 \times 10^{-3}$ M

$pOH = -\log [OH^-] = -\log (3.7 \times 10^{-3}) = 2.43$

$pH = 14 - pOH = 14 - 2.43 = 11.57$

Common misconception: The most difficult part of performing calculations involving strong and weak acids and bases is recognizing the chemistry. Does the problem involve an acid or a base? Is the acid or base strong or weak? If you can first answer these questions based on the chemistry, then the calculations are relatively uncomplicated.

Your Turn 16.4

Classify the following as strong or weak acids or bases. Justify your answers.

chloric acid, ammonium chloride, calcium hydroxide, ethyl amine, sodium cyanide. Write your answer in the space provided.

Once the identity of a strong or weak acid or base solution is established, the calculations fall into a predictable pattern.

Example:

Measurements show that the pH of a 0.10 M solution of acetic acid is 2.87. What is Kb for potassium acetate, $KC_2H_3O_2$?

Solution:

The question gives information regarding a solution of a weak acid but asks for the Kb of its conjugate weak base. First calculate Ka for the weak acid and then convert it to Kb for the conjugate base.

$Ka = x^2/I$

If the pH is 2.87, the $[H^+] = 10^{-2.87} = 1.3 \text{ X } 10^{-3}$ *M.*

$[H^+] = [C_2H_3O_2^-] = x = 1.3 \text{ X } 10^{-3}$ *M*

Substituting:

$Ka = x^2/(0.10\text{-}x)$

$Ka = (1.3 \text{ X } 10^{-3})(1.3 \text{ X } 10^{-3})/(0.10 - 1.3 \text{ X } 10^{-3} \text{ M})$

$Ka = 1.7 \text{ X } 10^{-5}$

$Kb = Kw/Ka = 1.0 \text{ X } 10^{-14}/1.7 \text{ X } 10^{-5} = 5.9 \text{ x } 10^{-10}$

Section 16.9 Acid-Base Properties of Salt Solutions

Hydrolysis of salts refers to the reactions of salt ions with water. Recall that, except for the anions of strong acids, anions tend to be weak bases. Their negative charges tend to attract protons from water. Similarly, except for the cations of strong bases, cations are weakly acidic, either by attracting a pair of electrons from water, as with Lewis acids or by donating an available proton to water.

To classify a salt solution as acidic, basic, or neutral, disregard any neutral cations or anions listed in Table 16.2. If what is left is an anion, the salt is basic. If a cation remains, the salt is acidic. If both the cation and anion are listed in Table 16.2, the salt is neutral. If neither the cation nor anion are neutral, the acid or base character of the salt cannot be determined by examining its formula.

Example:

Classify the salt, sodium nitrite, $NaNO_2$ as acid, base, or neutral. Explain your reasoning. Write a chemical equation for its reaction with water.

Solution:

Sodium nitrite is a base. The sodium ion is a cation of a strong base so it is neutral. Nitrite is the conjugate base of the weak acid, nitrous acid, HNO_2. A solution of sodium nitrite in water will be basic because of the hydrolysis reaction of the nitrite ion with water. (We ignore the sodium ion because it is neutral.) The Kb of the nitrite ion can be calculated from the Ka of nitrous acid.

$$NO_2^-(aq) + H_2O(l) \leftrightarrows HNO_2(aq) + OH^-(aq)$$

Example:

Classify an aqueous solution of methyl ammonium chloride, CH_3NH_3Cl, as acidic, basic or neutral. Explain your reasoning. Write an equation to illustrate your answer.

Solution:

Methyl ammonium chloride is acidic because the chloride ion is neutral and the methyl ammonium ion is the conjugate weak acid of the base, methyl amine, CH_3NH_2. The Ka of methyl ammonium ion can be calculated from the Kb of methyl amine.

$$CH_3NH_3^+(aq) + H_2O(l) \leftrightarrows CH_3NH_2(aq) + H_3O^+(aq)$$

Acid-Base Behavior and Chemical Structure — Section 16.10

Acidity of a substance is directly related to the strength of attraction for a pair of electrons to a central atom. In general, acidity increases with stronger attractions for electrons.

Three factors affect attraction for electrons:

1. Ionic charge. When comparing similar ions, the more positive ions are stronger acids. (The more negative ions are more strongly basic.) For example:

Relative acid strengths: $Na^+ < Ca^{2+} < Cu^{2+} < Al^{3+}$

Metal cations of higher charge act as Lewis acids in water. The higher the charge, the greater attractions for electrons and the stronger is the acid. If the charges are equal, the smaller ion displays the stronger attraction for electrons.

Relative acid strengths: $PO_4^{3-} < HPO_4^{2-} < H_2PO_4^- < H_3PO_4$

Note that in this example, PO_4^{3-} is clearly a base because it has a negative charge, and H_3PO_4 is clearly an oxyacid. However, HPO_4^{2-} and $H_2PO_4^-$ are recognized as amphoteric, because they are protonated anions. The ionic charge generalization cannot predict whether a substance will be an acid (more acidic than water) or a base (less acidic than water) relative to water. It can predict only the acidity of substances relative to each other.

2. Oxidation number: When comparing similar formulas with the same central atom, the greater the oxidation number of the central atom, the stronger the acid. For example:

Relative strengths: $HClO < HClO_2 < HClO_3 < HClO_4$

Relative strengths: $H_2SO_3 < H_2SO_4$

3. Electronegativity: When comparing similar formulas with different central atoms, in general, the greater the electronegativity of the central atom, the stronger the acid.

Relative strengths: $H_3BO_3 < H_2CO_3 < H_2SO_3 < HNO_3$

(Recall that the electronegativity of elements generally increases from the lower left to the upper right of the periodic table.)

Some Common Acid-Base Reactions

Besides the hydroxides of Groups 1 and 2, strong bases include hydrides, nitrides, and carbides:

Hydride ion, H^- : $NaH(s) + H_2O \rightarrow H_2(g) + Na^+(aq) + OH^-(aq)$

Nitride ion, N^{3-} : $Mg_3N_2(s) + 6H_2O \rightarrow 2NH_3(g) + 3Mg^{2+}(aq) + 6OH^-(aq)$

Carbide ion, C_2^{2-} : $Ca_2C_2(s) + 2H_2O \rightarrow C_2H_2(g) + Ca^{2+}(aq) + 2OH^-(aq)$

$C_2H_2(g)$ is ethyne (acetylene), $HC{\equiv}CH$.

Strong bases also include oxides of Groups 1 and 2 like Li_2O, MgO and CaO. These "base anhydrides" react with water to give hydroxides:

$Li_2O(s) + H_2O(l) \rightarrow 2\,Li^+(aq) + 2OH^-(aq)$

$CaO(s) + H_2O(l) \rightarrow Ca^{2+}(aq) + 2OH^-(aq)$

Similarly, non-metal oxides, called acid anhydrides, give solutions of acids in water.

$SO_2(g) + H_2O(l) \leftrightarrows H_2SO_3(aq)$

$SO_3(g) + H_2O(l) \rightarrow H^+(aq) + HSO_4^-(aq)$

$CO_2(g) + H_2O(l) \leftrightarrows H_2CO_3(aq)$

$Cl_2O(g) + H_2O(l) \leftrightarrows 2HClO(aq)$

$Cl_2O_7(g) + H_2O(l) \rightarrow 2H^+(aq) + 2ClO_4^-(aq)$

$P_2O_5(s) + 3H_2O(l) \leftrightarrows 2H_3PO_4(aq)$

Notice that the non-metal tends to retain its oxidation number when going from the oxide to the acid. One notable exception is the reaction of nitrogen dioxide with water to form nitric acid and nitrous acid.

$2NO_2(g) + H_2O(l) \rightarrow H^+(aq) + NO_3^-(aq) + HNO_2(aq)$

Notice that in all the equations, whenever a strong acid is formed, it is written in its ionized form to reflect its strong electrolytic character.

Additional Practice in Chemistry the Central Science

For more practice working acid-base equilibrium problems in preparation for the Advanced Placement examination, try these problems in Chapter 16 of Chemistry the Central Science:

Additional Exercises: 16.110, 16.111, 16.112, 16.116, 16.113.

Integrated Exercises: 16.118, 16.120, 16.121, 16.124, 16.125, 16.126.

Multiple Choice Questions

1. According to the Lewis definition, an acid is a substance that

A) increases the hydrogen ion concentration in water.

B) can react with water to form H^+ ions.

C) can accept an electron pair to form a covalent bond.

D) *can donate a proton to a base.*

E) *can react with water to form hydronium ions.*

2. *Which pair of chemical species is not a conjugate acid-base pair?*

A) H_2CO_3 *and* CO_3^{2-}

B) OH^- *and* H_2O

C) HPO_4^{2-} *and* PO_4^{3-}

D) NH_3 *and* NH_4^+

E) CH_3NH_2 *and* CH_3NH^-

3. *Which oxide when mixed in equal molar amounts with water forms a solution with the lowest pH?*

A) CaO

B) CO_2

C) SO_2

D) SO_3

E) P_2O_5

4. *When 0.2 moles of each of these salts are dissolved in one liter of water, what is the order of increasing pH (lowest pH first) of the resulting soultions?*

A) $Na_2CO_3 < NaC_2H_3O_2 < NaCl < NH_4Cl$

B) $NaCl < Na_2CO_3 < NaC_2H_3O_2 < NH_4Cl$

C) $NH_4Cl < NaC_2H_3O_2 < NaCl < Na_2CO_3$

D) $NH_4Cl < NaCl < NaC_2H_3O_2 < Na_2CO_3$

E) $Na_2CO_3 < NaCl < NaC_2H_3O_2 < NH_4Cl$

5. *Aqueous solutions of equal molar concentrations of these salts are listed in order of increasing pH.* $NaBr < NaIO_3 < NaF < NaC_2H_3O_2 < Na_2SO_3$

Which acid is the weakest?

A) HBr

B) HIO_3

C) *HF*

D) CH_3COOH

E) $NaHSO_3$

6. *The net-ionic equation for the addition of 10.0 mL of 0.10 M sulfurous acid to 10.0 ml of 0.10 M aqueous sodium hydroxide is*

A) $H_2SO_3 + 2OH^- \leftrightharpoons 2H_2O + SO_3^{2-}$

B) $H_2SO_4 + OH^- \leftrightharpoons H_2O + HSO_4^-$

C) $H_2SO_3 + OH^- \leftrightharpoons H_2O + HSO_3^-$

D) $H^+ + OH^- \leftrightharpoons H2O$

E) $HSO_3^- + NaOH \rightarrow NaSO_3^- + H_2O$

7. *Which is the acid anhydride of chlorous acid,* $HClO_2$*?*

A) Cl_2O

B) ClO

C) ClO_2

D) Cl_2O_3

E) Cl_2O_5

8. *Equal molar aqueous solutions of the chloride salts of the following metals in their highest oxidation states were prepared and their respective pH's measured. Which is the correct order of pH?*

A) $Sn < Al < Cu < Ca < K$

B) $K < Ca < Cu < Al < Sn$

C) $Al < Cu < Ca$ $Sn < K$

D) $Sn < Al < Ca < Cu < K$

E) $K < Cu < Ca < Al < Sn$

9. *The acid dissociation constants for the diprotic acid, malonic acid,* $H_2C_3H_2O_4$, *are* $Ka_1 = 1.5 X 10^3$ *and* $Ka_2 = 2.0 X 10^6$.

What is Kb for $HC_3H_2O_4^-$*?*

A) $Kw \times Ka_1$

B) $Kw \times Ka_2$

C) Kw/Ka_1

D) Kw/Ka_2

E) $Ka_1 \times Ka_2$

10. *What is the percent ionization of a 0.10 M solution of hydroazoic acid, HN_3? $Ka = 1.9 \times 10^{-5}$.*

A) 1.9×10^{-3}

B) 1.9.

C) 0. 19

D) 1.4

E) 0.14

Free Response Questions

1. *Sulfurous acid, H_2SO_3 is a diprotic acid. $Ka_1 = 1.7 \times 10^{-2}$. $Ka_2 = 6.4 \times 10^{-8}$.*

a. *Write an ionic equation for the aqueous ionization that corresponds to Ka_1.*

Write an ionic equation for the aqueous ionization that corresponds to Ka_2.

Identify the conjugate acid-base pairs in each of your two equations.

b. *Identify any amphoteric species, other than water, in your equations.*

c. *Assume the amphoteric species you identified in Part b is a base. Write an ionic equation for its aqueous ionization and calculate the corresponding Kb.*

d. *Is an aqueous solution of $NaHSO_3$ acidic or basic? Explain your reasoning.*

e. *Calculate the pH of a 0.50 M solution of Na_2SO_3.*

2. *It is found that 0.30 M solutions of the three salts XCl_3, YCl_2 and ZCl have pH's of 5.5, 7.0 and 3.7, not necessarily in order. (X, Y and Z are metal ions. Cl is chloride ion)*

a. *i. Explain how metal ions can act as acids in aqueous solution.*

ii. Which pH goes with which salt? Explain.

iii. What is the approximate pH of a 0.010 M solution of ZOH? Explain your reasoning.

b. If a 0.5 M solution of copper(II) sulfate is added to a 1.0 M solution of sodium hydrogen carbonate, significant effervescence is observed. If the same copper(II) sulfate solution is added to a 1.0 M solution of sodium carbonate, little or no effervescence in observed. Explain these observations.

Answers and explanations for multiple choice questions.

1. *C. A Lewis acid is an electron pair acceptor. A Brønsted Lowry acid is a proton donor. An Arrhenius acid increases [H^+] when dissolved in water.*

2. *A. An acid-base conjugate pair consists of two chemical species that differ in formula by an H^+. The species of the pair with the more positive charge (and hence the extra H) is the acid.*

3. *D. A lower pH implies a stronger acid. Non-metal oxides tend to be acidic (acid anhydrides) while metal oxides tend to be basic (base anhydrides). When comparing oxides having different central atoms, the more electronegative atom forms the strongest acid. Sulfur is the most electronegative atom represented. When comparing oxides having the same central atom, the one with the greater number of oxygen atoms is more acidic.*

4. *D. A lower pH implies a stronger acid. In general, cations (except for the cations of strong bases), tend to be acidic and anions (except for the anions of strong acids) tend to be basic. The more positive the charge the more acidic the cation, and the more negative the charge the more basic the anion. Sodium ion is the neutral ion of the strong base, NaOH, and chloride is the neutral ion of the strong acid, HCl, so sodium chloride is neutral. Ammonium ion is slightly acidic, and acetate and carbonate ion are weak bases.*

5. *E. A higher pH implies a stronger base. The strongest base, Na_2SO_3, has the weakest conjugate acid, $NaHSO_3$. The sodium ions are neutral because Na^+ is the cation of the strong base, NaOH.*

6. *C. Sulfurous acid, H_2SO_3, is a weak acid and should not be confused with sulfuric acid, H_2SO_4, a strong acid. Sodium hydroxide is a strong*

electrolyte so the sodium ion is a spectator ion. Only one millimole each of the acid and base are present so only one of the two ionizable protons on sulfurous acid will react.

7. *D. The oxidation number of chlorine in $HClO_2$ is + 3. The acid anhydride of an acid is generally the non-metal oxide with the same oxidation number of the non-metal.*

8. *A. Charged metal cations in aqueous solutions are Lewis acids. The higher the charge of the metal cation, the more acidic the solution. When charges are equal, the smaller the cation, the more acidic the solution. The salts are: $SnCl_4$, $AlCl_3$, $CuCl_2$, $CaCl_2$ and KCl and their respective metal cations are: Sn^{4+}, Al^{3+}, Cu^{2+}, Ca^{2+} and K^{+}. Cu^{2+} is smaller than Ca^{2+} so Cu^{2+} is more acidic.*

9. *C. The chemical reactions are:*

   1. $H_2C_3H_2O_4 + H_2O \leftrightarrows HC_3H_2O_4^- + H_3O^+$
   $Ka_1 = [HC_3H_2O_4^-][H_3O^+]/[H_2C_3H_2O_4]$

   2. $HC_3H_2O_4^- + H_2O \leftrightarrows C_3H_2O_4^{2-} + H_3O^+$
   $Ka_2 = [C_3H_2O_4^{2-}][H_3O^+]/[HC_3H_2O_4^-]$

   3. $HC_3H_2O_4^- + H_2O \leftrightarrows H_2C_3H_2O_4 + OH^-$
   $Kb = [H_2C_3H_2O_4][OH^-]/[HC_3H_2O_4^-]$

   4. $H_2O + H_2O \leftrightarrows H_3O^+ + OH^-$
   $kw == [H_3O^+][OH^{-1}]$

   The correct equation is Kb = Kw/Ka. Which Ka is the correct one to use in this case?

   Substituting $Kb = Kw/Ka_1$ we get the correct solution:

   $[H_2C_3H_2O_4][OH^-]/[HC_3H_2O_4^-] = [H_3O^+][OH^{-1}]/([HC_3H_2O_4^- H_2C_3H_2O_4])$

   Alternatively: Reaction 3 includes the same conjugate pair as Reaction 1, so $Kb = Kw/Ka_1$ is correct.

10. *D. $Ka = x^2/I$ and % ionization = (x/I)(100) where X is the moles per liter ionized and I is the initial concentration.*

    $1.9 \times 10^{-5} = X^2/0.10$

    $x = 1.4 \times 10^{-3}$

    % ionization $= (1.4 \times 10^{-3} / 0.10)(100) = 1.4$

Answers to free response questions

1a. $H_2SO_3(aq) + H_2O(l) = H_3O^+ (aq) + HSO_3^-(aq)$

acid *base* *acid* *base*

$HSO_3^-(aq) + H_2O(l) = H_3O^+(aq) + SO_3^{2-}(aq)$

acid *base* *acid* *base*

b. $HSO_3^-(aq)$

c. $HSO_3^-(aq) + H_2O(l) = OH^-(aq) + H_2SO_3(aq)$

$Kb = Kw/Ka_1 = 1.0 \times 10^{-14}/1.7 \times 10^{-2} = 5.9 \times 10^{-13}$

d. $HSO_3^-(aq)$ *forms an acidic solution because its Ka is much greater than its Kb.*

$Ka_2 = 6.4 \times 10^{-8} > Kb = 5.9 \times 10^{-13}$

e. $Kb = Kw/Ka_2 = 1.00 \times 10^{-14}/6.4 \times 10^{-8} = 1.56 \times 10^{-7}$

$Kb = y^2/I$

$1.56 \times 10^{-7} = y^2/0.50$

$y = 2.8 \times 10^{-4} = [OH^-]$

$pOH = -\log [OH^-] = -\log 2.8 \times 10^{-4} = 3.55$

$pH = 14.00 - pOH = 14.00 - 3.55 = 10.45$

2a. i. *Highly charged metal cations are Lewis acids. They attract non-bonding pairs on water molecules. This attraction polarizes the H-O bond making it weaker causing* $H^+(aq)$ *to enter the solution.*

$Al^{3+}(aq) + H_2O{:}(l) \leftrightarrows Al{:}OH_2^{3+}(aq) \leftrightarrows AlOH^{2+}(aq) + H^+(aq)$

ii. $XCl_3 = pH\ 3.7$, $YCl_2 = pH\ 5.5$, $ZCl = pH\ 7.0$

The higher the positive charge the more acidic the cation. Aqueous chloride ion, Cl^- *is neutral so does not affect the pH.* X^{3+} *is the most acidic ion and forms the solution with the lowest pH.*

iii. $pH = 12.00$. Z^+ *is a neutral ions because it forms a solution of pH 7.0 with chloride ion, also neutral. Therefore ZOH is a strong base.* $pOH = -\log 0.01 = 2$. $pH = 14 - pOH = 14 - 2 = 12$.

b. *Both carbonate and hydrogen carbonate ions effervesence producing gaseous carbon dioxide in the presence of acids according to the following equations:*

$CO_3^{2-}(aq) + 2H^+(aq) \rightarrow CO_2(g) + H_2O(l)$

$HCO_3^-(aq) + H^+(aq) \rightarrow CO_2(g) + H_2O(l)$

However, copper(II) ion is a weak Lewis acid which can provide only enough hydrogen ion to cause hydrogen carbonate to effervesence but not enough to cause carbonate ion to produce carbon dioxide. The stoichiometry requires each mole of carbonate ion to react with two moles of hydrogen ion.

Sulfate ion is nearly neutral and has no affect on the reaction.

Your Turn Answers

16.1. *Aqueous solutions of the salts NaCl, KNO_3, $LiClO_4$, $BaBr_2$ and CsI are neutral because they contain neutral cations and neutral anions. The cations are all cations of strong bases, and the anions are all anions of strong acids.*

16.2. *Of the acids listed in table 16.4, HIO is the weakest because it has the smallest Ka value. In Table 16.5, fluoride ion is the weakest base listed because it has the smallest value of Kb.*

16.3. *BF_3 is a Lewis acid because it accepts a lone pair of electrons from the Lewis base, $:NH_3$. BF_3 has a non-octet Lewis structure where boron has only six electrons and readily accepts ammonia's non-bonding pair.*

16.4. *Chloric acid is listed in Table 16.1 as a strong acid.*

Chloride is listed as a neutral ion in Table 16.2. Ammonium ion, NH_4^+, is the conjugate acid of the weak base ammonia, NH_3.

Calcium hydroxide is listed in Table 16.1 as a dibasic strong base.

Ethyl amine is a neutral nitrogen compound and a weak base.

Cyanide ion, CN^-, is the weak conjugate base of hydrocyanic acid, HCN. Sodium ion is listed in Table 16.2 as a neutral ion.

TOPIC

17

ADDITIONAL ASPECTS OF AQUEOUS EQUILIBRIA

Calculations involving buffer solutions and solubility equilibria are in many ways the most important single topics on the AP exam, and are also among the most difficult to master. Students must also master the quantitative relationships that govern various points during the titrations of weak acids and bases, and be able to match suitable indicators to particular titrations. Reactions involving complex ions are also included on the AP exam.

The Common Ion Effect

Section 17.1

The **common ion effect** decreases the ionization of a weak electrolyte when a common ion is added to the solution.

Consider the ionization of acetic acid in water.

$HC_2H_3O_2(aq) + H_2O(l) \leftrightarrows C_2H_3O_2^-(aq) + H_3O^+(aq)$

Adding sodium acetate, $NaC_2H_3O_2$, to this solution will increase the concentration of acetate ion, $C_2H_3O_2^-(aq)$. The effect will be to shift the equilibrium to the left in accordance with Le Chatelier's principle decreasing the ionization of acetic acid. The resulting solution is a buffer containing both a weak acid and a conjugate weak base.

Buffer Solutions

Section 17.2

A **buffer solution** is a solution that resists a change in pH upon addition of small amounts of strong acid or base. A mixture of a weak acid and its conjugate base, both existing in significant quantities in the same solution, constitutes an effective buffer. The solution's resistance to pH change arises because the weak acid can react with small quantities of strong base, and the weak base can react with small quantities of strong acid.

Your Turn 17.1

To make a buffer solution what weak acid must be added to an amount of sodium acetate? Write a chemical equation and its corresponding equilibrium expression.

Write your answer in the space provided.

Understanding the reactions of acids with bases is the key to understanding how a buffer solution works both qualitatively and quantitatively. It is important to know how acids and bases react with each other. For example, consider the reactions of equal molar amounts of the following acids and bases.

1. A strong acid reacts completely with a strong base to yield a neutral solution.

Complete equation: $HCl(aq) + NaOH(aq) \rightarrow H_2O(l) + NaCl(aq)$

Net ionic equation: $H^+(aq) + OH^-(aq) \rightarrow H_2O(l)$

The solution is neutral because both the sodium and chloride ions are neutral ions.

2. A strong acid reacts completely with a weak base to yield a weak acid.

Complete equation: $HCl(aq) + NaC_2H_3O_2(aq) \rightarrow NaCl(aq) + HC_2H_3O_2(aq)$

Net ionic equation: $H^+(aq) + C_2H_3O_2^-(aq) \rightarrow HC_2H_3O_2(aq)$

3. A strong base reacts completely with a weak acid to yield a weak base.

Complete equation: $NaOH(aq) + HC_2H_3O_2(aq) \rightarrow NaC_2H_3O_2(aq) + H_2O(l)$

Net ionic equation: $OH^-(aq) + HC_2H_3O_2(aq) \rightarrow C_2H_3O_2^-(aq) + H_2O(l)$

4. A weak acid does not react appreciably with a conjugate weak base.

Your Turn 17.2

Write and balance complete and net ionic equations for the reaction of aqueous ammonia solution with nitric acid. Write your answer in the space provided.

Calculating the pH of a buffer

The pH of a buffer solution is calculated just like the pH of a weak acid solution.

Example:

What is the pH of an aqueous mixture containing 0.20 M acetic acid and 0.10 M sodium acetate?

Solution:

Employ an ICE table for the ionization of acetic acid and include the initial concentrations of both the weak acid and the weak base:

$HC_2H_3O_2(aq) + H_2O(l) \leftrightharpoons C_2H_3O_2^-(aq) + H+(aq)$

	$HC_2H_3O_2$	$C_2H_3O_2^-$	H^+
I	0.20	0.10	0
C	-x	+x	+x
E	0.20-x	0.10+x	x

$Ka = [C_2H_3O_2^-][H^+]/[HC_2H_3O_2] = (0.10+x)(x)/(0.20-x)$

If x is small compared to 0.10, then $0.10 + x \simeq 0.10$ *and* $0.20-x \simeq 0.20$

So $Ka = 0.10x/0.20$

$1.8 \times 10^{-5} = 0.10x/0.20$

$x = [H_3O^+] = 3.6 \times 10^{-5}$

$pH = -log[H_3O^+] = -log(3.6 \times 10^{-5}) = 4.44$

Because the ICE tables for all buffer solutions are the same, the equation used in all buffer calculations can be generalized as:

Ka = x[base]/[acid]

or its logarithmic form:

$$pH = pKa + \log([base]/[acid])$$

where Ka is the ionization constant of the weak acid,

x is the molar concentration of hydrogen ion, $[H^+]$

[base] is the initial concentration of the weak base and

[acid] is the initial concentration of the weak acid.

Common misconception: Although calculations involving acid-base equilibria can be reduced to a few simple equations, the key to solving acid-base equilibria problems lies in understanding the chemistry and how the equations apply to the chemical reactions involved. Although the math may seem simple, the chemistry can be complex.

Table 17.1 reviews the equations used in solving quantitative acid-base problems.

Table 17.1. Common equations useful in solving acid-base equilibria problems.

Strong acid	$x = I$	$x = [H^+]$, I = initial acid concentration
Strong base	$y = I$ ($y = 2I$ for dibasic)	$y = [OH^-]$, I = initial base concentration
Weak acid	$Ka = x^2/I$	Ka = weak acid ionization constant
Weak base	$Kb = y^2/I$	Kb = weak base ionization constant
Buffer	$Ka = x[base]/[acid]$ or $pH = pKa + \log([base]/[acid])$	[base] = initial base concentration [acid] = initial acid concentration $pKa = -\log Ka$

Before the equations in Table 17.1 can be applied, the chemistry of the solution must be identified and understood. Table 17.2 summarizes what happens when an acid and base are mixed.

Table 17.2. Solutions resulting from various acid-base reactions.

If the solution contains:	And the base is in excess, the resulting solution is a:	And the acid is in excess, the resulting solution is a:	And neither the acid nor the base are in excess, the resulting solution is a:
Strong acid + strong base	Strong base	Strong acid	Neutral solution pH = 7
Strong acid + weak base	*buffer	Strong acid	Weak acid
Weak acid + strong base	Strong base	*buffer	Weak base
Weak acid + *buffer	weak base	*buffer	*buffer

* Whenever a buffer solution results, the volume change need not be considered when calculating pH.

Common misconception: A buffer can be prepared from more than just a weak acid and the salt of the acid or a weak base and the salt of the base. A buffer is formed whenever a limiting amount of strong base is added to an excess amount of weak acid because the strong base reacts completely converting some, but not all of the weak acid to its conjugate base. Similarly, a limiting amount of strong acid added to an excess amount of weak base forms a buffer solution.

Your Turn 17.3

Explain how nitric acid and sodium acetate can be used to make a buffer solution. Illustrate your answer using a chemical equation. Write your answer in the space provided.

Examples:

a. What is the pH of a solution made by mixing 1.0 L of 0.10 M HCl with 2.0 L of 0.060 M NaOH?

1. Write the net ionic equation.

$H^{+}(aq) + OH\text{-}(aq) \rightarrow H_2O(l)$

2. Calculate the amount of moles of acid and of base.

mol HCl = 1.0 L x 0.10 mol/L = 0.10 mol HCl

mol NaOH = 2.0 L x 0.06 mol/L = 0.12 mol NaOH

3. Subtract the limiting reactant from the excess reactant to obtain the amount of excess reactant that remains. (A strong acid and a strong base will react completely to the extent of the limiting reactant.)

0.12 mol NaOH - 0.10 mol HCl = 0.02 mol NaOH remains.

4. Calculate the total volume of the solution and the concentration(s) of the acid or base in solution.

Volume = 1.0 L + 2.0 L = 3.0 L.

$[OH^{-}] = 0.02 \text{ mol}/3.0 \text{ L} = 0.0067 \text{ M}$

5. Calculate the pH.

$pOH = -\log [OH\text{-}] = -\log (.0067) = 2.18.$

$pH = 14 - pOH = 14 - 2.18 = 11.82$

6. Is the answer reasonable? Yes. This pH is consistent for a solution containing an excess of strong base.

b. What is the pH of a solution made by mixing 1.0 L of 0.11 M HCl with 3.0 L of 0.080 M NaF?

1. Write the net ionic equation

$H^{+}(aq) + F^{-}(aq) \rightarrow HF(aq))$

2. Calculate the amount of moles of acid and of base.

mol HCl = 1.0 L x 0.11 mol/L = 0.11 mol HCl

mol NaF = 3.0 L x 0.080 mol/L = 0.24 mol NaF

3. All of the limiting strong acid will react with the base to form an amount of conjugate acid equal to the amount of limiting acid. Additionally, there is some weak base left over.

0.24 mol NaF - 0.11 mol HCl = 0.13 mol NaF plus 0.11 mol HF.

4. This is a buffer because it contains weak acid and weak base and the volume change is not important.

5. Calculate the pH.

pH = pKa + log [base]/[acid]

Ka for HF = 6.8 x 10^{-4} (from Table D-1 of Chemistry the Central Science)

pKa = -log Ka = -log 6.8 x 10-4 = 3.17

pH = 3.17 + log (0.13/0.11) = 3.24.

6. Is the answer reasonable? Yes. This pH is consistent for a buffer solution containing slightly more weak base than weak acid and whose weak acid has a pKa = 3.17.

c. What is the pH of a solution made by mixing 1.0 L of 0.20 M HNO_3 and 2.0 L of 0.10 M NaCN?

1. Write the net ionic equation.

$H^+(aq) + CN^-(aq) \leftrightarrows HCN(aq)$

2. Calculate the amount of moles of acid and of base.

Mol HNO_3 = 1.0 L x 0.20 mol/L = 0.20 mol HNO_3.

Mol NaCN = 2.0 L x 0.10 mol/L = 0.20 mol NaCN.

3. Neither strong acid nor weak base is in excess. The strong acid will convert all of the weak base into 0.20 moles of HCN.

4. Calculate the total volume of the solution and the concentration(s) of the species in solution.

Volume = 1.0 L + 2.0 L = 3.0 L.

[HCN] = 0. 20 mol/3.0 L = 0.067 M

5. Calculate the pH.

This is a 0.067 M solution of the weak acid, HCN

Ka = x^2/I

4.9 X 10^{-10} = x^2/0.067

x = 5.7 X 10^{-6} = [H^+]

pH = -log [H^+] = -log (5.7 X 10-6) = 5.24

6. Is the answer reasonable? Yes. A pH of 5.24 is consistent for a solution containing a weak acid.

d. What is the pH of a solution made by mixing 2.5 L of 0.20 M $HC_2H_3O_2$ with 1.0 L of 0.30 M $KC_2H_3O_2$?

Because this is a mixture of a weak acid and its conjugate weak base, no reaction will occur. This is a buffer solution.

1. Calculate the amount of moles of the acid and of the base.

mol $HC_2H_3O_2$ *= 2.5 L x 0.20 mol/L = 0.50 mol* $HC_2H_3O_2$.

mol $KC_2H_3O_2$ *= 1.0 L x 0.30 mol/L = 0.30 mol* $KC_2H_3O_2$.

2. Calculate the pH.

pH = pKa + log [base]/[acid]

pKa of $HC_2H_3O_2$ *= -log Ka = -log* (1.8×10^{-5}) *= 4.74*

pH = 4.74 + log(0.30/0.50) = 4.74 - 0.22 = 4.52

3. Is the answer reasonable? Yes. A pH of 4.52 is consistent with a buffer whose weak acid has a Ka of 4.74. Because the buffer contains more weak acid than weak base, the pH is slightly lower than the pKa.

Section 17.3 Acid-Base Titrations

An **acid-base titration** is a method to determine an unknown concentration of an acid or a base. A titration determines the volume of a standard solution of base of known concentration that is required to completely react with the acid sample. Similarly, a standard solution of an acid is used to measure an unknown concentration of base.

An **acid-base indicator** changes color at the **end point** of the titration.

The indicator signals the **equivalence point**, the point at which there are equal molar amounts of acid and base.

Alternatively, a pH meter can be used to monitor a titration from start to finish.

A **titration curve** is a graph of pH vs. mL of titrant. Various acid-base titrations produce distinctive titration curves.

Figure 17.1 shows a typical titration curve for a strong acid and a strong base. A strong acid-strong base titration curve is typified by the following properties:

The pH before the titration begins is that of a strong acid.

As strong base is added, the pH rises slightly because the concentration of the strong acid decreases as it becomes neutralized by strong base.

Near the equivalence point the pH rises dramatically. The pH at the equivalence point is always 7 for a strong acid-strong base titration.

After the equivalence point the pH is that of a strong base. It rises slightly because of the increasing amount of strong base.

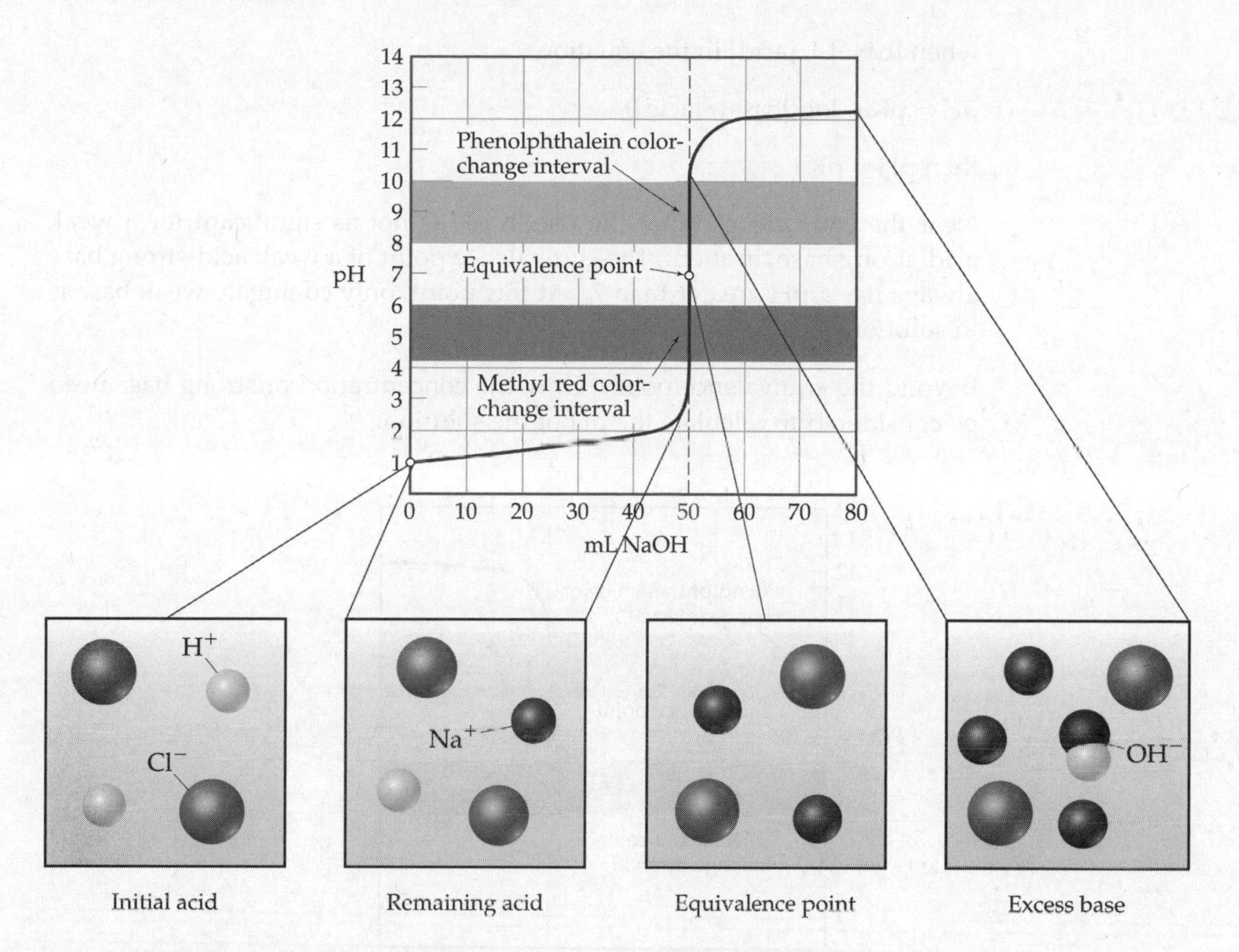

Figure 17.1. Typical acid-base titration curve for a strong acid and a strong base.

Figure 17.2 shows a titration curve of a weak acid and a strong base. Although a weak acid-strong base titration curve looks similar to a titration curve for a strong acid and strong base, there are significant differences:

The pH before the titration begins is higher because it is that of a weak acid in water.

The "buffered region" of the titration curve shows that the pH changes slightly as more and more strong base is added which replaces *some, but not all*, of the weak acid with conjugate base.

$$HA(aq) + OH^{-}(aq) \rightarrow A^{-}(aq) + H_2O(l)$$

$$pH = pKa - \log([base]/[acid])$$

The point where the amount of titrant is equal to half the amount required to reach the equivalence point defines the pKa of the weak acid because when [base] = [acid] in the equation,

pH = pKa -log([base]/[acid]),

then pH = pKa.

Near the equivalence point the rise in pH is not as significant for a weak acid-strong base titration. The equivalence point of a weak acid-strong base always has a pH greater than 7. At this point, only conjugate weak base is in solution.

Beyond the equivalence point. Only the concentration of strong base need be considered to calculate the pH of the solution.

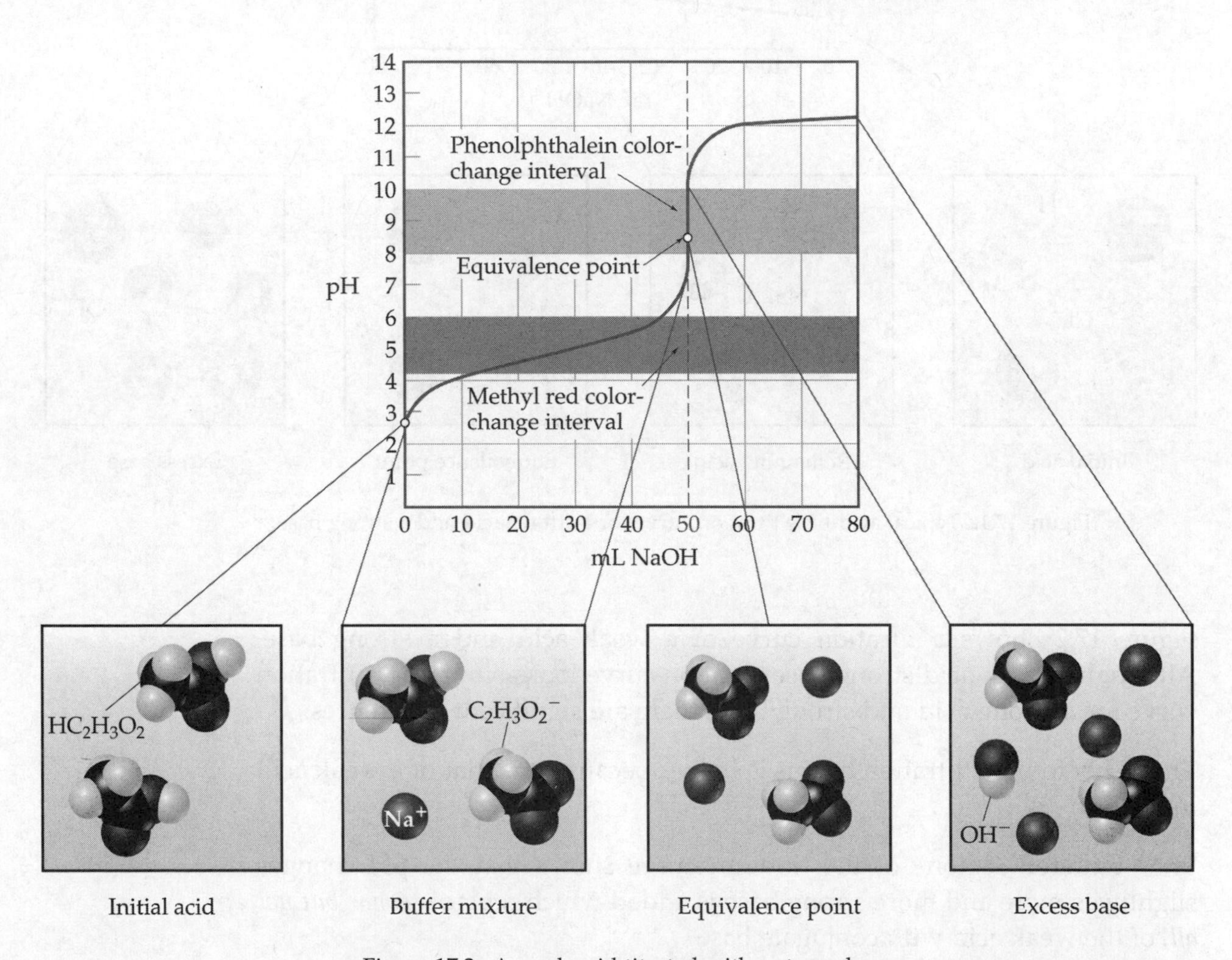

Figure 17.2. A weak acid titrated with a strong base.

In order for an indicator to accurately signal the equivalence point of a titration, the pH at which it changes color must match the pH of the equivalence point. Table 17.3 show the pKa's and pH range in which common indicators change color.

Table 17.3. Color changes of acid-base indicators and their pKa's.

Indicator	Color change	pH range	pKa
Bromphenol blue	yellow to blue	3-4.5	4
Bromcresol green	yellow to blue	4-5	4.5
Methyl red	red to yellow	4.5-6	5
Bromthymol blue	yellow to blue	6-7.5	7
Phenol red	yellow to red	7-8	7.5
Phenolphthalein	colorless to pink	8-10	9
Alizarin yellow R	yellow to red	10-12	11

Solubility Equilibria

Section 17.4

When a precipitate forms from the mixing of two solutions, an equilibrium is established between the solid precipitate and its dissolved ions. For example, consider the slightly soluble salt, silver chloride, AgCl. In a saturated solution, one that has dissolved the maximum amount of solute, the ions are in equilibrium with the solid:

$$AgCl(s) \leftrightarrows Ag^{+}(aq) + Cl^{-}(aq)$$

The equilibrium expression for this reaction is:

$$Ksp = [Ag^{+}][Cl^{-}]$$

The equilibrium constant, **Ksp**, is another special case of Kc and is called the **solubility product constant**.

For silver chloride, Ksp = 1.8×10^{-10} (From Table D-3, Appendix D in Chemistry the Central Science)

The ICE table is:

	$AgCl(s) =$	$Ag^{+}(aq) +$	$Cl^{-}(aq)$
I	I	0	0
C	-x	+x	+x
E	I - x	x	x

The **molar solubility** of silver chloride is the number of moles of silver chloride that dissolve in a liter of water. That is, the molar solubility = x. For every mole of AgCl that dissolves, one mole of Ag^+ and one mole of Cl^- is formed so:

$x = [Ag^+] = [Cl^-]$ and

$Ksp = [Ag^+] [Cl^-]$

Substituting:

$Ksp = x^2$

$1.8 \times 10^{-10} = x^2$

$x = 1.3 \times 10^{-5}$ M = the molar solubility of silver chloride.

Consider the slightly soluble salt, lead(II) chloride, $PbCl_2$. In a saturated solution, the ions are in equilibrium with the solid:

$PbCl_2(s) \leftrightarrows Pb^{2+} + 2\ Cl^-$

The equilibrium expression for this reaction is:

$Ksp = [Pb^{2+}] [Cl^-]^2$

For lead(II) chloride, $Ksp = 1.6 \times 10^{-5}$.

For every mole of $PbCl_2$ that dissolves, one mole of Pb^{2+} and two moles of Cl^- are formed so:

$x = [Pb^{2+}]$ and $[Cl^-] = 2x$

Substituting:

$\mathbf{Ksp} = (x)(2x)^2 = \mathbf{4x^3}$

$(1.6 \times 10^{-5})/4 = x^3$

$x = 0.016$ M = the molar solubility of lead(II) chloride

When a solid precipitate establishes an equilibrium with its ions in solution, the resulting Ksp expression, and its relationship to the molar solubility, x, is derived from the ICE table. It is dependent on the stoichiometry of the dissolution reaction. Table 17.4 shows the relationship of the Ksp expression and the molar solubility in water expression to various reactions.

Table 17.4. The Ksp expressions and molar solubility in water expressions for various solubility equilibria reactions.

Dissolution equilibrium	Ksp expression	molar solubility, x
$AgCl(s) \leftrightarrows Ag^{+}(aq) + Cl^{-}(aq)$	$Ksp = [Ag^{+}]\,[Cl^{-}]$	$Ksp = (x)(x) = x^2$
$PbCl_2(s) \leftrightarrows Pb^{2+}(aq) + 2\,Cl^{-}(aq)$	$Ksp = [Pb^{2+}]\,[Cl^{-}]^2$	$Ksp = (x)(2x)^2 = 4x^3$
$LaCl_3(s) \leftrightarrows La^{3+}(aq) + 3F^{-}(aq)$	$Ksp = [La^{3+}][F^{-}]^3$	$Ksp = (x)(3x)^3 = 27x^4$
$Ca_3(PO_4)_2(s) \leftrightarrows 3Ca^{2+}(aq) + 2PO_4^{3-}(aq)$	$Ksp = [Ca^{2+}]^3[PO_4^{3-}]^2$	$Ksp = (3x)^3(2x)^2 = 108x^5$

Factors that Affect Solubility — Section 17.5

The molar solubility of a precipitate is always greater in pure water than it is in an aqueous solution containing a common ion.

Common misconception: The molar solubility expression in water does not apply to solutions containing common ions. For common ions the molar solubility must be derived from an ICE table.

Example:

Calculate the molar solubility of cerium fluoride, CeF_3, in

a. pure water

b. a solution containing 0.2 M $CeCl_3$.

c. a solution containing 0.2 M NaF.

$Ksp = 8 \times 10^{-16}$.

Solution:

a. The reaction is: $CeF_3(s) \leftrightarrows Ce^{3+}(aq) + 3F^{-}(aq)$

The Ksp expression is

$Ksp = [Ce^{3+}]\,[F^{-}]^3$

$Ksp = (x)(3x)^3 = 27x^4$

$Ksp = 8 \times 10^{-16} = 27x^4$

$x = 7 \times 10^{-5}\ M$

b. The molar solubility expression in water does not apply to solutions containing common ions. It must be derived from the ICE table. Here cerium ion is the common ion. (Chloride is a soluble spectator ion.)

	$CeF_3(s) \leftrightarrows$	$Ce^{3+}(aq)$ +	$3F^-(aq)$
I	I	0.2	0
C	-x	+x	+3x
E	I - x	0.2 + x	3x

$Ksp = [Ce^{3+}][F^-]^3$

$Ksp = 8 \times 10^{-16} = (0.2 + x)(3x)^3$

If $x <<<< 0.2$ then $0.2 + x \simeq 0.2$

$8 \times 10^{-16} = (0.2)(3x)^3$

$x = 5 \times 10^{-6}$ M

c. The ICE table is similar to Part b except that here, fluoride is the common ion. (Sodium ion is a soluble spectator ion.)

	$CeF_3(s) \leftrightarrows$	$Ce^{3+}(aq)$ +	$3F^-(aq)$
I	I	0	0.2
C	-x	+x	+3x
E	I - x	x	0.2 + 3x

$Ksp = [Ce^{3+}][F^-]^3$

$Ksp = 8 \times 10^{-16} = (x)(0.2 + 3x)^3$

If $3x <<<< 0.2$ then $0.2 + 3x \simeq 0.2$

$8 \times 10^{-16} = (x)(0.2)^3$

$x = 1 \times 10^{-13}$ M

Notice than in both Parts b and c where a common ion is involved, the molar solubility of cerium fluoride is less than in pure water.

The **reaction quotient ,Q**, is used to predict whether a precipitate will form under a set of given conditions.

Example:

Consider a solution made by adding 175 mL of 0.10 M $BaCl^2$ to 40 mL of 0.50 M NaOH. Assume the volumes are additive. Ksp for barium hydroxide is 5.0×10^{-3}.

a. Write the net ionic equation for the dissolution of barium hydroxide.

b. What is the value of Q?

c. Ksp for $Ba(OH)_2$ is $5.0 X 10^{-3}$. Will a precipitate form?

d. What must be the concentration of OH^- for a precipitate to form in 0.10 M $BaCl_2$?

Solution:

a. $Ba(OH)_2(s) \leftrightarrows Ba^{2+}(aq) + 2OH^-(aq)$

b. $Q = [Ba^{2+}][OH^-]^2$

$Q = (0.10M)(175/215)[(0.50)(40/215)]^2 = 7.0 x 10^{-4}$

c. Ksp > Q. No precipitate will form because the concentrations of the ions are too small. Q must be at least as large as Ksp.

d. $Ksp = [Ba^{2+}][OH^-]^2$

$5.0 X 10^{-3} = [0.10][OH^-]^2$

[OH-] = 0.22M

The solubility of a hydroxide precipitate is affected by the pH of a solution.

Example:

Calculate the molar solubility of $Cd(OH)_2$ in a pH 9.5 buffer.
$Ksp = 2.5 X 10^{-14}$

Solution:

$Cd(OH)_2(s) = Cd^{2+}(aq) + 2OH^-(aq)$

$Ksp = 2.5 X 10^{-14} = [Cd^{2+}][OH^-]^2$

When the pH = 9.5, the pOH = 14 - 9.5 = 4.5 so $[OH^-] = 10^{-4.5}$.

$Ksp = 2.5 x 10^{-14} = [Cd^{2+}][10^{-4.5}]^2$

$[Cd^{2+}] = 2.5 X 10^{-5}$

Complex ions

A **complex ion** is a metal ion bonded to one or more Lewis bases. We have seen that a characteristic property of metal ions is their ability to act as Lewis acids by attracting non-bonding electron pairs of water molecules. Some metal ions also commonly attract the non-boding pairs of other Lewis bases, such as ammonia molecules and hydroxide ions, to form complex ions. Table 17.5 illustrates the formation of some common complex ions. A rough rule of thumb that works about 3/4 of the time is that the number of Lewis bases (called ligands) that a given metal ion attracts is equal to double its positive charge. The common exceptions to this rule are shown in table 17.4 with an asterisk, *.

Table 17.5. The formation of some common complex ions.

Complex ion	**Equilibrium reaction**
$Ag(NH_3)_2^+(aq)$	$Ag^+(aq) + 2NH_3(aq) \leftrightharpoons Ag(NH_3)_2^+(aq)$
$Ag(CN)_2^-(aq)$	$Ag^+ (aq) + 2CN^-(aq) \leftrightharpoons Ag(CN)_2^-(aq)$
$Ag(S_2O_3)_2^{3-}(aq)$	$Ag^+ (aq) + 2S_2O_3^{2-}(aq) \leftrightharpoons Ag(S_2O_3)_2^{3-}(aq)$
$Cu(NH_3)_4^{2+}(aq)$	$Cu^{2+} (aq) + 4NH_3(aq) \leftrightharpoons Cu(NH_3)_4^{2+}(aq)$
$Cu(CN)_4^{2-}(aq)$	$Cu^{2+} (aq) + 4CN^-(aq) \leftrightharpoons Cu(CN)_4^{2-}(aq)$
*$Fe(CN)_6^{4-}(aq)$	$Fe^{2+} (aq) + 6CN^-(aq) \leftrightharpoons Fe(CN)_6^{4-}(aq)$
$Fe(CN)_6^{3-}(aq)$	$Fe^{3+} (aq) + 6CN^-(aq) \leftrightharpoons Fe(CN)_6^{3-}(aq)$
*$Al(OH)_4^-(aq)$	$Al^{3+} (aq) + 4OH^-(aq) \leftrightharpoons Al(OH)_4^-(aq)$
*$Cr(OH)_4^-(aq)$	$Cr^{3+} (aq) + 4OH^-(aq) \leftrightharpoons Cr(OH)_4^-(aq)$
$Zn(OH)_4^{2-}(aq)$	$Zn^{2+} (aq) + 4OH^-(aq) \leftrightharpoons Zn(OH)_4^{2-}(aq)$
$Sn(OH)_4^{2-}(aq)$	$Sn^{2+} (aq) + 4OH^-(aq) \leftrightharpoons Sn(OH)_4^{2-}(aq)$
$Pb(OH)_4^{2-}(aq)$	$Pb^{2+} (aq) + 4OH^-(aq) \leftrightharpoons Pb(OH)_4^{2-}(aq)$

The hydroxides of the cations, Al^{3+}, Cr^{3+}, Zn^{2+}, Sn^{2+} and Pb^{2+}, are amphoteric because they can act as either an acid or a base. These hydroxides dissolve in both acidic and basic solutions. Each hydroxide acts as a base when it is neutralized by an acid.

$$Zn(OH)_2(s) + 2H^+(aq) \leftrightharpoons Zn^{2+}(aq) + 2H_2O(l)$$

Each hydroxide can act as an acid in the presence of excess hydroxide to form a complex ion.

$$Zn(OH)_2(s) + 2OH^-(aq) \leftrightharpoons Zn(OH)_4^{2-}(aq)$$

Multiple Choice Questions

1. *A solution is known to contain ions of Ag^+, Cu^{2+} and Pb^{2+}. What will happen when 1.0 M HCl is added to the solution?*

A) *No precipitate will form.*

B) *Only AgCl will precipitate.*

C) *Both AgCl and $PbCl_2$ will precipitate.*

D) *Both $PbCl_2$ and $CuCl_2$ will precipitate.*

E) *All three ions will form precipitates.*

2. *The Ksp of CaF_2 is 4.3×10^{-11} in pure water. In acidic solution the solubility of CaF_2 is expected to:*

A) *increase because Ca^{2+} ion is acidic.*

B) *decrease because Ca^{2+} ion is acidic.*

C) *increase because F^- ion is basic.*

D) *decrease because F^- ion is basic.*

E) *remain the same as in pure water.*

3. *Which has the greatest molar solubility in water at 25°C?*

A) *ammonium nitrate*

B) *barium sulfate*

C) *iron(II) sulfide*

D) *lead(II) carbonate*

E) *copper(II) phosphate*

4. *When mixed, which pair of aqueous solutions will form a precipitate?*

A) *KNO_3 and $CaBr_2$*

B) *$NaNO_3$ and $(NH_4)_2CO_3$*

C) *$CaCl_2$ and K_2CO_3*

D) *K_2SO_4 and $(NH_4)_2S$*

E) *$LiClO_4$ and $CuSO_4$*

5. *Which of these slightly soluble salts would show an increased solubility in a 1.0 M aqueous solution of HCl?*

I. $PbCl_2$ *II.* $CuCO_3$ *III.* $Ba_3(PO_4)_2$

A) I only

B) II only

C) I and II only

D) II and III only

E) I, II and III

6. *If equal volumes of the following pairs are mixed, which of the resulting solutions will NOT make a buffer solution?*

A) 0.20 M NaCN and 0.40 M HCN

B) 0.20 M HF and 0.10 M KOH

C) 0.20 M KF and 0.05 M HCl

D) 0.15 M NaCN and 0.20M HCl

E) 0.20 M NaOH and 0.30 M HF

7. *0.10 M hydrofluoric acid (Ka = 6.8 X* 10^{-4}*) is titrated with 0.10 M sodium hydroxide solution. Which indicator is the most appropriate for signaling the endpoint of the titration? The approximate pH range for the color change of each indicator is given.*

A) bromphenyl blue pH = 3 - 4.5

B) phenolphthalein pH = 8 - 10

C) thymol blue pH = 1.5 - 2.5

D) alizarin yellow R pH = 11 – 12

E) bromthymol blue pH = 6 - 7

8. *Which acid is best when preparing a buffer of pH = 9.0?*

A) hydrazoic acid, Ka = 1.9 x 10^{-5}

B) hydrofluoric acid, Ka = 6.8 x 10^{-4}

C) sulfuric acid, strong acid, Ka2 = 1.2 x 10^{-2}

D) hypobromous acid, Ka = 2.5 x 10^{-9}

E) carbonic acid, $Ka_1 = 4.3 \times 10^{-7}$, $Ka_2 = 5.6 \times 10^{-11}$

9. *Which of the following hydroxides is not amphoteric?*

A) *KOH*

B) *Al(OH)*$_3$

C) *Zn(OH)*$_2$

D) *Pb(OH)*$_2$

E) *Sn(OH)*$_2$

10. *When dissolved in aqueous solution which pair would behave as a buffer?*

A) *HCl and NaCl*

B) *KOH and KCl*

C) HNO_2 *and* $NaNO_2$

D) HNO_3 *and* NH_4NO_3

E) $Ca(OH)_2$ *and* $CaSO_4$

Additional Practice in Chemistry the Central Science

For more practice working buffer and solubility equilibria problems in preparation for the Advanced Placement examination try these problems in Chapter 17 of Chemistry the Central Science:

Additional Exercises: 17.73, 17.75, 17.76, 17.77, 17.78, 17.80, 17.81, 17.83, 17.89.

Integrated Exercises: 17.96, 17.97, 17.98, 17.99.

Multiple Choice Answers and Explanations

1. *C. Chlorides are soluble except those of mercury(I), lead(II), and silver ions.*

2. *C. Most anions are basic. Their negative charges attract protons. The only neutral anions are those anions of strong acids. HF is a weak acid.*

3. *A. All common salts of ammonium, sodium, and potassium ions are soluble in water.*

4. *C. Most carbonates, phosphates, hydroxides, and sulfides are insoluble. The exceptions include salts of the ammonium ion and the ions of Group 1.*

5. *D. The anions of weak acids are basic and will show increased solubility in acids. The anions of strong acids are neutral and the solubility of neutral salts will not be affected. Sulfate ion is the anion of a strong acid and is only slightly basic.*

6. *D. A buffer consists of a weak acid and a conjugate weak base. It can be formed by mixing the conjugates directly, or in this case by mixing an excess of weak base with a limiting strong acid or by mixing excess weak acid by limiting strong base. Excess strong or weak acid will not result in a buffer solution.*

7 *B. For an indicator to correctly signal the endpoint of the titration, the color change must closely match the pH of the equivalence point of the tiration. A titration of a weak acid with a strong base will produce only a weak base at the equivalence point. Therefore the pH of the equivalence point will be higher than 7, but not in the strong base range (pH >11).*

8. *D. The pH of a buffer will be close to the pKa of the weak acid from which it is made.*

   pH = pka + log[base]/[acid].

   The pKa of hydrobromous acid = –log Ka = -log (2.5 x 10^9) = 8.6

9. *A. Amphoteric hydroxides are those that form complex ions in the presence of excess hydroxide ions. All those listed are amphoteric except KOH.*

10. *C. A buffer consists of a weak acid and its conjugate weak base. The only conjugate pair listed is nitrous acid and sodium nitrite.*

Free Response Questions

1. *Solid NaCl is added slowly to a solution containing 0.10 M $AgNO_3$ and 0.20 M $Pb(NO_3)_2$. Ksp for AgCl is $1.8 X 10^{10}$. Ksp for $PbCl_2$ is $1.6 X 10^{5}$.*

a. *Write a net ionic equation and corresponding Ksp expression for the dissolution of solid*

i *silver chloride*

ii. *lead(II) chloride*

b. *Calculate the $[Cl^-]$ required to form each precipitate.*

c. *Which precipitate forms first? Explain your answer.*

d. *What is the concentration of the first metal ion to precipitate when the second one just begins to precipitate?*

e. *If 100 mL of 0.05 M NaCl is added to 200 mL of solution containing 0.005 M $AgNO_3$ and 0.10M M $PbCl_2$, does a precipitate form? If so, which one(s) form? Explain.*

2. *0.450 moles of hydrazoic acid, HN_3, ($Ka = 1.9 X 10^{5}$) are added to enough water to make 1.55 liters of solution.*

a. *Write a chemical equation for the reaction of hydrazoic acid with water and write the corresponding equilibrium expression.*

b. *Calculate the pH of the solution.*

c. *Calculate the pH of the solution after 0.350 moles of sodium azide, NaN_3, is added. Assume no volume change.*

d. *Calculate the pH of the solution in Part c after 0.0150 moles of HCl are added. Assume no volume change.*

e. *What is the pH of a 0.350 M solution of NaN_3?*

Free Response Answers

1a. i. $AgCl(s) \leftrightarrows Ag^+(aq) + Cl^-(aq)$ $\qquad Ksp = [Ag^+][Cl^-]$

ii. $PbCl_2(s) \leftrightarrows Pb^{2+}(aq) + 2Cl^-(aq)$ $\qquad Ksp = [Pb^{2+}][Cl^-]^2$

b. $Ksp = [Ag^+][Cl^-]$

$1.8 X 10^{10} = (0.10) [Cl^-]$

$[Cl^-] = 1.8 X 10^{9}$ *M required to precipitate AgCl.*

$Ksp = [Pb^{2+}][Cl^-]^2$

$1.6 \times 10^{-5} = (0.20)^{+}][Cl^-]^2$

$[Cl^-] = 8.9 \times 10^{-3}$ M required to precipitate $PbCl_2$.

c. AgCl precipitates first because it requires less $[Cl^-]$ to form a precipitate.

d. The second metal ion, Pb^{2+}, just begins to precipitate when $[Cl^-]$ reaches 8.9×10^{-3} M. At this chloride ion concentration, the silver ion can be calculated from:

$Ksp = [Ag^+][Cl^-]$

$1.8 \times 10^{-10} = [Ag^+](8.9 \times 10^{-3}\ M)$

$[Ag^+] = 2.0 \times 10^{-8}\ M.$

e. Calculate Q for each equilibrium in Part a and compare each Q to the corresponding Ksp. In each case, if Ksp < Q, a precipitate will form.

$Q = [Ag^+][Cl^-] = [(0.005M)(200/300)][(0.05)(100/300)] = 5.6 \times 10^{-5}.$

Ksp < Q for AgCl so a precipitate does form.

$Q = [Pb^{2+}][Cl^-]^2$

$1.6 \times 10^{-5} = (0.10M)(200/300)[(0.05)(100/300]^2 = 1.9 \times 10^{-5}$. *Ksp < Q for $PbCl_2$ so a precipitate forms.*

2a. $HN_3 + H_2O(l) \leftrightarrows N_3^-(aq) + H_3O^+(aq)$

$Ka = [N_3^-][H_3O^+]/[HN_3]$

b. $Ka = [N_3^-][H_3O^+]/[HN_3]$

$[HN_3] = 0.450\ mol/1.55\ L = 0.290\ M$

$x = [N_3^-] = [H_3O^+]$

$1.9 \times 10^{-5} = (x)(x)/(0.290\ M)$

$x = [H_3O^+] = 2.35 \times 10^{-3}\ M$

$pH = -log[H_3O^+] = -log(2.35 \times 10^{-3}) = 2.63$

c. $pH = pKa + log([N_3^-]/[HN_3])$

$[N_3^-] = 0.350\ moles/1.55\ L = 0.229\ M$

$pKa = -log\ Ka = -log(1.9 \times 10^{-5}) = 4.72$

$pH = 4.72 + log(0.350/0.450) = 4.72 - 0.11 = 4.61$

d. 0.015 mol of HCl will convert 0.0150 mol of azide to hydrazoic acid:

moles of $N_3^- = 0.350 - 0.015 = 0.335$ mol

moles of $HN_3 = 0.450 + 0.015 = 0.465$ mol

$pH = 4.72 + log(.335/.465) = 4.72 - 0.14 = 4.58$

e. $Kb = Kw/Ka = (y)(y)/0.350 = 1 x 10^{-14}/1.9 x 10^{-5} = y^2/0.350$

$y = [OH^-] = 1.36 x 10^{-5} M$

$pOH = -log (1.36 x 10^{-5}) = 4.87$

$pH = 14 = pOH = 14 – 4.87 = 9.13$

Your Turn Answers

17.1. *Acetic acid added to sodium acetate will form a buffer solution.*

$HC_2H_3O_2(aq) + H_2O(l) \leftrightarrows C_2H_3O_2^-(aq) + H_3O^+(aq)$

$Ka = [C_2H_3O_2^-][H_3O^+]/[HC_2H_3O_2]$

17.2. $NH_3(aq) + HNO_3(aq) -> NH_4NO_3(aq)$

$NH_3(aq) + H^+(aq) -> NH_4^+(aq)$

17.3. *A limiting amount of nitric acid added to an excess of sodium acetate will form a buffer because the strong acid will convert some, but not all of the weak base to acetic acid, a weak acid. The resulting solution will contain both a weak acid and its conjugate weak base.*

$H^+(aq) + C_2H_3O_2^-(aq) \rightarrow HC_2H_3O_2(aq)$

TOPIC 18

CHEMISTRY OF THE ENVIRONMENT

Chapter 18 of *Chemistry the Central Science* is perhaps the most interesting chapter even though very little of its content is found on the Advanced Placement Chemistry exam. Expect no more than one or two general knowledge multiple choice questions from this chapter. In preparation for the exam, do little more than read this summary.

The four major gases that make up the composition of dry air (in mole fraction) are:

Nitrogen, N_2, 0.781

Oxygen, O_2, 0.209

Argon, Ar, 0.00934

Carbon dioxide, CO_2, 0.000375

Water and carbon dioxide are the gases principally responsible for the natural greenhouse effect, the trapping of heat in the Earth's atmosphere.

The combustion of fossil fuels like coal and petroleum contribute to the increasing amount of carbon dioxide in the atmosphere.

$$2C_8H_{18}(l) + 25O_2(g) \rightarrow 16CO_2(g) + 18H_2O(g)$$

Sulfur dioxide, SO_2, is a principal air pollutant resulting from the combustion of various forms of sulfur in coal and oil.

$$S(s) + O_2(g) \rightarrow SO_2(g)$$

Acid rain is formed when sulfur dioxide is oxidized to sulfur trioxide in the atmosphere and then combines with water.

$$SO_3(g) + H_2O(l) \rightarrow H_2SO_4(aq)$$

Powdered limestone ($CaCO_3$) injected into the furnace of a power plant decomposes to lime (CaO) and carbon dioxide.

$$CaCO_3(s) + heat \rightarrow CaO(s) + CO_2(g)$$

The lime prevents the escape of sulfur dioxide by reacting to form calcium sulfite.

$$CaO(s) + SO_2(g) \rightarrow CaSO_3(s)$$

The high heats generated by internal combustion engines cause the chief components of air to react to form nitrogen monoxide.

$$N_2(g) + O_2(g) + heat \rightarrow 2NO(g)$$

Nitrogen monoxide oxidizes in air to nitrogen dioxide.

$$2NO(g) + O_2(g) \rightarrow 2NO_2(g)$$

Nitrogen dioxide undergoes photodissociation in sunlight to form atomic oxygen.

$$NO_2(g) + h\nu \rightarrow NO(g) + O(g)$$

Atomic oxygen produces ozone as well as other products collectively referred to as photochemical smog.

$$O(g) + O_2(g) \rightarrow O_3(g)$$

Municipalities treat water by adding lime (CaO) and alum, $Al_2(SO_4)_3$. Lime in water forms calcium hydroxide.

$$CaO(s) + H_2O(l) \rightarrow Ca^{2+}(aq) + 2OH^-(aq)$$

Aluminum ions react with hydroxide ions to produce a spongy, gelatinous precipitate that absorbs suspended particles and bacteria as it settles removing them from the water.

$$Al^{3+}(aq) + 3OH^-(aq) \rightarrow Al(OH)_3(s)$$

Chlorine added to water produces hydrochloric acid, a strong acid and hypochlorous acid, a weak acid, which is deadly to any remaining bacteria.

$$Cl_2(g) + H_2O(l) \rightarrow H^+(aq) + Cl^-(aq) + HClO(aq)$$

Notice that in the above disproportionation, chlorine atoms are both oxidized and reduced. Also the strong acid is written in ionic form and the weak acid is written in molecular form.

TOPIC

19

CHEMICAL THERMODYNAMICS

Chemical thermodynamics deals with energy and its transformations in chemical systems. Students must recognize, in both a qualitative and a quantitative sense, the complex relationship between the spontaneity of a chemical reaction and its free energy, enthalpy, and entropy. Calculations involving changes in entropy, enthalpy, and free energy and their relationship to the equilibrium constant of a chemical reaction must be mastered. The important quantitative and qualitative relationships between thermodynamics, kinetics, and equilibrium are presented in Chapter 19 of *Chemistry the Central Science*.

Section 19.1

Chemical thermodynamics answers a fundamental question. Why does change occur?

The driving influences for any chemical and physical change are:

1. **Change in enthalpy**, ΔH, (heat transferred between the system and the surroundings)

2. **Change in entropy**, ΔS, (randomness or disorder of the system.)

In general, chemical and physical systems tend to change in a direction that moves toward lower enthalpy (they release heat to the environment) and higher entropy (they become more random or disordered.)

Sometimes these two influences are in direct conflict with each other and the reaction seeks a balance between moving toward lower enthalpy and higher entropy. Systems reach this balance at equilibrium.

Section 19.1

Spontaneous Reactions

A **spontaneous change** is one that proceeds on its own without any outside assistance. A spontaneous change occurs in a definite direction. Processes that are spontaneous in one direction are non-spontaneous in the opposite direction. For example, a rock falls to the ground spontaneously. The opposite process, a rock rising from the ground, is non-spontaneous. Similarly, a campfire burns spontaneously. The "un-burning" of a campfire is non-spontaneous.

Review Section 5.3

Enthalpy

Recall from Topic 5 that **enthalpy** is the heat transferred between a chemical or physical system and its surroundings during a constant-pressure process.

The **change in enthalpy**, ΔH, is the **heat absorbed** by a system at constant pressure.

An **endothermic** process is one that absorbs heat from the environment. ΔH is positive.

An **exothermic** process is one that releases heat to the environment. ΔH is negative.

The **first law of thermodynamics**, also called the law of conservation of energy, states that in all cases, energy is conserved. It can neither be created nor destroyed. This means that the amount of energy gained by a system must equal the amount of energy lost by the environment and vice-versa.

A major driving influence for a chemical and physical change is the tendency for systems to move toward lower enthalpy by releasing energy to the environment. Thus exothermic processes, those that release energy to the environment, usually but not always, are spontaneous. They proceed on their own without any outside assistance.

Recall from Section 5.7 of Chemistry the Central Science that the enthalpy change for a given reaction can be calculated from the enthalpies of formation for products minus the enthalpies of formation for reactants:

$$\Delta H^\circ_{rxn} = \Sigma\Delta H^\circ_{f\ products} - \Sigma\Delta H^\circ_{f\ reactants}$$

Entropy and the Second Law of Thermodynamics

Section 19.2

Entropy is best considered to be the extent of randomness or disorder in a chemical or physical system. For example, gases are more random than are liquids.

Kinetic molecular theory states that gases consist of a large number of atoms and/or molecules having a high kinetic energy so they are in continuous, random motion. The attractive forces between gas particles are negligible compared to their kinetic energy. (See Section 10.7.) In contrast, liquids have lower kinetic energy and the attractive forces between the particles are sufficiently significant to hold the molecules together but still allow them to freely move past one another. Solids have relatively low kinetic energy and the attractive forces work to lock the molecules into place, often in a very ordered crystal lattice. (See Section 11.1.) Gases, because of their relative disorder, have higher entropy than do liquids. Solids generally have lower entropy than either liquids or gases.

The **change in entropy,** ΔS, for any process is a measure of the change in randomness or disorder of the system. In general, as the phase of a given system changes from solid to liquid to gas, the entropy of the system increases and ΔS is positive for such changes. Phase changes from gas to liquid, gas to solid or liquid to solid all happen with a decrease in entropy and ΔS is negative. Table 19.1 illustrates the

Table 19.1. Changes that increase entropy: ΔS is positive.

In general, entropy increases when:	**Example:**
solids change to liquids or gases	$CO_2(s) \rightarrow CO_2(g)$
liquids change to gases	$H_2O(l) \rightarrow H_20(g)$
solids dissolve in liquid solutions	$NaCl(s) \rightarrow Na^+(aq) + Cl^-(aq)$
the number of gas molecules increases	$2NH_3(g) \rightarrow N_2(g) + 3H_2(g)$
temperature increases	Water increases in temperature.
volume increases	A gas expands.
the number of particles increase	A rock is crushed.
two or more pure substances are mixed	Sugar dissolves in water

Your Turn 19.1

When helium is released from a toy balloon, does the entropy of the system increase or decrease? Explain your answer. Write your answer in the space provided.

Section 19.3

The Molecular Interpretation of Entropy

Entropy also increases for substances increases with increasing molecular complexity. For example, under the same conditions, O_3 has a higher entropy than O_2 because it is more complex.

Your Turn 19.2

Under the same conditions of pressure and temperature, which has greater entropy, gaseous methane or gaseous propane? Explain. Which has greater entropy, steam or ice? Explain. Write your answers in the space provided.

The **second law of thermodynamics** states that any spontaneous change is always accompanied by an overall increase in entropy.

The **third law of thermodynamics** states that the entropy of a pure crystalline substance at absolute zero is zero. Upon heating, the entropy of a pure crystalline substance gradually increases. Continued heating brings upon a sharp increase in entropy upon melting and another marked increase upon boiling as illustrated in Figure 19.1.

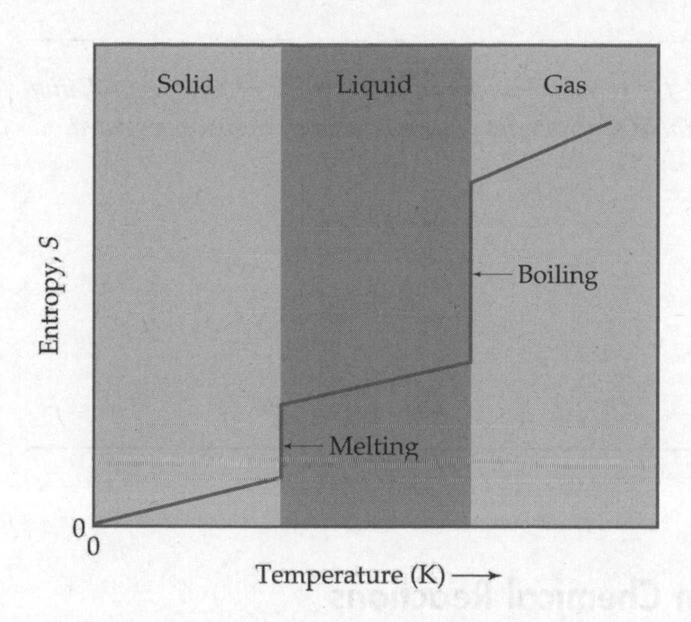

Figure 19.1. As the temperature of a substance increases, its entropy increases. Large changes in entropy are associated with phase changes.

Standard molar entropies, S°, are values for pure substances at one atmosphere pressure and 298 K. The units most often used for S° are J/mol K. Table 19.2 lists some examples. Notice than S° is comparatively larger for gases than for liquids or solids. Also, substances having the same phase but having of molecular mass and/or higher molecular complexity, have higher entropy values. A more complete list of standard entropy values is given in Appendix C of Chemistry the Central Science.

Table 19.2. Standard molar entropies of selected substances at 298K

Substance	S°, kJ/mol K	Substance	S°, kJ/mol K
I_2(s)	116.73	NO(g)	210.62
I_2(l)	180.66	NO_2(g)	240.45
I_2(g)	260.57	N_2O_4(g)	304.30
O(g)	161.0	H_2O(l)	69.91
O_2(g)	205.0	H_2O(g)	188.83
O_3(g)	237.6		

Your Turn 19.3

Rationalize the difference in the standard molar entropy values of solid, liquid, and gaseous iodine as listed in Table 19.2. Write your answer in the space provided.

Section 19.4

Entropy Changes in Chemical Reactions

The change in entropy for a chemical reaction can be calculated by the following equation:

$$\Delta S^{o}_{rxn} = \Sigma S^{o}_{products} - \Sigma S^{o}_{reactants}$$

Example:

What is the entropy change for the reaction, $2NO(g) + O_2(g) \rightarrow 2NO_2(g)$?

Solution:

Substitute the entropy values listed in Table 19.2 into the equation. Don't forget to also use the coefficients that balance the equation!

$$\Delta\Delta S^{o}_{rxn} = \Sigma S^{o}_{products} - \Sigma S^{o}_{reactants}$$

$$\Delta\Delta S^{o}_{rxn} = 2(240.45 \text{ J/mol K}) - 205.0 \text{ J/mol K} - 2(210.62 \text{ J/mol K})$$

$$\Delta\Delta S^{o}_{rxn} = -145.34 \text{ J/mol K.}$$

The answer is negative and makes sense because the system moves from three moles of gas to two moles of gas with a decrease in entropy.

Section 19.5

Gibbs Free Energy

Free energy (or **change in free energy**, ΔG) represents the amount of useful work that can be obtained from a process at constant temperature and pressure. Free energy, enthalpy, and entropy are related by the equation:

$$\Delta G = \Delta H - T\Delta S$$

ΔG is change in free energy measured in kJ/mol.

ΔH is change in enthalpy measured in kJ/mol.

ΔS is change in entropy measured in J/mol K.

T is absolute temperature in K.

The sign of ΔG tells if any given process is spontaneous as illustrated by Table 19.3.

Table 19.3. The sign of ΔG and spontaneity of reactions

If the ΔG is:	the process is:
Negative (-)	spontaneous
Positive (+)	non-spontaneous (The reverse reaction is spontaneous.)
Zero (0)	at equilibrium

The change in free energy of a reaction can be calculated by the following equation: $\Delta G^\circ_{rxn} = \Sigma \Delta G^\circ_{f\ products} - \Sigma \Delta G^\circ_{f\ reactants}$

The standard free energies of formation, ΔG°_f, of various substances are listed in Appendix C of Chemistry the Central Science.

Example:

What is the standard free energy change for the reaction,
$2NO(g) + O_2(g) \rightarrow 2NO_2(g)$?

Solution:

Obtain the values of standard free energy of formation for each of the reactants and products from Appendix C of Chemistry the Central Science and substitute them into the equation. Don't forget to take into account the number of moles of each substance as defined by the coefficients of the balanced equation.

$\Delta G^\circ_{rxn} = \Sigma G^\circ_{products} - \Sigma G^\circ_{reactants}$

$\Delta G^\circ_{rxn} = 2(+51.58\ kJ/mol) - 0 - 2(+86.71\ kJ/mol)$

$= -70.26\ kJ/mol$

The sign of $\Delta\Delta G^\circ$ *for the reaction is negative indicating that this is a spontaneous process at 298K.*

Section 19.6 Free Energy and Temperature

What causes a process to be spontaneous? What influences chemical and physical change? The spontaneity of any process involves the two thermodynamic concepts, enthalpy and entropy. Often a balance exists between enthalpy and entropy, and temperature determines in which direction a reversible reaction will proceed spontaneously.

Phase change provides a good example. Consider the melting of ice and freezing of liquid water. Ice melts spontaneously at temperatures above 0°C and liquid water freezes spontaneously at temperatures below 0°C. Whether ice melts spontaneously or liquid water freezes spontaneously depends on the temperature.

Ice melts spontaneously at temperatures higher than 0°C because liquid water is more random than is solid water. The increase in entropy (the positive ΔS) as ice melts, is the driving influence for the change from solid to liquid.

Liquid water freezes spontaneously at temperatures lower than 0°C because freezing is exothermic (ΔH is negative) and the move to lower enthalpy as liquid changes to solid influences the change.

$$H_2O(l) \rightarrow H_2O(s) + \text{heat}$$

At exactly 0°C, an equilibrium is established. At 0°C ice changes to liquid water at the same rate that liquid water changes to ice. At equilibrium, the drive toward lower enthalpy exactly balances the drive toward higher entropy.

The equation, $\Delta G = \Delta H - T\Delta S$ and the signs of ΔH and ΔS are useful qualitatively to predict the spontaneity of any given reaction. It provides a model to guide our thinking in determining whether a reaction will be spontaneous or nonspontaneous. Table 19.4 illustrates.

Table 19.4. The effect of temperature on the spontaneity of reactions.

ΔH	ΔS	$-T\Delta S$	ΔG	The reaction is:
-	+	-	-	Spontaneous at all temperatures
+	-	+	+	Non-spontaneous at all temperatures
-	-	+	+/-	Spontaneous only at low temperatures
+	+	-	+/-	Spontaneous only at high temperatures

When ΔH is negative and ΔS is positive, ΔG is always negative and the reaction is spontaneous at all temperatures because the reaction proceeds toward lower enthalpy and higher enthalpy.

If ΔH is positive and ΔS is negative, ΔG is always positive and the reaction is non-spontaneous at all temperatures. (However, the reverse reaction is spontaneous at all temperatures.)

Table 19.5 illustrates that often the change in free energy represents a balance between systems tending toward decreasing enthalpy and increasing entropy. If ΔH and ΔS both are positive, ΔG is negative only when the temperature is high enough for the entropy term to outweigh the enthalpy term. On the other hand, if ΔH and ΔS both are negative, ΔG is negative only when the temperature is low enough for the enthalpy term to outweigh the entropy term.

Table 19.5. A Summary of Important Thermodynamic Quantities

Quantity	**Change in Enthalpy**	**Change in Entropy**	**Change in Free Energy**
Symbol	ΔH	ΔS	ΔG
Unit	kJ/mol	J/mol K k	J/mol
Definition	Heat gained by a system	Change in randomness of a system	Available useful work
Comments	+ for endothermic - for exothermic	+ for increasing randomness - for decreasing randomness	+ for non-spontaneous - for spontaneous $\Delta G = 0$ at equilibrium

Your Turn 19.4

Is ΔG for a burning campfire positive or negative? Explain citing the positive or negative signs of ΔH and ΔS for the process. Is there a temperature at which ΔG for a burning campfire will change signs? Explain. Write your answers in the space provided.

Section 19.7 Free Energy and the Equilibrium Constant

The free energy change for any reaction under nonstandard conditions, ΔG, can be calculated from the standard free energy change, ΔG°, by the following equation:

$$\Delta G = \Delta\Delta G^{\circ} + RT \ln Q$$

ΔG = the free energy change under nonstandard conditions

ΔG° = the free energy change under standard conditions: 25°C and one atmosphere partial pressure of all gases and 1 M concentration for all solutes.

R = 8.314 J/mol K, the ideal gas constant.

T = the absolute temperature in Kelvin.

Q = the reaction quotient.

Example:

Using standard free energies of formation for the reaction,

$2H_2S(g) + 3O_2(g) \rightarrow 2H_2O(g) + 2SO_2(g)$,

calculate ΔG° and ΔG for a mixture at 25°C with the composition below.

$P_{H_2S} = 1.00\ atm \quad P_{H_2O} = 0.500\ atm \quad P_{O_2} = 2.00\ atm$
$P_{SO_2} = 0.750\ atm$

Solution:

Using the standard free energies of formation from Appendix C of Chemistry the Central Science, calculate ΔG° for the reaction.

$$\Delta\Delta G^{o}_{rxn} = \Sigma\Delta G^{o}_{products} - \Sigma\Delta G^{o}_{reactants}$$

$$\Delta\Delta G^{o}_{rxn} = 2(-300.4) + 2(-228.57) - 2(-33.01) - 0$$

$$\Delta\Delta G^{o}_{rxn} = -991.9\ kJ$$

Evaluate Q and use the equation, ΔG = ΔGo + RT ln Q, to calculate ΔG.

$$Q = P^2_{SO_2} P^2_{H_2O}/P^2_{H_2S} P^3_{O_2} = (0.750)^2(0.500)^2/(1.00)^2(2.0\text{-})^3 = 0.0176$$

$$\Delta G = \Delta G^{o} + RT \ln Q =$$

$$-991.9\ kJ + (8.314\ J/mol\ K)(298\ K) \ln (0.0176)(1\ kJ/1000\ J) = -1002\ kJ$$

Common misconception: When using the equation, ΔG = ΔG° + RT ln Q, keep in mind that ΔG° is usually given in kJ whereas R is in units of J/mol K. Be sure to convert J to kJ using 1 kJ = 1000 J.

The standard free energy for any reaction is related to the equilibrium constant. At equilibrium, $\Delta G = 0$ and $Q = K$, the equilibrium constant. At equilibrium,

$$\Delta G^{\circ} = -RT \ln K$$

Example:

The standard free-energy change for the following reaction at 25°C is -118.4 kJ/mol:

$$KClO_3(s) \rightarrow KCl(s) + 3/2\ O_2(g)$$

Calculate Kp for the reaction at 25°C and the equilibrium pressure of O_2 gas.

Solution:

$$\Delta G^{\circ} = -RT \ln K$$

$$-118.4 \text{ kJ/mol} = -(8.314 \text{ J/mol K})(298 \text{ K})(1 \text{ kJ}/1000 \text{ J}) \ln K_p$$

$$\ln Kp = 47.8$$

$$Kp = e^{47.8} = 5.68 \times 10^{20}$$

$$Kp = P^{3/2}\ O_2$$

$$PO_2 = Kp^{2/3} = (5.68 \times 10^{20})^{2/3} = 6.9 \times 10^{13} \text{ atm}$$

Multiple Choice Questions.

1. *The standard free energy of formation, ΔG°_f, for sodium chloride is the free energy change for which of the following reactions?*

A) $Na^+(g) + Cl^-(g) \rightarrow NaCl(g)$

B) $Na^+(g) + Cl^-(g) \rightarrow NaCl(s)$

C) $2Na(s) + Cl_2(g) \rightarrow 2NaCl(s)$

D) $Na(s) + 1/2Cl_2(g) \rightarrow NaCl(s)$

E) $2NaCl(s) \rightarrow 2Na(s) + Cl_2(g)$

2. *For which of these processes is the value of ΔS expected to be negative?*

I. *NaCl is dissolved in water.*

II. *Steam condenses to liquid water.*

III. *$MgCO_3$ is decomposed into MgO and CO_2.*

A) I only

B) I and III only

C) II only

D) II and III only

E) I and II only

3. *What are the signs of the enthalpy change and entropy change for the melting of ice?*

$H_2O(s) \rightarrow H_2O(l)$

	ΔH^o	ΔS^o
A)	+	+
B)	-	-
C)	-	+
D)	+	-
E)	0	0

4. *Under which set of conditions does a temperature exist at which equilibrium can be established with all reactants and products in standard states?*

I. ΔS is +, ΔH is + II. $\Delta\Delta S$ is -, ΔH is - III. $\Delta\Delta S$ is +, ΔH is -

A) I only

B) II only

C) II only

D) I and II only

E) I and III only

5. *Which of the following substances is expected to have the largest standard molar entropy?*

A) $H_2O(s)$

B) $H_2O(l)$

C) $H_2O(g)$

D) $H_2O_2(l)$

E) $H_2O_2(g)$

6. *In which process would ΔG be expected to be positive?*
   A) *melting ice at -10°C and 1.0 atm pressure*
   B) *water evaporating at 25°C into dry air*
   C) *cooling hot water to room temperature*
   D) *sublimation of dry ice (solid carbon dioxide) at 25°C*
   E) *a burning campfire*

7. *Which is not expected to have a value of zero?*
   A) ΔH°_f *for* $O_2(g)$
   B) ΔG°_f *for* $K(s)$
   C) *S for Ca(s) at 0 K*
   D) ΔH°_f *for* $Br_2(l)$
   E) ΔG°_f *for* $O(g)$

8. *Which process is exothermic and occurs with a decrease in entropy?*
   A) $H_2O(l) \rightarrow H_2O(s)$
   B) $H_2O(s) \rightarrow H_2O(l)$
   C) $H_2O(s) \rightarrow H_2O(g)$
   D) $2H_2O(l) \rightarrow 2H_2(g) + O_2(g)$
   E) $H_2O(l) \rightarrow H_2O(g)$

9. *In order to calculate the entropy change for the following reaction at 25°C and one atmosphere, what additional thermodynamic information is needed?*

   C(solid graphite) + $O_2(g) \rightarrow CO_2(g)$

   A) ΔH°_f *for* $CO_2(g)$
   B) ΔG°_f *for* $CO_2(g)$
   C) ΔH°_f *and* ΔG°_f *for* $CO_2(g)$
   D) *So for* $CO_2(g)$
   E) ΔH°_f *and* ΔG°_f *for both C(solid graphite) and* $O_2(g)$

10. *For a certain reaction, ΔH° = - 150 kJ/mol and ΔS° = - 50 J/mol K. Which of the following statements is true about the reaction?*

A) *It is spontaneous at high temperatures only.*

B) *It is spontaneous at low temperatures only.*

C) *It is spontaneous at all temperatures.*

D) *It in non-spontaneous at all temperatures.*

E) *There is no temperature at which the reaction will reach equilibrium.*

Free Response Questions

1. *At 25°C the equilibrium constant, Kp, for the reaction below is 0.281 atm.*

   $Br_2(l) \rightarrow Br_2(g)$

a. *Calculate ΔG° for this reaction.*

b. *It takes 193 joules to vaporize 1.00 gram of Br_2 (l) at 25°C and 1.00 atmosphere pressure. Calculate ΔH° at 298 K.*

c. *Calculate ΔS° at 298 K for this reaction.*

d. *The normal boiling point of a liquid is the temperature at which the liquid and its vapor are at equilibrium at 1 atm pressure. Calculate the normal boiling point of bromine. Assume that ΔH° and ΔS° remains constant as the temperature is changed.*

e. *What is the equilibrium vapor pressure of bromine in torr at 25°C?*

2. *In principle, ethanol can be prepared by the following reaction.*

$$2\ C(s) + 2\ H_2(g) + H_2O\ (l) \rightarrow C_2H_5OH(l)$$

	2 C(s)	2 $H_2(g)$	H_2O (l)	$C_2H_5OH(l)$	
ΔH° =	0	0	-285.85	-277.69	kJ/mol
ΔS° =	5.740	130.6	69.91	160.7	J/molK

a. *Calculate the standard enthalpy change, ΔH°, for the reaction above.*

b. *Calculate the standard entropy change, ΔS°, for the reaction above.*

c. *Calculate the standard free energy change, ΔG°, for the reaction above at 298 K.*

d. *Calculate the value of the equilibrium constant, Kp at 25°C for the reaction above.*

e. *Calculate the partial pressure of $H_2(g)$ at equilibrium.*

Additional Practice in Chemistry the Central Science

For more practice answering questions in preparation for the Advanced Placement examination try these problems in Chapter 19 of Chemistry the Central Science:

Additional Exercises: 19.82, 19.84, 19.87, 19.90, 19.92.

Integrative Exercises: 19.98, 19.99, 19.101, 19.102, 19.103, 19.104.

Multiple Choice Answers and Explanations

1. *D. The standard free energy of formation is the free energy change of a formation reaction, one that forms one mole of a substance from its component elements in their most stable forms. Solid sodium and gaseous diatomic chlorine are the most stable forms of those elements.*

2. *C. A negative ΔS denotes a change from a more random state to one of more order. In general, solids are more ordered than liquids which are more ordered than gases. Liquid water is more ordered than is steam. When solid NaCl dissolves in liquid water, it dissociates into more random aqueous ions in the liquid phase. Carbon dioxide gas is more random than solid magnesium carbonate.*

3. *A. All solids melt endothermically (where ΔH° is positive) because solids must absorb energy to overcome the strong attractive forces that hold the particles together. All solids melt with an increase in entropy (where ΔS° is positive) because liquids are more random than solids.*

4. *D. The fundamental thermodynamic condition at equilibrium is when ΔG is zero. At equilibrium, the equation $\Delta G = \Delta H - T\Delta S$ becomes $\Delta H = T\Delta S$. Because T is the absolute temperature in Kelvin, it must be positive. A positive temperature can only exist when the signs of ΔH and ΔS are both negative or both positive.*

5. *E. Entropy is the degree of randomness or disorder of a system. Gases have higher entropies than liquids which have higher entropies than solids. Additionally, when comparing substances having the same phase, the more complex the substance the higher is its entropy because complex substances have more possible rotational and vibrational modes of motion.*

6. *A. ΔG is positive for non-spontaneous processes. When determining qualitatively the sign of ΔG, ask the question, left on its own, does it happen? If the answer is yes, the process is spontaneous and the sign of ΔG is negative. If the answer is no, the process in non-spontaneous and the sign of ΔG is positive. Ice will not melt spontaneously at -10°C unless it's is under pressure. ΔG is negative for all the other processes listed because left on their own, they will happen spontaneously.*

7. *E. O(g) is not the most stable form of oxygen so its standard free energy is non-zero. Standard heats of formation, ΔH^o_f, and standard free energies of formation, ΔG^o_f, are zero for elements in their most stable form at 25°C. The most stable forms of oxygen, potassium and bromine are O_2(g), K(s) and Br_2(l), respectively. The third law of thermodynamics states that the entropy, S, of a pure crystalline substance at absolute zero is zero. Ca(s) at 0 K has S = 0.*

8. *A. In general, gases have more entropy than liquids and liquids have more entropy than solids. Phase changes from gas to liquid, gas to solid and liquid to solid all are exothermic (release heat to the environment) and occur with a decrease in entropy (result in less random, more ordered products). The decomposition of liquid water to form gaseous products is endothermic and takes place with a large increase in entropy.*

9. *C. The given reaction is the formation reaction for CO_2(g). Therefore, the values for ΔH^o_f and ΔG^o_f for CO_2(g) are the same as the values for ΔH^o_{rxn} and and ΔG^o_{rxn} for CO_2(g). ΔS_{rxn} can be calculated using the equation, $\Delta G^o_{rxn} = \Delta H^o_{rxn} - T\Delta S^o_{rxn}$. The given Celsius temperature must be converted to Kelvin. ΔH^o_f and ΔG^o_f for neither C(solid graphite) nor O_2(g) are necessary because both are elements in their most stable forms so their values are each, by definition, zero.*

10. *B. Use the equation, $\Delta G = \Delta H - T\Delta S$ to guide your thinking. To be spontaneous requires a negative ΔG and that can only be achieved if the absolute value of the negative ΔH term is larger than the absolute value of the $T\Delta S$ term. At low temperatures ΔG will be negative. At high temperatures it will be positive. There is a temperature at which ΔG will be zero and the system will reach equilibrium.*

Answers to Free Response Questions

1a. $\Delta G^{\circ} = -RT \ln K = -(8.314 \text{ J/mol K})(25 + 273 \text{ K}) \ln (0.281)$

$\Delta G^{\circ} = 3.15 \text{ kJ/mol}$

b. $x \text{ kJ/mol} = (193 \text{ J/g})(160 \text{ g/mol})(1 \text{ kJ}/1000 \text{ J}) = 30.9 \text{ kJ/mol}$

c. $\Delta G^{\circ} = \Delta H^{\circ} - T\Delta S^{\circ}$

$\Delta S^{\circ} = (\Delta H^{\circ} - \Delta G^{\circ})/T = (30.9 - 3.15)/298 \text{ K} = .0931 \text{ kJ/mol K} = 93.1 \text{ J/mol K}$

d. *The boiling point is the temperature at which the equilbrium vapor pressure of bromine equals the external pressure. Because the system is at equilibrium, $\Delta G = 0$.*

$\Delta G = \Delta H - T\Delta S$

$0 = 30.9 \text{ kJ/mol} - T(.0931 \text{ kJ/mol K})$

$T = 30.9/.0931 = 332 \text{ K}$

e. $Kp = P_{Br_2(g)}$

$Kp = (0.281 \text{ atm})(760 \text{ torr/atm}) = 214 \text{ torr}$

2a. $\Delta H^{\circ} rxn = \Sigma \Delta H^{\circ}_{f\,products} - \Sigma \Delta H^{\circ}_{f\,reactants} = -(-285.85) + (-227.69) = +8.16 \text{ kJ/mol}$

b. $\Delta S^{\circ} rxn = \Sigma S^{\circ}_{products} - \Sigma S^{\circ}_{reactants}$ –

$+160.7 - 69.91 - 2(130.6) - 2(5.740) = -181.9 \text{ J/mol K}$

c. $\Delta G^{\circ}_{rxn} = \Delta H^{\circ}_{rxn} - T\Delta S^{\circ}_{rxn}$

$\Delta G^{\circ}_{rxn} = -50.8 \text{ kJ/mol} - (298)(118.9 \text{ J/mol K})/(1 \text{ kJ}/ 1000 \text{ J}) = +62.7 \text{ kJ/mol}$

$\Delta G^{\circ} = -RT \ln K$

$\ln K = -(-86.2 \text{ kJ/mol})/(8.314 \text{ J/mol K})(1 \text{ kJ}/1000 \text{ J})(298 \text{ K}) = -25.2$

$K = e^{34.8} = 1.14 \times 10^{-11}$

e. $Kp = P^{-2}_{H_2}$

$P_{H_2} = Kp^{-1/2} = (1.14 \times 10^{-11})^{-1/2} = 2.96 \times 10^{5} \text{ atm}$

Answers to Your Turn

19.1 *Entropy of the system increases when helium is released from a toy balloon because the gas from the balloon mixes with the gases in the atmosphere causing an increase in randomness or disorder.*

19.2. *Propane, $CH_3CH_2CH_3$, has greater entropy than methane, CH_4, because the molecules of propane are much more complex than the molecules of methane.*

Steam has more entropy than ice because gases are more random and disordered than are solids.

19.3. *Gaseous iodine has a higher standard molar entropy than either liquid or solid iodine because gases are more random than liquids or solids. Liquid iodine has a higher value than solid iodine because liquids are more random than solids.*

19.4. *A burning campfire is a spontaneous process where ΔG is negative. For the process, ΔH is negative and ΔS is positive so ΔG is negative at all temperatures. Thus, a burning campfire will never become nonspontaneous.*

TOPIC

20

ELECTROCHEMISTRY

Students need to understand the concept of electrochemical cells, their structures, the chemical processes that take place in electrochemical cells and the differences between voltaic and electrolytic cells. They must be able to calculate cell EMF and quantitatively determine spontaneity. Calculations using the Nernst equation relating cell EMF to concentrations need to be mastered as well as quantitative electrolysis determinations using Faraday's constant, time, amperage,, and moles of substances. Pay particular attention to these sections in *Chemistry the Central Science.*

- 20.1 **Oxidation-Reduction Reactions**
- 20.2 **Balancing Oxidation-Reduction Equations**
- 20.3 **Voltaic Cells**
- 20.4 **Cell EMF**
- 20.5 **Spontaneity of Redox Reactions**
- 20.6 **The Effect of Concentration on EMF**
- 20.9 **Electrolysis**

Oxidation-Reduction Reactions

Section 20.1 and Review Section 4.4

Electrochemistry refers to the interchange of electrical and chemical energy. In electrochemical processes, chemical energy is transformed to electrical energy or electricity is used to cause chemical change.

An **oxidation state**, also called oxidation number, is a positive or negative number assigned to an element in a molecule or ion according to a set of rules.

Common misconception: Oxidation state or oxidation number is often incorrectly confused with charge. Oxidation number is not charge. Oxidation number and charge are equivalent only when considering a monatomic ion.

Oxidation numbers are a way to keep track of electrons in redox reactions. Table 20.1 lists a set of simplified rules for determining oxidation numbers. For a more complete list of rules, see Section 4.4 of *Chemistry the Central Science*.

Table 20.1. Simplified rules for determining oxidation numbers

Rules
1. The oxidation number of combined oxygen is usually 2-, except in the peroxide ion, O_2^{2-} where the oxidation number of oxygen is 1-. Examples: In H_2O_2 and BaO_2 O is 1-.
2. The oxidation number of combined hydrogen is usually 1+, except in the hydride ion, H-, were it is 1-. Examples: In NaH and CaH_2 H is 1-.
3. The oxidation numbers of all individual atoms of a formula add to the charge on that formula. When in doubt, separate ionic compounds into common cation-anion pairs.

Examples:

Na	K^+	O_2	Ca^{2+}	H_2SO_4	NO_3	-Mg_3 $(PO_4)_2$	=	Mg^{2+} $PO4^{3-}$
0	1+	0	2+	1+ 6+ 2-	5+ 2-	2+ 5+ 2-		2+ 5+ 2-

Your Turn 20.1

What is the oxidation number of sulfur in each of the following atoms, molecules and ions? In which case(s) is oxidation number equal to charge?

Sulfate, thiosulfate, sulfide, sulfite, thiocyanate, sulfur, sulfur dioxide, sulfur trioxide. Write your answer in the space provided.

An **oxidation-reduction reaction**, also called a **redox reaction** is a reaction where electrons are transferred between reactants. The oxidation numbers of some of the elements change as they become products.

Oxidation is the loss of electrons. Substances that lose electrons are said to be oxidized.

Reduction is the gain of electrons. Substances that gain electrons are reduced. Both an oxidation and a reduction occurs in all redox reactions.

In a redox reaction, one substance loses electrons and another gains electrons. For example, the chemical reaction in a common alkaline flashlight cell is complex but it can be approximately represented by the equation below. The oxidation numbers of each element have been assigned according to the rules given in Table 20.1.

$$2\ MnO_2(s) + Zn(s) + 2H_2O(l) \longrightarrow 2MnO(OH)(s) + Zn(OH)_2(s)$$

4+ 2-　　0　　1+ 2-　　3+ 2- 2-1+　　2+ 2-1+

Notice that Mn changes oxidation numbers from 4+ to 3+ as it goes from MnO_2 to MnO(OH). Also Zn changes from 0 to 2+. MnO_2 is reduced because it has gained electrons. (The oxidation number of Mn has been reduced.) Zn is oxidized because it has lost electrons. (The oxidation number of Zn has increased.)

In the alkaline cell reaction, there is a transfer of electrons from Zn to MnO_2. The cell is designed to take advantage of this transfer of electrons by allowing the electrons to flow as electricity through an external circuit

Zn is called the **reducing agent** because it is the agent that provides electrons for MnO_2 to be reduced.

MnO_2 is the **oxidizing agent** because it is the agent that takes electrons away from Zn causing Zn to be oxidized.

The substance oxidized is the reducing agent and the substance reduced is the oxidizing agent.

Your Turn 20.2

In the following reaction, identify the oxidizing agent, the reducing agent, the substance oxidized and the substance reduced: $N_2(g) + H_2(g) \rightarrow NH_3(g)$*. Explain your reasoning. Write your answer in the space provided.*

Section 20.2

Balancing Oxidation-Reduction Equations

A **half-reaction** is an equation that shows either a reduction or an oxidation. Aqueous redox equations are conveniently balanced by the method of half-reactions. For example, we can represent the alkaline cell reaction shown above as separate oxidation and reduction half-reactions.

$2\ MnO_2(s) + 2H_2O(l) + 2e^- \rightarrow 2MnO(OH)(s) + 2OH^-(aq)$ reduction

$Zn(s) + 2OH^-(aq) \rightarrow 2e^- + Zn(OH)_2(s)$ oxidation

$2\ MnO_2(s) + Zn(s) + 2H_2O(l) \rightarrow 2MnO(OH)(s) + Zn(OH)_2(s)$

The reduction half-reaction shows electrons as reactants. The oxidation half-reaction shows electrons as products. The sum of the half-reactions yields the complete equation.
Table 20.1 illustrates some simplified rules for balancing half-reactions. For a more complete set of rules see Section 20.2 of *Chemistry the Central Science*.

Table 20.2. Simplified rules for balancing half reactions

If in acidic solution stop at Step 4:	
1.	Balance all elements other than H or O
2.	Balance O by adding H_2O as required.
3.	Balance H by adding H^+ as required.
4.	Balance charge by adding e- to more positive side.
If in basic solution continue through Step 7:	
5.	Count H^+ and add equal numbers of OH^- to both sides.
6.	Combine each $H^+ + OH^-$ on one side to yield water. ($H^+ + OH^- = H_2O$)
7.	Combine or cancel water molecules as needed.

Example:

Balance the following equation in acidic solution.

$HOCl(aq) + NO(g) \rightarrow Cl_2(g) + NO_3^-(aq)$

Solution:

First separate the reactants and products into two half-reactions. HOCl and Cl_2 go together because they both contain Cl. Also NO and NO_3^- are a pair because they both have N. Once the reactants and products are separated, just follow the rules.

$HOCl(aq) \rightarrow Cl_2(g)$ $\quad\quad$ $NO(g) \rightarrow NO_3^-(aq)$

1. $2HOCl(aq) \rightarrow Cl_2(g)$
2. $2HOCl(aq) \rightarrow Cl_2(g) + 2H_2O$
3. $2H^+ + 2HOCl(aq) \rightarrow Cl_2(g) + 2H_2O$
4. $2e- + 2H^+ + 2HOCl(aq) \rightarrow Cl_2(g) + 2H_2O$

$NO(g) \rightarrow NO_3^-(aq)$

1. $NO(g) \rightarrow NO_3^-(aq)$
2. $2H_2O + NO(g) \rightarrow NO_3^-(aq)$
3. $2H_2O + NO(g) \rightarrow NO_3^-(aq) + 4H^+$
4. $2H_2O + NO(g) \rightarrow NO_3^-(aq) + 4H^+ + 3e-$

Now equalize the number of electrons transferred in the half-reactions by multiplying the Cl reaction by 3 and the N reaction by 2 and add the half reactions. The goal is to cancel all of the electrons in the final balanced equation.

$3(2e- + 2H^+ + 2HOCl(aq) \rightarrow Cl_2(g) + 2H_2O) = 6e- + 6H+ + 6HOCl(aq) \rightarrow 3Cl_2(g) + 6H_2O$

$2(2H_2O + NO(g) \rightarrow NO_3^-(aq) + 4H^+ + 3e-) = 4H_2O + 2NO(g) \rightarrow 2NO_3^-(aq) + 8H^+ + 6e-$

$6HOCl(aq) + 2NO(g) \rightarrow 3Cl_2(g) + 2NO_3^-(aq) + 2H_2O + 2H^+$

Example:

Balance the above reaction assuming it also proceeds in basic solution.

Solution:

First apply Rules 1 through 4 as if the reaction were in acidic solution. Then continue with Rules 5 through 7 to balance it in basic solution:

$6HOCl(aq) + 2NO(g) \rightarrow 3Cl_2(g) + 2NO_3^-(aq) + 2H_2O + 2H^+$

5. $2OH^- + 6HOCl(aq) + 2NO(g) \rightarrow 3Cl_2(g) + 2NO_3^-(aq) + 2H_2O + 2H^+ + 2OH^-$

6. $2OH^- + 6HOCl(aq) + 2NO(g) \rightarrow 3Cl_2(g) + 2NO_3^-(aq) + 2H_2O + 2H_2O$

7. $2OH^- + 6HOCl(aq) + 2NO(g) \rightarrow 3Cl_2(g) + 2NO_3^-(aq) + 4H_2O$

Voltaic Cells

Section 20.3

A **voltaic** (or galvanic) cell is a device that spontaneously transforms chemical energy into electrical energy. The transfer of electrons of a redox reaction takes place through an external pathway.

Figure 20.1 illustrates a typical voltaic cell. Two compartments, called half-cells, physically separate the reactants. Each half-cell consists of a metal electrode immersed in an aqueous solution.

Oxidation, the loss of electrons, takes place in the anode compartment. In Figure 20.1, the anode compartment contains a zinc metal anode and a zinc nitrate solution.

Reduction, the gain of electrons, takes place in the cathode compartment, consisting in Figure 20.1, of a copper metal cathode and a copper(II) nitrate solution.

Because the reactants are separated, the reaction can occur only when the transfer of electrons takes place through an external circuit. Electrons always flow spontaneously from the anode to the cathode.

The voltaic cell uses a salt bridge to complete the electrical circuit. As oxidation and reduction take place, ions from the half-cell compartments migrate through the salt bridge to maintain the electrical neutrality of the respective solutions. Cations always move toward the cathode and anions move toward the anode. (To examine an atomic view of an operating voltaic cell, see Figures 20.7 and 20.8 of *Chemistry the Central Science*.)

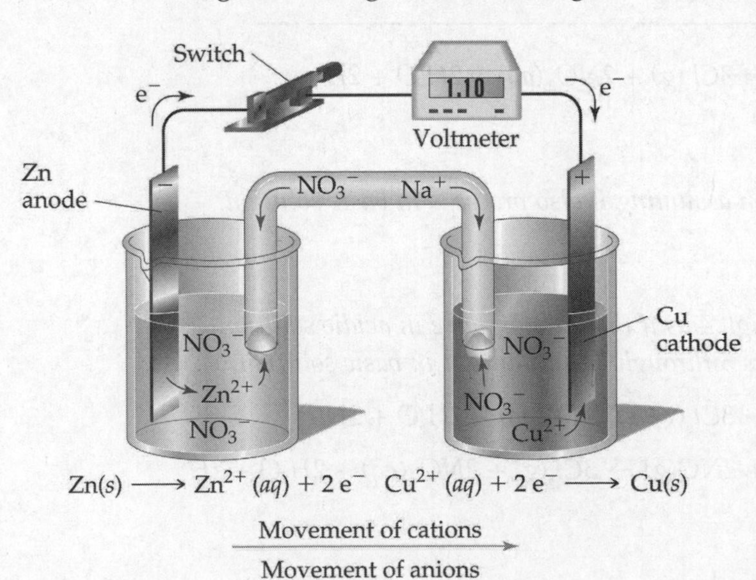

Figure 20.1. A voltaic cell uses a redox reaction to spontaneously generate an electric current . Electrons flow from the anode to the cathode. Anions migrate toward the anode and cations migrate toward the cathode.

Anode compartment oxidation half-reaction: $Zn(s) \rightarrow Zn^{2+}(aq) + 2e^-$

Cathode compartment reduction half-reaction: $Cu^{2+}(aq) + 2e^- \rightarrow Cu(s)$

Overall reaction: $Zn(s) + Cu^{2+}(aq) \rightarrow Zn^{2+}(aq) + Cu(s)$

Cell EMF

Section 20.4

A potential difference exists between anode and cathode of a voltaic cell. As the redox reaction takes place this cell potential, also called electromotive force (emf), pushes electrons through the external circuit. The **cell potential, E_{cell}**, (also called **electromotive force, EMF**) is measured in volts.

One **volt** is the potential difference required to impart one joule of energy to one coulomb of charge. (1 volt = 1 joule per coulomb, 1 v = 1 J/C)

Cell voltage, E_{cell} is positive for a spontaneous reaction.

A standard reduction potential for a given half-reaction is the electric potential when that half-cell is coupled with a reference half cell. The standard conditions are 25°C, 1 atm pressure and 1 M solutions. The reference half-cell is the standard hydrogen electrode whose potential is assigned 0.000 volts. The reduction half-reaction of the hydrogen electrode is:

$$2H^+(aq) + 2e^- \rightarrow H_2(g) \qquad E^\circ_{red} = 0.000 \text{ volts}$$

Table 20.3 lists the standard reduction potentials of several half-cells (Appendix E of *Chemistry the Central Science* has a more complete list of standard reduction potentials.) For convenience, all half-reactions are written as reductions. The reverse reactions (oxidations) have the same cell voltage but with the opposite signs.

Table 20.3. Selected Standard Reduction Potentials

oxidizing/reducing agents	**half-reaction**	**E°_{red} (volts) Forward reaction**	**E°_{ox} (volts) Reverse reaction**
F_2/F^-	$F_2(g) + 2e^- \rightarrow 2F^-(aq)$	+2.87	-2.87
Cl_2/Cl^-	$Cl_2(g) + 2e^- \rightarrow 2Cl^-(aq)$	+1.359	-1.359
Ag^+/Ag	$Ag^+(aq) + 1e^- \rightarrow Ag(s)$	+0.799	-0.799
Cu^{2+}/Cu	$Cu^{2+}(aq) + 2e^- \rightarrow Cu(s)$	+0.337	-0.337
H^+/H_2	$2H^+(aq) + 2e^- \rightarrow H_2(g)$	0.000	0.000
Zn^{2+}/Zn	$Zn^{2+}(aq) + 2e^- \rightarrow Zn(s)$	-0.762	+0.762
Al^{3+}/Al	$Al^{3+}(aq) + 3e^- \rightarrow Al(s)$	-1.66	+1.66

A table of standard reduction potentials is useful in determining the spontaneous reaction that will take place in a voltaic cell and what its voltage will be. When comparing two reduction potentials, a more positive voltage signifies a greater tendency for reduction. Therefore, when coupling two half reactions, the one with the more positive voltage will be the reduction and it will force the other half reaction to occur in reverse and become the oxidation.

To determine the spontaneous reaction for any cell made up of two half-cells, reverse the half reaction with the less positive E^{o}_{red}, and add it to the half reaction with the more positive voltage. Similarly, to determine the cell potential of any two coupled half reactions, change the sign of the potential for the reversed half reaction and add it to the potential of the reduction half reaction.

Example:

What is the reaction that occurs when a Ag half cell is coupled with a Cu half cell?

Solution:

From Table 20.3, the half-reactions are:

	half-reaction	E^{o}_{red} *(volts)*	E^{o}_{ox} *(volts)*
Ag^{+}/Ag	$Ag^{+}(aq) + 1e^{-} \rightarrow Ag(s)$	+0.799	-0.799
Cu^{2+}/Cu	$Cu^{2+}(aq) + 2e^{-} \rightarrow Cu(s)$	+0.337	-0.337

The Ag half-reaction has the more positive voltage. Reverse the Cu half-reaction and add it to the Ag half reaction. (Notice that to cancel the electrons, the Ag half-reaction is multiplied by a factor of 2! This factor does not change the value of E^{o}_{red}!)

$$2Ag^{+}(aq) + 2e^{-} \rightarrow 2Ag(s) \qquad E^{o}_{red} = +0.799$$

$$Cu(s) \rightarrow Cu^{2+}(aq) + 2e^{-} \qquad E^{o}_{ox} = -0.337$$

$$2Ag^{+}(aq) + Cu(s) \rightarrow 2Ag(s) \qquad E^{o}_{cell} = +0.462$$

To determine the voltage, E^{o}_{cell}, for the cell, add the voltages of the half reactions. (Remember to change the sign of the reduction potential for the oxidation half reaction.)

Common misconception: Changing the coefficient in a balanced half reaction does not affect the value of the standard reduction potential. Although in the example, the Ag half-reaction is multiplied by a factor of 2 to make the electrons balance, the value of E° is not proportional to the balanced equation and should not be multiplied by 2!

Table 20.4 summarized summarizes the process.

Table 20.4. Simplified rules for using a standard reduction potentials table to determine the overall reaction and the voltage, E°_{cell}, for a cell consisting of any two half cells.

1. Select the half-reaction with the more positive voltage as the reduction.
2. Reverse the half-reaction with the less positive voltage and add it to the reduction half-reaction.
3. Calculate the cell voltage, E°_{cell}, by adding E°_{red} for the reduction half-reaction to $-E^\circ_{red}$ for the oxidation half-reaction.
$E^\circ_{cell} = E^\circ_{red} + E^\circ_{ox}$ ($E^\circ_{ox} = -E^\circ_{red}$ for the reversed reaction.)

Your Turn 20.3

Using the data in Table 20.3, choose the half cells that, when coupled, will give the most spontaneous reaction. Write the balanced equation for the reaction and calculate the voltage. Explain your answer. Write your answer in the space provided.

How to recognize common oxidizing and reducing agents.

Oxidizing agents are substances that take electrons away from another reactant in a redox reaction. In the process, oxidizing agents are reduced. They gain electrons. Any table of standard reduction potentials lists the relative strengths of oxidizing agents. In general, strong oxidizing agents have relatively high positive values of E°_{red}. The more positive the E°_{red} value for a substance listed in a table of standard reduction potentials, the stronger the oxidizing agent. For example, $F_2(g)$ is the strongest oxidizing agent listed in Table 20.3.

Reducing agents are substances that provide electrons to another reactant. Reducing agents lose electrons. They are oxidized in redox reactions. The relative strengths of reducing agents are indirectly listed in tables of standard reduction potentials. The more negative the standard reduction potential, the stronger reducing agent is the *product* of the listed half-reaction.

If all the half reactions in a table of standard reduction potentials are reversed, the signs of the accompanying E° values are reversed as well. Such a "reverse table" is a table of oxidation potentials. The more positive the E°_{ox} value, the stronger the oxidation potential. For example, Al(s) is the strongest oxidizing agent listed in Table 20.3.

Whether a substance acts as an oxidizing agent or a reducing agent, depends on the relative values of E°_{red} for their respective reduction half-reactions. For example, half-reactions for the reduction of Mn^{2+} to Mn and Fe^{2+} to Fe both have negative reduction potentials. Fe^{2+} has the more positive E°_{red} so when coupled with Mn, Fe^{2+} will act as the oxidizing agent and Mn will be the reducing agent. However, if Fe and Ni are coupled, Ni^{2+} will be the oxidizing agent and Fe will be the reducing agent.

	E°_{red} (volts)	E°_{ox} (volts)
$Fe^{2+}(aq) + 2e^- \rightarrow Fe(s)$	-0.440	+0.440
$Ni^{2+}(aq) + 2e^- \rightarrow Ni(s)$	-0.28	+0.28
$Mn^{2+}(aq) + 2e^- \rightarrow Mn(s)$	-1.18	+1.18

Recognizing redox reactions, predicting their products and writing their equations.

Given reactants in word form, the AP exam requires students to predict products and write the formulas of reactants and products in net ionic form. Redox reactions are one class of equations that are included.

Recognize a redox reaction if:

1. One reactant is a metal and the other is an aqueous metal ion. The metal will be oxidized to form an aqueous ion, usually but not always having a charge of 2+, and the aqueous ion will be reduced to the corresponding metal.

Example:

A piece of solid zinc is placed in an aqueous solution of copper(II) sulfate.

Solution:

$Zn(s) + Cu^{2+}(aq) \rightarrow Zn^{2+}(aq) + Cu(s)$

2. One reactant is a polyatomic anion with a metallic element displaying its highest oxidation number and the other is an anion displaying an oxidation number lower than its maximum. The polyatomic ion reduces to an ion displaying the metal in a lower oxidation state. The anion is oxidized to a higher oxidation state.

Example:

Acidic aqueous sodium dichromate is mixed with a solution of potassium bromide.

Solution:

$14H^{+}(aq) + Cr_2O_7^{2-}(aq) + 6Br^{-}(aq) \rightarrow 3Br_2(aq) + 2Cr^{3+} + 7H_2O.$

Remember: The equation must account for each reactant and product. "Acidic solution" implies that $H^{+}(aq)$ is a reactant and H_2O is often a product.

3. An organic compound burned in air (or oxygen) produces carbon dioxide and water.

Example:

Ethanol is burned in air.

Solution:

$CH_3CH_2OH(l) + 3O_2(g) \rightarrow 2CO_2(g) + 3H_2O(g)$

If you cannot convert the organic name to a formula, for partial credit, re-write the name and finish the equation with O_2 as a reactant and carbon dioxide and water as products.

4. A metal reacts with a non-metal to produce a binary salt.

Example:

Solid sodium is mixed with chlorine gas.

Solution:

$2Na(s) + Cl_2(g) \rightarrow 2NaCl(s)$

5. An active metal reacts with water to produce hydrogen gas and a hydroxide base.

Example:

Solid lithium is placed in water.

Answer:

$2Li(s) + 2H_2O \rightarrow 2Li^+(aq) + 2OH^-(aq) + H_2(g)$

Remember that lithium must be oxidized because the metal is in its lowest oxidation state. Hydrogen is reduced because it can change oxidation states from 1+ to 0. Oxygen cannot be reduced because in water it already displays its lowest oxidation number (2-).

Section 20.5 Spontaneity of Redox Reactions

A negative value for free energy, ΔG indicates a spontaneous process. A spontaneous reaction also has a positive value for cell potential, E_{cell}.

Free energy and cell potential are related by the following equation:

$\Delta G^o = -nFE^o$

ΔG is free energy in kJ/mol

n is the number of moles of electrons transferred

F is Faraday's constant. One Faraday is the electrical charge carried by one mole of electrons. 1 F = 96,485 coulombs/mol e- = 96,485 J/v-mol e-

E^o is the cell potential

Example:

Assume that aluminum and zinc half cells are suitably connected at 298 K and standard conditions and that both aqueous solutions are 1.00 M concentrations.

a. Write the half-reaction for the cathode.

b. Write the half-reaction for the anode.

c. Write the overall cell reaction.

d. Calculate the E^o for the voltaic cell.

e. Calculate the free energy change for the cell. Is the reaction spontaneous or non-spontaneous? Explain.

f. Calculate the equibrium constant for this reaction.

Solution:

a. From Table 20.3 the reduction half reactions and their corresponding E^o values are:

		E^o_{red}	E^o_{ox}
Zn^{2+}/Zn	$Zn^{2+}(aq) + 2e^- \rightarrow Zn(s)$	-0.762	+0.762
Al^{3+}/Al	$Al^{3+}(aq) + 3e^- \rightarrow Al(s)$	-1.66	+1.66

The reduction half-reaction is the Zn^{2+}/Zn half reaction because it has the more positive E^o_{red}.

$Zn^{2+}(aq) + 2e^- \rightarrow Zn(s)$ $\quad E^o_{red} = -0.762$ *volts*

b. The oxidation half-reaction is the reverse of the listed Al^{3+}/Al half-reaction:

$Al(s) \rightarrow Al^{3+}(aq) + 3e^-$ $\quad E^o_{ox} = +1.66$ *volts*

c. Multiply each half reaction by coefficients that balance the electrons and add the half reaction.

$3(Zn^{2+}(aq) + 2e^- \rightarrow Zn(s)) \quad = 3Zn^{2+}(aq) + 6e^- \rightarrow 3Zn(s)$

$2(Al(s) \rightarrow Al^{3+}(aq) + 3e^-) \quad = 2Al(s) \rightarrow 2Al^{3+}(aq) + 6e^-$

$3Zn^{2+}(aq) + 2Al(s) \rightarrow 3Zn(s) + 2Al^{3+}(aq)$

d. Add the cell potentials of the two half reactions. Do not multiply the cell voltages by the coefficients that balance the equation.

$E^o_{cell} = E^o_{red} + E^o_{ox} = +1.66 + (-0.762) = +0.90$ *volts.*

e. In the equation, $\Delta G^o = -nFE^o$, the number of moles of electrons transferred is 6, as evidenced by the balanced half reactions.

$\Delta G^o = -nFE^o$

$\Delta G^o = -(6)(96{,}484\ 482\ J/v\text{-}mol\ e^-)(+0.90\ v) = -520{,}000\ J/mol = -520\ kJ/mol.$

The positive value of E^o and the corresponding negative value of ΔG^o both indicate that the reaction is spontaneous, which we would expect in a voltaic cell.

f. From Topic 19, $\Delta G = -RT \ln K$

$R = 8.314\ J/K\ mol$

$T =$ *absolute temperature*

$K =$ *the equilibrium constant*

$\Delta G^{\circ} = -RT \ln K$

$-520{,}000 J/mol = -(8.314 J/K mol)(298 K) \ln K$

$\ln K = 210$

$K = e^{210} = 1.6 \times 10^{91}$

The equilibrium constant is very large which means that the reaction goes nearly to completion. This is consistent with the very large negative free energy indicating a spontaneous reaction.

Section 20.6 The Effect of Concentration on EMF

As a voltaic cell discharges, reactants are consumed and products are generated. The concentrations of reactants and products change. As the cell operates, the conditions become nonstandard and the voltage drops. The voltage of the cell is dependent on the concentrations of reactants and products, and can be calculated using the Nernst equation:

$E = E^{\circ} - (0.0592/n) \log Q$ when $T = 298$ K

E is the cell voltage under nonstandard conditions

E° is the standard cell voltage (calculated from a table of reduction potentials)

Q is the reaction quotient

n is the number of electrons transferred in the balanced equation

0.0592 is a constant, which incorporates Faraday's constant, F, the universal gas constant, R, and the absolute temperature at 298 K.

Example:

What is the voltage of the Zn/Al cell in the previous example when $[Al^{3+}] = 1.60$ M and $[Zn^{2+}] = 0.10$ M?

Solution:

The balanced equation is: $3Zn^{2+}(aq) + 2Al(s) \rightarrow 3Zn(s) + 2Al^{3+}(aq)$

Q is defined in the same way as the equilibrium constant and its value can be calculated from the given data:

$Q = [Al^{3+}]^2/[Zn^{2+}]^3 = (1.60)^2/(0.10)^3 = 2560$

From the previous example, the standard cell voltage, $E^{\circ} = +0.90$ volts. Substituting into the Nerst equation we get:

E = Eo – (0.0592/n) log Q

E = +0.90 volts – (0.0592/6) log (2560)

E = +0.90 volts - 0.034 volts = 0.87 v

The cell voltage drops as the concentration of reactant decreases from 1 M standard conditions.

Common misconception: Remember that E is the voltage of a cell at non-standard conditions and that E° is the voltage at standard conditions.

Electrolysis

Section 20.9

Electrolysis is the application of an electric current to a chemical system.

An **electrolytic cell** is a device that uses electrical energy to cause a non-spontaneous chemical reaction to occur. An electrolytic cell converts electrical energy into chemical energy. Figure 20.2 illustrates an electrolytic cell.

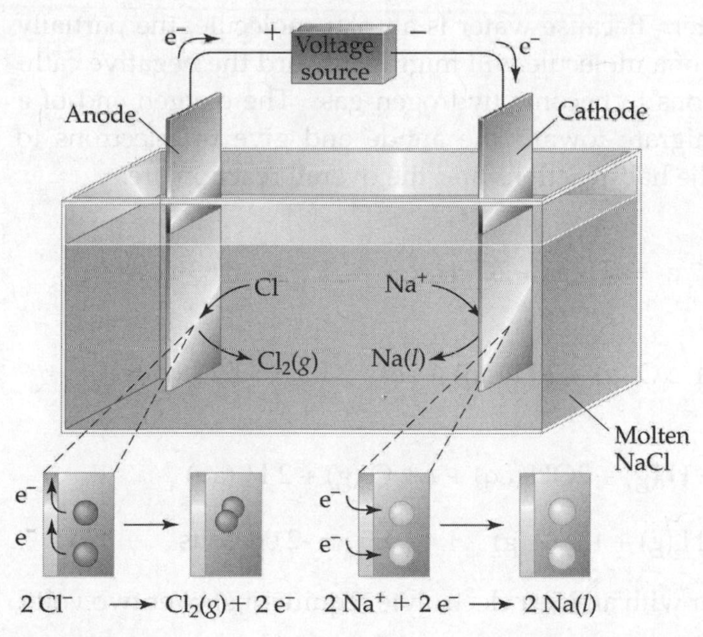

Anode oxidation:	$Cl^- + \rightarrow 1/2\ Cl_2(g) + 1\ e^-$
Cathode reduction:	$Na^+ + 1\ e^- \rightarrow Na(l)$
Overall reaction:	$Na^+ + Cl^- \rightarrow Na(l) + 1/2\ Cl_2(g)$

Figure 20.2. Diagram of an electrolytic cell for the electrolysis of molten NaCl.

The Electrolysis of Molten Sodium Chloride

Solid sodium chloride does not conduct electricity and cannot be electrolyzed because its ions are locked in place in a strong ionic lattice. Molten sodium chloride does conduct electricity because the energy required to melt the solid overcomes the attractive forces that hold the ions in the lattice. The ions are free to migrate to the electrodes. Negative chloride ions migrate to the anode, give up their electrons to become chlorine gas. Positive sodium ions move to the cathode, collect electrons, and become liquid sodium. The half reactions, the overall equation and the corresponding voltages are:

Cathode Anode
reduction: $Cl^- \rightarrow 1/2\ Cl_2(g) + 1\ e^-$ $E^\circ_{ox} = -1.359$ volts

Anode Cathode
oxidation: $Na^+ + 1\ e^- \rightarrow Na(l)$ $E^\circ_{red} = -2.71$ volts

Overall reaction: $Na^+ + Cl^- \rightarrow Na(l) + 1/2\ Cl_2(g)$ $E^\circ_{red} = -4.07$ volts

Notice that the overall reaction has a negative voltage so it is non-spontaneous. It requires just over 4 volts to electrolyze molten sodium chloride. All electrolysis reactions are non-spontaneous.

The Electrolysis of Water Using an Inert Electrolyte

Water, in the presence of an inert electrolyte like sodium sulfate will undergo electrolysis. While pure water does not conduct electricity, inert ions in water can migrate toward the electrodes and carry an electric current sufficient to electrolyze water. Because water is a polar molecule, the partially positive hydrogen end of a molecule will migrate toward the negative cathode and pick up electrons to become hydrogen gas. The oxygen end of a water molecule will migrate toward the anode and give up electrons to become oxygen gas. The half-reactions and the overall reaction are:

Cathode
reduction: $2H_2O(l) + 2\ e^- \rightarrow H_2(g) + 2OH^-(aq)$ $E^\circ_{red} = -0.83$ volts

Anode
oxidation: $H_2O(l) \rightarrow 1/2\ O_2(g) + 2\ H^+(aq) + 2e^-$ $E^\circ_{ox} = -1.23$ volts

Overall
reaction: $3H_2O(l) \rightarrow H_2(g) + 2OH^-(aq) + 1/2\ O_2(g) + 2\ H^+(aq)$

or $H_2O(l) \rightarrow H_2(g) + 1/2\ O_2(g)$ $E^\circ_{cell} = -2.06$ volts

The electrolysis of water with an inert electrolyte requires just over two volts.

The Electrolysis of an Aqueous Sodium Chloride Solution

Electrolysis of an aqueous solution of sodium chloride produces hydrogen gas at the cathode and chlorine gas at the anode. The half-reactions and the overall process are given below.

$2H_2O(l) + 2\ e- \rightarrow H_2(g) + 2OH^-(aq)$ $E^\circ_{red} = -\ 0.83$ volts

$2Cl^- \rightarrow Cl_2(g) + 2\ e-$ $E^\circ_{ox} = -\ 1.359$ volts

$2H_2O(l) + 2Cl^- \rightarrow H_2(g) + 2OH^-(aq) + Cl_2(g)$ $E^\circ_{cell} = -\ 2.19$ volts

In the case of the electrolysis of aqueous sodium chloride, there are two possible reductions that can take place at the cathode. Either water will gain electron and produce hydrogen gas or sodium ions pick up electrons to become sodium metal.

Possible cathode reductions:

$Na^+(aq) + 1\ e- \rightarrow Na(l)$ $E^\circ_{red} = -\ 2.71$ volts

$2H_2O(l) + 2\ e- \rightarrow H_2(g) + 2OH^-(aq)$ $E^\circ_{red} = -\ 0.83$ volts *

It is usually possible to predict which of the possible half-reactions will occur when solutions are electrolyzed on the basis of their relative E°_{red} values. The reduction potential for water is more favorable than the reduction potential for sodium ion because water has the more positive value for E°_{red}. Therefore the water reduction, not the sodium ion reduction will occur at the cathode.

The same analysis can be made for the possible oxidation reactions. In general, the oxidation with the more positive E°_{ox} value will predominate.

The Chloride Ion is an Uncommon Exception.

The oxidation potential for chloride ion is less positive that the oxidation potential for water so, from a thermodynamic argument, water should oxidize in the presence of aqueous chloride ion. And, indeed, it does. However, the oxidation of water is kinetically very slow. By comparison, the oxidation of chloride ion is rapid. Additionally, both half-reactions require similar voltages to carry out. Usually in most electrolyses, there is sufficient voltage to drive both reactions. Because of the relative rates of reaction, chlorine gas, not oxygen gas is produced at the anode.

Possible anode oxidations:

$Cl^-(aq) \rightarrow 1/2\ Cl_2(g) + 1\ e-$ $E^\circ_{ox} = -\ 1.359$ volts *

$H_2O(l) \rightarrow 1/2\ O_2(g) + 2H^+(aq) + 2e-$ $E^\circ_{red} = -\ 1.23$ volts

Electroplating

Electrolysis is commonly used in electroplating, the process of depositing a thin coating of one metal onto another metal for ornamental purposes or for corrosion resistance. For example, steel is commonly plated with nickel or chromium and eating utensils are often plated with silver or gold.

To plate nickel on a steel surface, for example, an electrolysis apparatus illustrated in Figure 20.3 consists of a nickel anode and a steel cathode. Both electrodes are immersed in an aqueous solution of nickel nitrate. Voltage applied to the cell forces electrons from the anode to the cathode. Reduction of Ni^{2+} ions occurs at the cathode where nickel metal plates the steel. The nickel anode oxidizes to nickel ions, which replace the ions in solution.

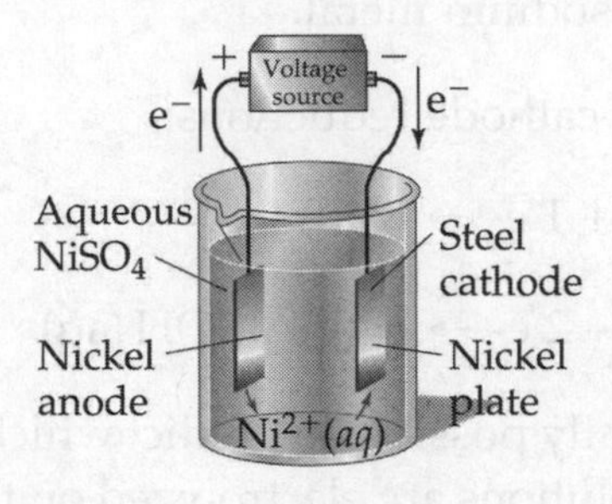

Figure 20.3. Electrolysis device for plating nickel onto a piece of steel.

The half-reactions are:

Reduction at the cathode: $Ni^{2+}(aq) + 2e- \rightarrow Ni(s)$ $E^\circ_{red} = -0.28$ v

Oxidation at the anode: $Ni(s) \rightarrow Ni^{2+}(aq) + 2e-$ $E^\circ_{ox} = +0.28$ v

The other possible anode and cathode reactions are the oxidation and reduction of water, both of which have less favorable potentials than do the nickel reactions.

Cathode reduction: $H_2O(l) + 2\ e- \rightarrow H_2(g) + 2OH^-(aq)$ $E^\circ_{red} = -0.83$ volts

Anode oxidation: $H_2O(l) \rightarrow 1/2\ O_2(g) + 2\ H^+(aq) + 2e-$ $E^\circ_{ox} = -1.23$ volts

The nickel reactions predominate and the net effect is to move nickel atoms from the anode to the surface of the steel cathode.

Quantitative Aspects of Electrolysis

A balanced half-reaction tells how many moles of electrons are involved in a redox reaction. For example, when silver ion is reduced to silver metal, one mole of electrons is transferred. When water is reduced to hydrogen gas and hydroxide ion, two moles of electrons are involved.

$Ag^+(aq) + 1\ e- \rightarrow Ag(s)$

$2H_2O(l) + 2\ e- \rightarrow H_2(g) + 2OH-(aq)$

The amount of a substance that is oxidized or reduced in an electrolytic cell is directly proportional to the number of electrons passed through the cell. The balanced half-reaction is a way to convert electrical quantities into moles of chemical quantities and vice-versa.

Example:

How many liters of dry hydrogen gas at 30°C and 720 torr can be obtained by the electrolysis of water for one hour at 2.50 amps?

Solution:

The key ideas in this quantitative electrolysis problem are:

The ***balanced half-reaction*** *relates moles of electrons transferred to moles of chemical reactants or products:*

2 mol e- = 1 mol $H_2(g)$

The ***ampere*** *is the measure of electric current. One ampere is one coulomb of electric charge passing a point in an electrical circuit in one second:*

2.50 amps = 2.50 Coulombs per second = 2.50 C/s

Faraday's constant *tells the quantity of electric charge carried by a mole of electrons.*

1 mol e- = 96485 C (Faraday's constant)

The ***ideal gas equation*** *calculates volume in liters from number of moles, n, absolute temperature, T, pressure in atmospheres and the universal gas constant R. R = 0.0821 Latm/mol-K.*

$V = nRT/P$

x mol $H_2(g)$ = 1 hr(60 min/hr)(60 s/min)(2.50 C/s)(1 mole e-/96485 C) (1 mol H_2/(2 mol e-) = 0.0466 mol H_2.

V = nRT/P = (0.0466 mol)(0.0821 Latm/mol-K)(303 K)/(720/760)atm = 1.22 L

Multiple Choice Questions

1. *When the following species are listed in order of increasing oxidation number of the sulfur atoms (most negative to most positive oxidation number), the correct order is*

A) $CuS, K_2S_2O_3, SO_2, Na_2SO_4$

B) $CuS, SO_2, K_2S_2O_3, Na_2SO_4$

C) $CuS, SO_2, Na_2SO_4, K_2S_2O_3$

D) $Na_2SO_4, CuS, K_2S_2O_3, SO_2$

E) $K_2S_2O_3, SO_2, Na_2SO_4, CuS$

2. *Use the reduction potentials to determine which one of the reactions below is spontaneous.*

Reduction Potentials, E°

1.	$Cd^{2+} + 2e \rightarrow Cd$	-0.403 V
2.	$Mn^{2+} + 2e \rightarrow Mn$	-1.18 V
3.	$Cu^{+} + 1e^- \rightarrow Cu$	+0.521 V
4.	$Fe^{3+} + 1e \rightarrow Fe^{2+}$	+0.771 V

A) $Cd^{2+} + 2Cu \rightarrow Cd + 2Cu^{+}$

B) $Mn^{2+} + 2Cu \rightarrow Mn + 2Cu^{+}$

C) $Cd^{2+} + Mn \rightarrow Cd + Mn^{2+}$

D) $Cu^{+} + Fe^{2+} \rightarrow Cu + Fe^{3+}$

E) $Cd^{2+} + 2Fe^{2+} \rightarrow Cd + 2Fe^{3+}$

3. *What is the coefficient for MnO_4^- when the following redox equation is balanced in acidic solution using the smallest whole-number coefficients?*

$$H^+ + Cr_2O_7^{2-} + Mn^{2+} \rightarrow H_2O + Cr^{3+} + MnO_4^-$$

A) 1

B) 2

C) 5

D) 6

E 10

4. *A cell is set up using the following reactions: What is the voltage of the cell?*

$Mg^{2+} + 2e- \rightarrow Mg$ $E^{\circ} = -2.37$ *volts*

$Ag^{+} + 1e- \rightarrow Age$ $E^{\circ} = +0.80$ *volts*

A) +1.57 volts

B) +3.17 volts

C) +3.97 volts

D) -0.77 volts

E) +0.77 volts

5. *All of these substances act as oxidizing agents except:*

A) H_2O_2

B) $Cr_2O_7^{2-}$

C) MnO_4^-

D) IO_3^-

E) Br^-

6. *Which element can have the highest positive oxidation number?*

A) O

B) C

C) Cl

D) N

E) Cr

7. *Magnesium reacts with dilute hydrochloric acid to produce hydrogen gas. Silver does not react in dilute hydrochloric acid. Based on this information, which of the following reactions will occur spontaneously?*

A) $H_2(g) + Mg^{2+}(aq) \rightarrow 2H^+(aq) + Mg(s)$

B) $2Ag(s) + Mg^{2+}(aq) \rightarrow 2\,Ag^+(aq) + Mg(s)$

C) $2Ag^+(aq) + Mg(s) \rightarrow 2Ag(s) + Mg^{2+}(aq)$

D) $2Ag + 2H^+(aq) \rightarrow H_2(g) + 2Ag^+(aq)$.

E) $H_2(g) + 2Ag\,(aq) \rightarrow 2Ag^+ + 2H^+(aq)$

8. *Which species can act as both an oxidizing agent and a reducing agent?*

A) $Cr_2O_7^{2-}$

B) S^{2-}

C) H_2O_2

D) *Fe*

E) MnO_4^-

As an oxidizing agent: $H_2O_2(aq) + 2H^+(aq) + 2e^- \rightarrow 2H_2O(l)$

As a reducing agent: $H_2O_2(aq) \rightarrow O_2(g) + 2H^+(aq) + 2e^-$

$Cr_2O_7^{2-}$ *and* MnO_4^- *can oly be oxidizing agents because Cr and Mn, respectively, have their maximum oxidation numbers. They can only gain electrons. Similarly, Fe and* S^{2-} *can only lose electrons because those atoms display their minimum oxidation number.*

9. *Of the following, which is the strongest oxidizing agent?*

A) IO_3^-

B) Cl_2

C) O_2

D) PbO_2

E) Co^{3+}

10. *Of the following, which is the strongest reducing agent?*

A) F^-

B) *Ca*

C) *Mn*

D) H_2O_2

E) N_2H_4

Free Response Questions

1. *Suppose that gold and silver half cells are suitably connected.*

$Au^{3+} + 3e \rightarrow Au(s) \quad E^\circ = +1.50\ v$

$Ag^+ + 1e^- \rightarrow Ag(s) \quad E^\circ = +0.80\ v$

a. i. Indicate the cathode and anode half reactions.

ii. Write the overall cell reaction.

iii Calculate E° for the cell.

b. Calculate ΔG° for the cell.

c. Calculate the equilibrium constant for the cell reaction 25°C.

d. Calculate E for the cell if $[Ag^{+}]$= 0.15 M and $[Au^{3+}]$= 0.75 M.

e. In a separate experiment, how many grams of gold can be plated onto a piece of jewelry if a 1.00 M solution of gold(III) nitrate is suitably electrolyzed for 45 minutes at 2.50 amps?

2. Consider the following half reactions and their standard reduction potentials at 25°C.

$PbO_2(s) + H_2O(l) + 2\,e^- \rightarrow PbO(s) + 2OH^-(aq)$ $\quad E^{\circ} = +0.28\ v$

$IO_3^- + 2\,H_2O + 4\,e^- \rightarrow IO^- + 4OH^-$ $\quad E^{\circ} = +0.56\ v$

$PO_4^{3-} + 2\,H_2O + 1/2\ O_2 + 2e^- \rightarrow HPO_4^{2-} + 3OH^-$ $\quad E^{\circ} = -1.05\ v$

a. i. Which two half reactions when combined will give the voltaic cell with the largest E^{0} for the cell?

ii. Write the overall cell reaction that would take place.

iii. Calculate E° for the cell.

b. Indicate how the E_{cell} will be affected by the following changes. Justify your answers.

i. If the pH of both the anode and cathode compartments is decreased.

ii. If $[OH^-]$ is fixed at 1.0 M, $O_2(g)$ is fixed at one atmosphere and all other ion concentrations are changed to 0.10 M.

iii. If the temperature is increased while all concentrations remain at 1.0 M.

Additional Practice in *Chemistry the Central Science*

For more practice working electrochemistry problems in preparation for the Advanced Placement examination try these problems in Chapter 20 of Chemistry the Central Science:

Additional Exercises: 20.91, 20.92, 20.93, 20.95, 20.96, 20.99.

Integrated Exercises: 20.104, 20.105, 20.108, 20.109, 20.110, 20.112, 20.113, 20.114, 20.115, 20.116.

Multiple Choice Answers and Explanations

1. *A. The oxidation numbers or sulfur in CuS, $K_2S_2O_3$, SO_2, Na_2SO_4 are -2, +2, +4 and +6 respectively. As a monatomic ion in CuS, sulfur has a -2 charge so its oxidation number is equal to its charge. The oxidation number of each sulfur adds to the oxidation numbers of each oxygen to equal the charge in the polyatomic ion, $S_2O_3^{2-}$. 3(2- for each oxygen) + 2(2+ for each sulfur) (note: in $S_2O_3^{2}$, one sulfur carries a 4+ oxidation number where the other sulfur SO, an average of 2+ for each sulfur). The oxidation number of sulfur in SO_2 is 4+ because it adds to the oxidation numbers of each oxygen to equal the zero charge on SO_2. 4 + 2(2-) = 0. A 6+ oxidation number for sulfur in SO_4^{2-} adds to the four 2- oxidation numbers of oxygen to give a charge of 2-. 6 + 4(2-) = 2-.*

2. *C. For an overall reaction to be spontaneous the values of E^o for the two half reactions must add to yield a positive E^o. Half reaction 1 adds to the reverse of half reaction 2 to give a positive E^o.*

1:	$Cd^{2+} + 2e \rightarrow Cd$	*-0.403 V*
plus reverse of 2:	$Mn^{2+} \rightarrow 2e- + Mn^{2+}$	*+1.18 V*
yields	$Cd^{2+} + Mn \rightarrow Cd + Mn^{2+}$	*+0.78 V*

The other combinations listed each yield a negative E^o.

3. *D. Taking apart the unbalanced equation provides two skeletal half-reactions and each is balanced according to the rules given in Table 20.2:*

   1. *Balance all elements other than H or O*
   2. *Balance O by adding H_2O as required.*
   3. *Balance H by adding H^+ as required.*
   4. *Balance charge by adding e^- to more positive side.*

   $Mn^{2+} \rightarrow MnO_4^-$

   1. $Mn^{2+} \rightarrow MnO_4^-$
   2. $4H_2O(l) + Mn^{2+} \rightarrow MnO_4^-$
   3. $4H_2O(l) + Mn^{2+} \rightarrow MnO_4^- + 8H^+(aq)$
   4. $4H_2O(l) + Mn^{2+} \rightarrow MnO_4^- + 8H^+(aq) + 5e-$

$Cr_2O_7^{2-} \rightarrow Cr^{3+}$

1. *$Cr_2O_7^{2-} \rightarrow 2Cr^{3+}$*
2. *$Cr_2O_7^{2-} \rightarrow 2Cr^{3+} + 7H_2O(l)$*
3. *$14H^+(aq) + Cr_2O_7^{2-} \rightarrow 2Cr^{3+} + 7H_2O(l)$*
4. *$6e- + 14H^+(aq) + Cr_2O_7^{2-} \rightarrow 2Cr^{3+} + 7H_2O(l)$*

Multiplication of the Mn and Cr half-reactions to make the electrons equal requires factors of 6 and 5, respectively. That means that in the balanced equation, MnO_4^- will have a coefficient of 6.

An alternate way to arrive at the answer is to assign oxidation numbers to all the elements and note that Mn changes by 5 electrons (from 2+ to 7+) and that Cr changes by 3 electrons (from 6+ to 3+) twice for a total of 6 electrons. The lowest common multiple of 5 and 6 is 30 so the Mn half reaction must be multiplied by 6 giving the MnO_4^- a coefficient of 6.

4. *B. The Ag half-reaction has the more positive voltage so it is the reduction. The Mg half-cell is the oxidation and it's voltage changes to +2.37 volts.*

   $E^\circ_{cell} = E^\circ_{red} + E^\circ_{ox} = 0.80 + 2.37 = 3.17$ volts.

   Note: The stoichiometry is not taken into account because voltage is independent of the number of electrons transferred.

5. *E. An oxidizing agent is the substance reduced in a chemical reaction. Therefore, an oxidizing agent must contain an element with an oxidation number greater than the lowest possible oxidation number for that element. Br- has an oxidation number of 1- for Br which is the lowest possible oxidation number for bromine.*

6. *C. Cl displays a 7+ oxidation number in the perchlorate ion, ClO_4^-. Cr is 6+ in CrO_4^{2-}. C is 4+ in CO_2. Oxygen is 2+ in OF_2.*

7. *C. On the basis of their behavior in the presence of hydrochloric acid, Mg is a more active metal than is Ag. Thus, Mg will react with Ag^+ ions, but Ag will not react with Mg^{2+}. Also Ag, because it does not react with HCl, does not react with H^+. E cannot be correct because both H_2 and Ag are oxidized.*

8. *C. An oxidizing agent is reduced. It gains electrons causing another species to lose electrons. A reducing agent is oxidized, losing electrons causing another species to gain electrons. Although hydrogen peroxide is often considered an oxidizing agent, it commonly acts as a reducing agent. The oxidation number of oxygen in H_2O_2 is 1^-. Oxygen can either lose an electron to become 0 or gain an electron to become 2-. The half reactions are:*

As an oxidizing agent: $H_2O_2(aq) + 2H^+(aq) + 2e^- \rightarrow 2H_2O(l)$

As a reducing agent: $H_2O_2(aq) \rightarrow O_2(g) + 2H^+(aq) + 2e^-$

$Cr_2O_7^{2-}$ and MnO_4^- can only be oxidizing agents because Cr and Mn, respectively, have their maximum oxidation numbers. They can only gain electrons. Similarly, Fe and S^{2-} can only lose electrons because those atoms display their minimum oxidation numbers.

9. *E. The correct answer requires access to a table of standard reduction potentials provided on the AP exam. The reactant with the most positive E°_{red} value is the strongest oxidizing agent.*

10. *B. The correct answer requires access to a table of standard reduction potentials provided on the AP exam. The reactant with the most positive E°_{ox} value is the strongest reducing agent. However, the answer is not given directly on a table of standard reduction potentials. To determine the correct answer, choose the product of the reduction half-reaction having the most negative E°_{red} value. The reverse of that half-reaction has the most positive E°_{ox} value.*

Free Response Answers

1a. *i. Cathode half cell reaction:*
$Au^{3+} + 3e^- \rightarrow Au(s)$ $E^\circ = +1.500\ v$

Anode half cell reaction:
$Ag(s) \rightarrow Ag^+ + 1e^-$ $E^\circ = -0.800\ v$

ii. Overall cell reaction:
$Au^{3+} + 3Ag(s) \rightarrow Au(s) \rightarrow + 3Ag^+$

iii. $E^\circ_{cell} = E^\circ_{ox} + E^\circ_{red} = +1.50\ v - 0.080\ v = 0.700\ volts$

b. $\Delta G^{\circ} = -nFE^{\circ} = -(3 \text{ mole e-})(96{,}500 \text{ Coul/mole e-})(0.700 \text{ J/Coul}) = -203 \text{ kJ}$

(1 volt = 1 J/Coul)

c. $\Delta G^{\circ} = -RT \ln K$

$-202 \text{ kJ} = (8.314 \text{ J/mol K})(1 \text{ kJ}/1000 \text{ J})(298 \text{ K}) \ln K$

$\ln K = +81.5$

$K = e^{+81.5} = 2.48 \times 10^{35}$

d. $E = E^{\circ} - (0.0592/n)\log Q$

$Q = [Ag^{+}]^{3}/[Au^{3+}] = (0.15)^{3}/0.75 = 0.0045$

$E = (0.700) - (0.0592/3)\log(0.0045) = 0.65$ volts

e. x g Au =

45 min (60 s/min)(2.50 C/s)(1 mol e-/96500 C)
(1 mol Au/3 mol e-)(197 g/mol) = 4.59 g

2a. i. $IO_3^- + 2H_2O + 4e^- \rightarrow IO^- + 4OH^-$ $E^{\circ} = +0.56$ v

$HPO_4^{2-} + 3OH^- \rightarrow PO_4^{3-} + 2H_2O + 1/2\, O_2 + 2e^-$ $E^{\circ} = +1.05$ v

ii. $2HPO_4^{2-} + 2OH^- + IO_3^- \rightarrow 2PO_4^{3-} + 2H_2O + IO^- + O_2$

iii. $E^{\circ} = E^{\circ}_{ox} + E^{\circ}_{red} = +1.05 \text{ v} + 0.56 \text{ v} = +1.61$ volts.

b. *If the pH of both the anode and cathode compartments is decreased, the net effect will be to decrease [OH⁻] and the voltage will decrease because, as evidenced by the Nernst equation.*

$E = E^{\circ} - (0.0592/n)\log([PO_4^{3-}]^2[IO^-][O_2]/[HPO_4^{2-}]^2[OH^-]^2[IO_3^-])$

ii. *If [OH⁻] is fixed at 1.0 M and all other ion concentrations are changed to 0.10 M the voltage will remain the same because, other than hydroxide, there are three moles of ions on both sides of the balanced equation.*

iii. *If the temperature is increased while all concentrations remain at 1.0 M the voltage will drop because the 0.0592 term will become larger.*

Also, the reaction is exothermic because the voltage is positive. Increasing temperature always favors the endothermic reaction (the reverse reaction).

Your Turn Answers

20.1. *SO_4^{2-}, 6+; $S_2O_3^{2-}$, 4+ and 0; S^{2-}, 2-; SO_3^{2-}, 4+; SCN^-, 2-; S, 0; SO_2, 4+; SO_3, 6+. Oxidation number and charge are equivalent only in sulfide, S^{2-}, and sulfur, S.*

20.2. *N_2 gains electrons and its oxidation number changes from 0 to 3-. N_2 is the substance reduced and is the oxidizing agent. H_2 loses electrons and its oxidation number changes from 0 to 1+. H_2 is the substance oxidized and is the reducing agent.*

20.3. *When coupled, the two half reactions with the largest difference in E° value will give the most spontaneous reaction, the one with the largest positive cell voltage. Reverse the half reaction with the most negative voltage and add it to the half reaction with the most positive voltage. Use coefficients to balance the electrons. Change the sign of the E° for the reversed reaction and add it to the E° of the forward reaction.*

$3F_2(g) + 6e^- \rightarrow 6F^-(aq)$	+2.87 volts
$2Al(s) \rightarrow 2Al^{3+}(aq) + 6e^-$	+1.66 volts
$3F_2(g) + 2Al(s) \rightarrow 6F^-(aq) + 2Al^{3+}(aq)$	+4.53 volts

TOPIC

21

NUCLEAR CHEMISTRY AND ORGANIC CHEMISTRY

Nuclear chemistry (Chapter 21) and organic chemistry (Chapter 25) comprise a small portion of the Advanced Placement Examination. Know what radioactivity is, the symbols for the particles involved, and how to write and balance nuclear equations. Because radioactive substances react according to first-order kinetics, know how to use the integrated first order rate equation and half-lives in calculations. Understand basic organic nomenclature, structural, geometrical and optical isomers and functional groups. Don't spend a lot of time on these chapters but focus on these sections:

- 21.1 **Radioactivity**
- 21.4 **Rates of Radioactive decay**
- 25.3 **Alkanes**
- 25.4 **Unsaturated Hydrocarbons**
- 25.5 **Functional Groups**

Radioactivity

Section 21.1

Nuclear chemistry is the branch of chemistry that deals with nuclear reactions, changes in the atomic nucleus.

Isotopes are atoms having the same atomic number (number of protons) but different mass numbers (numbers of neutrons).

A **radioactive isotope (radioisotope)** is an atom having an unstable nucleus.

Radioactivity (or radioactive decay) is the spontaneous disintegration of an unstable atomic nucleus with the accompanying emission of radiation.

Radiation associated with radioactive decay includes alpha and beta particles, positrons and gamma radiation. Table 21.1 summarizes the properties of each type of radiation.

Table 21.1 Properties of alpha, beta, gamma, and positron radiation.

Radiation	Symbols		Charge	Mass	Definition
alpha	α	$^{4}_{2}He$	2+	6.6 x 10^{-24} g	Helium nucleus
beta	β	$^{0}_{-1}e$	1-	9.11 x 10^{-28} g	electron
positron		$^{0}_{+1}e$	1+	9.11 x 10^{-28} g	Anti-electron
gamma	γ		0	0	High-energy electromagnetic radiation

Except for Technetium, atomic number 43, atoms having atomic numbers less than 84 have at least one stable (non-radioactive) isotope. All isotopes having atomic numbers greater than 83 are radioactive.

Nuclear equations represent the transformations of nuclei in nuclear reactions. Mass numbers and atomic numbers of reactants and products are balanced in nuclear equations.

Types of Radioactive Decay

Alpha decay: Emits an alpha particle (helium nucleus), $^{4}_{2}He$.

$$^{226}_{88}Ra \rightarrow ^{222}_{86}Rn + ^{4}_{2}He$$

Beta decay: Emits a beta particle (electron), $^{0}_{-1}e$. The atomic number increases by one.

$$^{14}_{6}C \rightarrow ^{14}_{7}N + ^{0}_{-1}e$$

Positron decay: Emits a positron, $^{0}_{+1}e$. The atomic number decreases by one.

$$^{11}_{6}C \rightarrow ^{11}_{5}B + ^{0}_{+1}e$$

Electron capture: The nucleus captures a 1s electron. The atomic number decreases by one.

$$^{81}_{37}Rb + ^{0}_{-1}e \rightarrow ^{81}_{36}Kr$$

Other Nuclear Reactions

Fission: A neutron splits a heavy nucleus producing smaller nuclei and more neutrons.

$$^{235}_{92}U + ^{1}_{0}n \rightarrow ^{142}_{56}Ba + ^{91}_{36}Kr + 3\,^{1}_{0}n$$

Fusion: Two light nuclei fuse to form a heavier one.

$${}^{2}_{1}H + {}^{3}_{1}H \rightarrow {}^{4}_{2}He + {}^{1}_{0}n$$

Rates of Radioactive Decay

Section 21.4

Radioactive decay is a first-order kinetic process. The half-life of a radioisotope is the time required for half of any given sample to react. The equations are:

$\ln N_o/N_t = +kt$ and $k = 0.0693/t_{1/2}$

N_o is the number of atoms present at time t = 0

N_t is the number of atoms left at an given time, t

$t_{1/2}$ is the half life

k is the rate constant is k.

Example:

Strontium-90 decays via beta emission and has a half life of 28.8 years.

a. *Into which isotope does strontium-90 decay?*

b. *Write a nuclear equation describing the radioactive decay of strontium-90.*

c. *How many grams of a 0.500 g sample of strontium-90 remain after 40.0 years?*

Solution:

a. *Strontium-90 has a mass number of 90 and an atomic number of 38: ${}^{90}_{38}Sr$.*

A beta particle is an electron: ${}^{0}_{-1}e$. If the nucleus loses an electron, the net effect is that a neutron is converted to a proton. Strontium-90 becomes Yttrium-90.

b. ${}^{90}_{38}Sr \rightarrow {}^{90}_{39}Y^{0} + {}_{-1}e$

c. $k = 0.693/t_{1/2} = 0.693/28.8\ yr = 0.0241\ yr^{-1}$

$$ln\ N_o/Nt = +kt$$

$$ln\ (0.500g/Nt) = (0.0241\ yr^{-1})(40.0\ yr)$$

$$0.500/Nt = e^{0.963}$$

$$N_t = 0.191\ g$$

Names of Simple Organic Molecules

Organic chemistry is the study of carbon compounds.

Hydrocarbons are compounds containing only carbon and hydrogen.

Alkanes are hydrocarbons containing C-C single bonds. Alkanes have the general formula, C_nH_{2n+2}. Carbons containing all single bonds have a tetrahedral geometry and are sp^3 hybridized. Alkanes are examples of saturated hydrocarbons.

Table 21.2 shows the names and formulas of some common alkanes. Notice in each name a prefix indicates the number of carbon atoms in the formula and it is followed by the suffix –ane to indicate that the formula is an alkane.

Table 21.2. Names and formulas of some simple alkanes. The prefix of the name tells how many carbon atoms are in the formula.

Number of carbon atoms	prefix	name	formula
1	meth-	methane	CH_4
2	eth-	ethane	CH_3CH_3
3	prop-	propane	$CH_3CH_2CH_3$
4	but-	butane	$CH_3CH_2CH_2CH_3$
5	pent-	pentane	$CH_3CH_2CH_2\ CH_2CH_3$
6	hex	hexane	$CH_3CH_2CH_2CH_2CH_2CH_3$
7	hept-	heptane	$CH_3CH_2CH_2CH_2CH_2CH_2CH_3$
8	oct-	octane	$CH_3CH_2CH_2CH_2CH_2CH_2CH_2CH_3$
9	non-	nonane	$CH_3CH_2CH_2CH_2CH_2CH_2CH_2CH_2CH_3$
10	dec-	decane	$CH_3CH_2CH_2CH_2CH_2CH_2CH_2CH_2CH_2CH_3$

Unsaturated hydrocarbons contain one or more multiple bonds.

Alkenes are hydrocarbons containing C=C double bonds. For example, the hydrocarbon having three carbons and one double bond is called propene, $CH_2=CHCH_3$.

Alkynes are hydrocarbons containing C≡C triple bonds. For example, the hydrocarbon having three carbons and one triple bond is called propyne, $CH≡CCH_3$.

Table 21.3 shows some characteristics of the three types of hydrocarbons.

Table 21.3. Characteristics of alkanes, alkenes and alkynes.

Hydrocarbon	**Alkane**	**Alkene**	**Alkyne**
Characteristic	C-C single bonds	C=C double bond	C≡C triple bond
General formula	C_nH_{2n+2}	C_nH_{2n}	C_nH_{2n-2}
Geometry	tetrahedral	trigonal planar	linear
Hybridization	sp^3	sp^2	sp
Bond angles	109.5°	120°	180°
isomers	structural and optical	structural, optical and geometrical	structural

Aromatic hydrocarbons are unsaturated hydrocarbons that have more than one resonance structure. Benzene, C_6H_6 is an example of an aromatic hydrocarbon. Figure 21.1 compares the geometries of various hydrocarbon compounds.

Structural isomers are compounds having the same molecular formula but different structural formulas.

Geometrical isomers are compounds having the same molecular formula and the same groups bonded to one another but they have different special spatial arrangement of these groups. Figure 21.2 shows four compounds, all of which are structural isomers. In addition, the two compounds on the right are geometrical isomers of each other because the methyl groups have different spatial arrangements.

Optical isomers (enantiomers) are isomers whose mirror images are not superimposable on each other. Figure 21.3 shows an example.

Functional groups are groups of atoms that give rise to the structure and properties of an organic compound. Table 21.4 lists the common organic functional groups together with example compounds. Notice that the name of each compound listed reflects the number of carbons it contains.

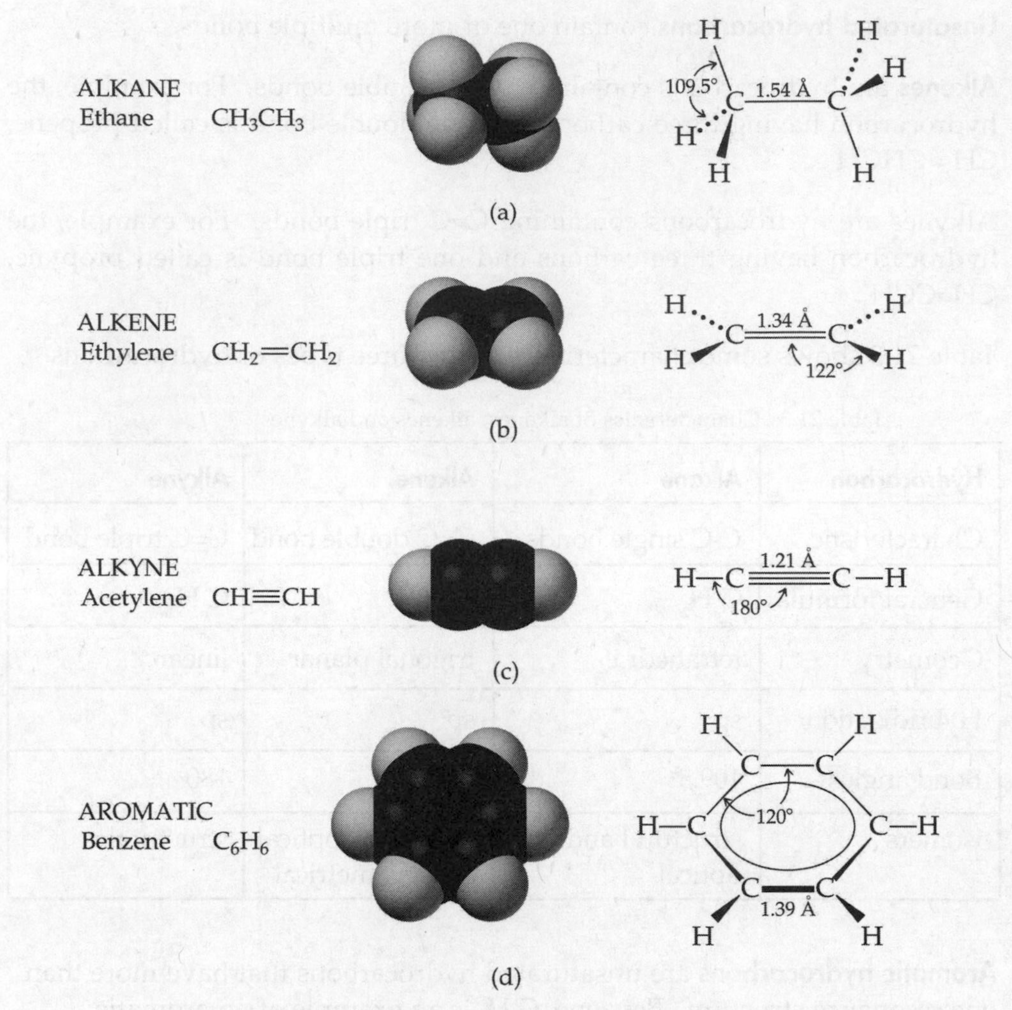

Figure 21.1. Names, formulas and geometrical structures of various hydrocarbons.

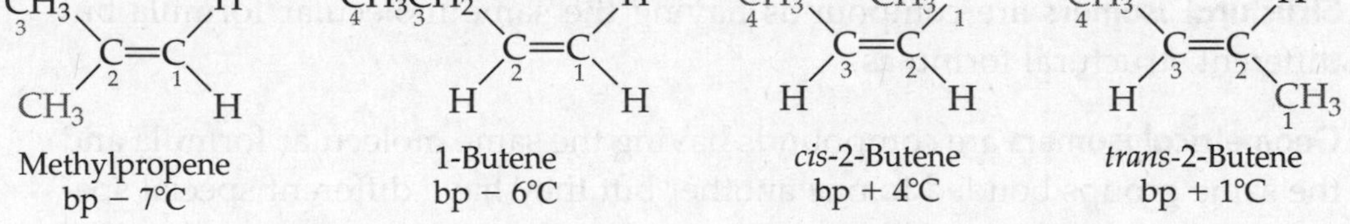

Methylpropene bp − 7°C

1-Butene bp − 6°C

cis-2-Butene bp + 4°C

trans-2-Butene bp + 1°C

Figure 21.2. Structural and geometrical isomers of C4H8.

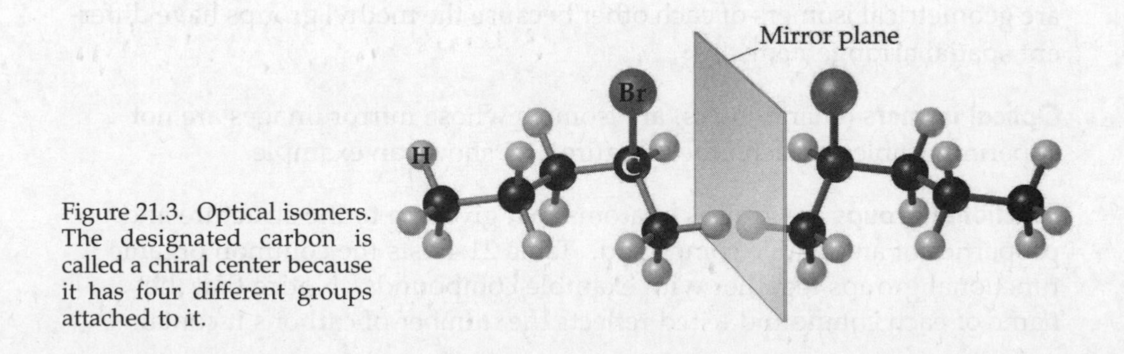

Figure 21.3. Optical isomers. The designated carbon is called a chiral center because it has four different groups attached to it.

TABLE 21.4 Common Functional Groups in Organic Compounds

Functional Group	Type of Compound	Suffix or Prefix	Example	Systematic Name (common name)
>C=C<	Alkene	*ene*	$H_2C=CH_2$	Ethene (Ethylene)
—C≡C—	Alkyne	*-yne*	H—C≡C—H	Ethyne (Acetylene)
—C—Ö—H	Alcohol	*-ol*	H_3C—Ö—H	Methanol (Methyl alcohol)
—C—Ö—C—	Ether	*ether*	H_3C—Ö—CH_3	Dimethyl ether
—C—X: (X = halogen)	Haloalkane	*halo-*	H_3C—Cl:	Chloromethane (Methyl chloride)
—C—N—	Amine	*-amine*	H_3C—CH_2—N—H_2	Ethylamine
—C(=O:)—H	Aldehyde	*-al*	H_3C—C(=O:)—H	Ethanal (Acetaldehyde)
—C—C(=O:)—C—	Ketone	*-one*	H_3C—C(=O:)—CH_3	Propanone (Acetone)
—C(=O:)—Ö—H	Carboxylic acid	*-oic acid*	H_3C—C(=O:)—Ö—H	Ethanoic acid (Acetic acid)
—C(=O:)—Ö—C—	Ester	*-oate*	H_3C—C(=O:)—Ö—CH_3	Methyl ethanoate (Methyl acetate)
—C(=O:)—N—	Amide	*-amide*	H_3C—C(=O:)—N—H_2	Ethanamide (Acetamide)

Table 21.4. Common Functional Groups in Organic Compounds

TEST 1

CHEMISTRY PRACTICE TEST 1

A few weeks before the Advanced Placement Exam you should approach the finish of your review of all the topics in this book. You should also be approaching the end of your coursework leading up to the Advanced Placement exam. When you are ready, take Practice Test 1 within the suggested time limits. When you are finished, look up the answers and calculate your score. The following scoring system is simplified and informal. Although it cannot predict what you will score on the real AP exam, it serves as a measure of your strengths and weaknesses a few weeks before the exam.

Self-score your practice exam this way:

Multiple-choice Questions. Section I

Number right: 33 x 0.667 = 22.011 A

Number wrong: 26 x 0.167 = 4.342 B

Sub-total for Section I (Line A minus Line B) = 17.7 C ≈18

Free Response Questions. Section II

For questions 1, 2, 3, 5 and 6, divide the maximum number of points for each question outlined below by the number of sub-questions contained in that question. Award yourself a corresponding number of points for each correct response for each sub-question but no more than the maximum number of points listed for each.

Question 1: maximum of 10 points = _____ D

Question 2: maximum of 10 points = _____ E

Question 3: maximum of 10 points = _____ F

Question 5: maximum of 7.5 points = _____ G

Question 6: maximum of 7.5 points = _____ H

Score each of the three parts of Question 4 like this:

Award +0.33 points for correctly written reactants.

Award +0.67 points for correctly written products.

Award +0.33 points for correctly balancing the equation.

Award +0.33 points for answering the question correctly.

(No more than +1.67 points for each part,

maximum of 5 points) = _____ I

Total points (Add lines C through I): = _____ = total score

A minimum score of about 66 to 70 might correspond to a 5.

A minimum score of about 53 to 57 might correspond to a 4.

A minimum score of about 40 to 44 might correspond to a 3.

When you finish Practice Test 1 and have analyzed your results, corrected your mistakes, and reviewed your weak areas, repeat the process with Practice Test 2.

CHEMISTRY PRACTICE TEST 1

Multiple Choice Questions Section I

90 minutes

You may not use a calculator for Section I.

Part A

Directions: Each set of lettered responses refers to the numbered statements or questions immediately below it. Choose the one lettered response that best fits each statement or question. You may use a response once, more than once, or not at all.

Questions 1-3 refer to the following terms.

A) normal boiling point
B) boiling point
C) critical temperature
D) critical pressure
E) triple point

1. The temperature at which the vapor pressure of the liquid equals 760 torr.

2 The temperature at which the vapor pressure of the liquid equals the barometric pressure.

3. The temperature above which, no distinct liquid phase can form.

Questions 4-7 refer to the following aqueous solutions.

A) 0.1 M NaF
B) 0.1 M $CaCl_2$
C) 0.1 M CH_3CH_2OH
D) 0.1 M HCN
E) 0.2 M $C_6H_{12}O_6$

4. Which solution is the most acidic?

5. Which solution has the highest pH?

6. Which solution has the highest freezing point?

7. Which solution has the highest boiling point?

Questions 8-12 refer to the following aqueous solutions.

A) 0.1 M Na_3PO_4
B) 0.1 M $CaCl_2$
C) 0.1 M CH_3CH_2OH
D) 0.1 M HNO_3
E) 0.2 M NH_4Cl

8. Which solution will form a precipitate with nickel(II) nitrate solution?
9. Which solution will cause a solution of sodium hydrogen carbonate to effervesce strongly?
10. Which solution is basic?
11. Which solution will react with an equal molar amount of hydrochloric acid to form a weakly basic solution?
12. Which solution will form a water-soluble gas upon reaction with sodium hydroxide?

Questions 13-17 refer to the following elements.

A. Cl B. Ba C. F D. Ne E. B.

13. Has the largest first ionization energy.
14. Has the largest difference between the second and third ionization energies.
15. Has the largest electron affinity.
16. Has the smallest atomic radius.
17. Has the most oxidation states.

Questions 18-22 refer to the following sets of quantum numbers.

(A) 4, 0, 0, - 1/2
(B) 3, 2, -1, - 1/2
(C) 3, 1, 1, +1/2
(D) 2, 2, 0, +1/2
(E) 2, 0, 0, +1/2

18. Is an electron in a 3p orbital
19. Is an impossible set of quantum numbers.
20. Is the last electron to fill a transition metal.
21. Can be the last electron to fill a ground state calcium atom.
22. Can be an electron in a ground state neon atom.

Questions 23-25 refer to the following data table

The table lists the solubilities, in grams of solute per 100 grams of H_2O, of various salts at two different temperatures.

	Salt	20ºC	60ºC
I	$Ce_2(SO_4)_3 \cdot 9H_2O$	9.16	3.73
II	KNO_3	31.6	10.0
III	NaCl	36.0	37.3
IV	$K_2Cr_2O_7$	13.1	50.5

23. For which of these salts will the solution process have a ΔH closest to zero?

A) I
B) II
C) III
D) IV
E) I and II

24. Which of these salts dissolve(s) exothermically?

A) I
B) II
C) III
D) IV
E) II, III and IV

25. For which of these salts is entropy, but not enthalpy, the driving force for dissolution?

A) I
B) II
C) III
D) IV
E) II, III and IV

Part B

Directions: For each of the following questions or incomplete statements select the letter of the best answer or completion directly below it.

26. The energy absorbed when dry ice sublimes is required to overcome which type of interaction?

A) covalent bonds.
B) ion-dipole forces
C) dipole-dipole forces
D) dispersion forces
E) hydrogen bonds

27. A container is half filled with a liquid and sealed at room temperature and atmospheric pressure. What happens inside the container?

A) Evaporation stops.
B) Evaporation continues for a time then stops.
C) The pressure in the container remains constant.
D) The pressure inside the container increases for a time and then remains constant.
E) The liquid evaporates until it is all in the vapor phase.

28. Acetone, $(CH_3)_2C{=}O$, is a volatile, flammable liquid. The central carbon is sp^2 hybridized. The strongest intermolecular forces present in acetone are

A) dipole-dipole forces.
B) London dispersions.
C) hydrogen bonds.
D) covalent bonds.
E) ion-dipole forces

29. Heating Br_2O_7 causes it to decompose to its gaseous elements. What is the ratio of bromine to oxygen molecules in the product?

A) 1 to 7
B) 2 to 7
C) 7 to 2
D) 7 to 1
D) 1 to 14

30. Complete combustion of a compound containing only carbon, hydrogen and oxygen yields data that allows for elemental analysis. The analysis of the combustion products relies upon which of these assumption?

I. The quantity of carbon dioxide formed relates directly to the amount of carbon present in the sample.
II. The quantity of water formed relates directly to the amount of hydrogen present in the sample.
III. The quantity of water formed is limited by the amount of oxygen present in the sample.
IV. The quantity of carbon dioxide formed is limited by the amount of air present.

A) I only
B) II only
C) I, II and III only
D) I and II only
E) I, II, III and IV.

31. A sample of hydrated copper(II) sulfate, $CuSO_4{\cdot}xH_2O$ weighs 24.95 g. When the water is driven off, the anhydrous form weighs 15.95 g. What is the value of x in the formula of the hydrated salt?

A) 1
B) 2
C) 3
D) 4
E) 5

32. A compound whose empirical formula is C_2H_4O has a molar mass that lies between 100 and 150 g/mol. What is the molecular formula of the compound?

A) C_2H_4O
B) $C_4H_8O_2$
C) $C_6H_{12}O_3$
D) $C_6H_{12}O_2$
E) $C_6H_8O_3$

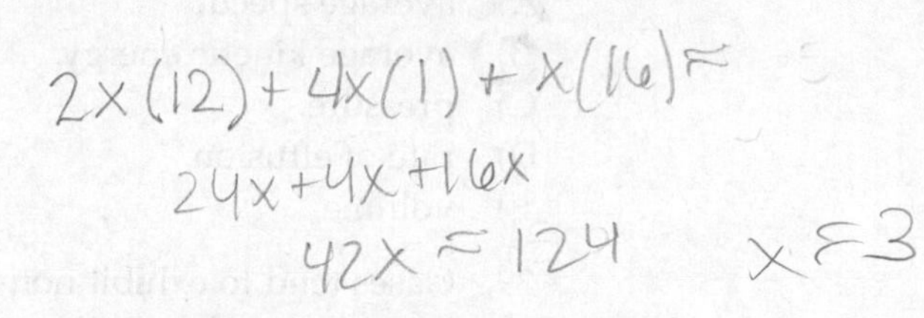

33. Atom Y has 3 valence electrons and atom Z has 6 valence electrons. What is the simplest formula expected for the binary ionic compound composed of Y and Z?

A) Y_2Z
B) YZ_2
C) Y_2Z_3
D) Y_3Z_2
E) YZ

34. Which contains the greatest mass of hydrogen?

A) 2.0 mol CH_4
B) 1.5 mol C_2H_4
C) 3.0 mol H_2O
D) 2.0 mol NH_3
E) 5.0 mol HCl

35. For the following reaction, ΔH = -400 kJ.

$$2K(s) + 2H_2O(l) \rightarrow 2KOH\ (aq) + H_2(g)$$

What is ΔH for this reaction?

$$KOH\ (aq) + 1/2\ H_2(g) \rightarrow K(s) + H_2O(l)$$

A) -400 kJ
B) +400 kJ
C) -200 kJ
D) +200 kJ
E) -100 kJ

36. A monatomic species that has 36 electrons and a net charge of 2- has

A) the same number of electrons as a neutral selenium atom.
B) 36 protons
C) more protons than electrons.
D) 2 unpaired electrons.
E) a noble gas configuration

37. The physical behavior of an ideal gas is dependent on all of the following except:

A) temperatue
B) volume
C) pressure
D) number of moles
E) chemical composition

38. The kinetic-molecular theory predicts that two different gases at the same temperature will have the same

A) average speed.
B) average kinetic energy.
C) pressure.
D) rate of effusion.
E) volume.

39. Gases tend to exhibit non-ideal behavior under conditions of

A) low temperature and low pressure.
B) low temperature and high pressure.
C) high temperature and high pressure.
D) high temperature and low pressure.
E) any temperatures above the critical temperature.

40. One liter of oxygen gas, O_2, and three liters of sulfur dioxide gas, SO_2, react to form gaseous sulfur trioxide, SO_3 at a given temperature and pressure. How many liters of $SO_3(g)$ can be produced at that same temperature and pressure?

A) 1
B) 2
C) 3
D) 4
E) 6

41. Which gas deviates the most from ideal behavior?

A) H_2
B) Ne
C) H_2O
D) CH_4
E) N_2

42. Based on effective nuclear charge, predict which element will have the smallest atomic radius.

A) H
B) He
C) Li
D) Be
E) B

43. Which element has the smallest first ionization energy?

A) Ca
B) Ga
C) Ge
D) As
E) Br

44. Which atom displays the greatest metallic character?

A) Be
B) Li

C) K
D Na
E) Mg

45. For which pair of atoms is the electronegativity difference the greatest?

A) Al, Si
B) Li, Br
C) Rb, Cl
D) Se, S
E) N, O

46. The neutral atom with the largest electronegativity is

A) Na
B) Al
C) P
D) Cl
E) S

47. Which one of the following compounds does not have a valid octet Lewis structure?

A) $SeCl_4$
B) CCl_4
C) SO_4^{2-}
D) PCl_3
E) NO_3^-

48. Which of the following diatomic molecules has the strongest bonds?

A) N_2
B) O_2
C) F_2
D) Cl_2
E) Br_2

49. The hybrid orbitals of carbon in CO_3^{2-} are

A) sp
B) sp^2
C) sp^3
D) dsp^3
E) d^2sp^3

50. Which molecules are polar?

I. H_2O II. CO_2 III. NO_2 IV. SO_2

A) I only
B) I and III only
C) I, II and III only
D) I, II, III and IV
E) I, III and IV only

51. The experimental data from the reaction A → products gives these three graphs. What is the most likely order for this reaction?

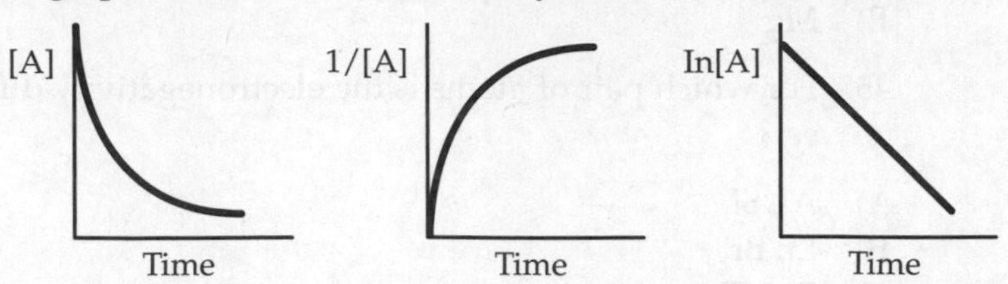

A) zero
B) first
C) second
D) third
E) not enough information to conclude

52. The rate of a chemical reaction between substances X and Y is found to follow the rate equation rate = k $[X][Y]^2$. If the concentration of Y is halved, what condition would result in keeping the reaction rate constant, assuming no temperature change?

A) If [X] remains constant.
B) If [X] is doubled.
C) If [X] is halved.
D) If [X] is tripled.
E) If [X] is quadrupled.

53. What can be correctly said about the energy diagram for a reversible endothermic reaction?

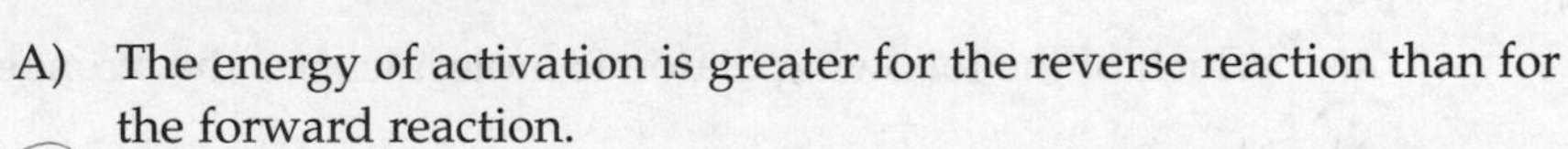

A) The energy of activation is greater for the reverse reaction than for the forward reaction.
B) The energy of activation is greater for the forward reaction than for the reverse reaction.

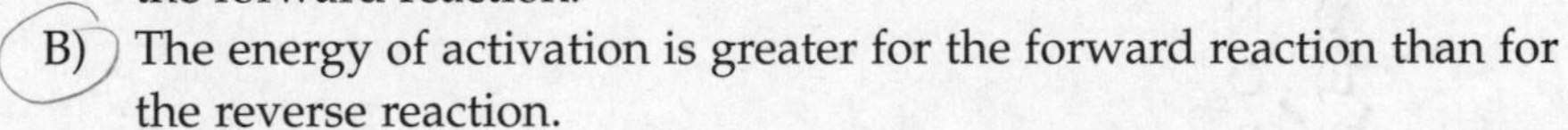

C) The energy of activation is the same for the reaction in both directions.
D) The enthalpy of reactants is greater than the enthalpy of products.
E) The forward reaction is faster than the reverse reaction.

54. Strontium-90 undergoes radioactive decay with a half-life of about 30 years. Approximately how many years will have elapsed before 97% of the ^{90}Sr in a sample will have decayed?

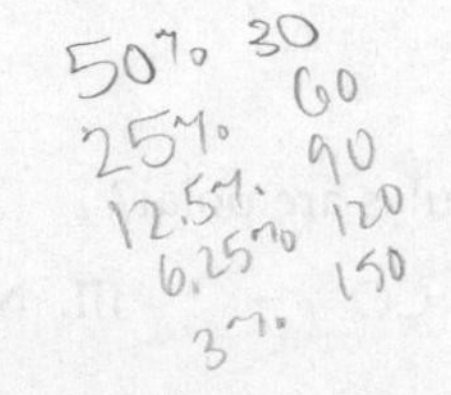

A) 30
B) 60
C) 120
D) 150
E) 180

55. When the following reaction in aqueous solution is balanced using whole number coefficients, what is the coefficient of H^+?

$$MnO_4^- + NO_2 \rightarrow MnO_2 + NO_3^-$$

A) 2
B) 4
C) 5
D) 6
E) 8

56. Which of these is the strongest oxidizing agent?

A) lithium
B) lithium ion
C) iodine
D) mercury (II) ion
E) bromine

57. What is the voltage of a cell consisting of the following half cells that are suitably connected?

$Cu^{2+}(aq) + 2e^- \rightarrow Cu(s)$ $E^\circ = 0.337$ volts

$Al^{3+}(aq) + 3e^- \rightarrow Al(s)$ $E^\circ = -1.66$ volts

A) 1.32 volts
B) 2.00 volts
C) 2.30 volts
D) -2.30 volts
E) 4.33 volts

58. Which of the following ions in aqueous solution is the strongest Lewis acid?

A) Ca^{2+}

B) NH_4^+
C) Na^+
D) Cu^{2+}
E) Fe^{3+}

59. Each of the following is amphoteric except

A) HSO_3^-
B) HPO_4^{2-}
C) NH_4^+
D) H_2O
E) HS^-

60. Which of the following species is in the greatest concentration in an aqueous 0.100 molar solution of H_3PO_4?

A) H_3O^+
B) H_3PO_4
C) $H_2PO_4^-$
D) HPO_4^{2-}
E) PO_4^{-3}

61. Which substance in aqueous solution is an electrolyte and is basic?

A) HCl
B) CH_3COOH
C) CH_3OH
D) KOH
E) NH_4NO_3

62. Which would you expect to be the weakest acid?
A) HClO
B) $HClO_2$
C) $HClO_3$
D) $HClO_4$
E) All have equal acid strength.

63. The solubility of which compound is pH dependent?
A) CaF_2
B) KNO_3
C) NaCl
D) CH_3OH
E) $LiClO_4$

64. 20 mL of 0.10 M solutions of each of the following acids are exactly neutralized with 20 mL of 0.10 M NaOH. Which of the resulting solutions has the highest pH?

A) $HC_2H_3O_2$ ($K_a = 1.8 \times 10^{-5}$)
B) HCN ($K_a = 4.9 \times 10^{-10}$)
C) HBrO ($K_a = 2.5 \times 10^{-9}$)
D) HIO ($K_a = 2.3 \times 10^{-11}$)
E) HN_3 ($K_a = 2.0 \times 10^{-5}$)

65. The Brønsted-Lowry theory of acids and bases would predict which of the following species would act as an acid?

A) NaH B) NH_4^+ C) Mg_3N_2 D) NH_2^- E) O_2^-

66. If the following system held at constant volume is at equilibrium

$PCl_3(g) + Cl_2(g) = PCl_5(g)$, $\Delta H = -92.6$ kJ,

the concentration of PCl_3 will decrease if

A) Cl_2 is removed from the system.
B) PCl_5 is removed from the system.
C) the temperature of the system is increased.
D) an inert gas is added to the system.
E) a catalyst is added

67. Which is the correct equilibrium constant expression for the reaction

$2\ NO_2(g) = N_2O_4(g)$?

A) $Kc = [NO_2]/[N_2O_4]$
B) $Kc = [N_2O_4]/[NO_2]$

C) $Kc = [N_2O_4]/[NO_2]^2$
D) $Kc = [N_2O_4]/2[NO_2]$
E) $Kc = [NO_2]^2/[N_2O_4]$

68. What is the value of the equilibrium constant for the following reaction when the equilibrium concentrations are:

$[N_2] = 2.0\ M \quad [H_2] = 2.0\ M \quad [NH_3] = 2.0\ M$

$N_2(g) + 3H_2(g) = 2\ NH_3(g)$

A) 0.25
B) 0.50
C) 0.75
D) 1.00
E) 2.00

69. Which change will cause the value of the equilibrium constant for the following reaction to decrease?

$X(g) + 2Y(g) = 2Z(g) \quad \Delta H = +120\ kJ$

A) adding a catalyst
B) decreasing the concentration of X
C) decreasing the concentration of Z
D) decreasing the temperature
E) decreasing the volume of the container

70. A mixture of 1.00 mole of $H_2(g)$ and 1.00 mole of $I_2(g)$ is placed in a 1.00 liter flask at a constant temperature and is allowed to come to equilibrium according to the equation

$$H_2(g) + I_2(g) = 2\ HI(g).$$

If the equilibrium constant at this temperature is Kc = 36.0, what is the molar concentration of $H_2(g)$ in the equilibrium mixture?

A) 0.500 M
B) 1.00 M
C) 0.750 M
D) 0.250 M
E) 0.125 M

71. What is the mole fraction of ethanol, CH_3CH_2OH, in a solution consisting of 92.0 g of ethanol, 116 g of acetone, $(CH_3)_2C{=}O$, and 108 g of water?

A) 0.10
B) 0.20
C) 0.33
D) 0.40
E) 0.60

72. What is the molar solubility of silver chloride ($Ksp = 1.8 \times 10^{-10}$) in 0.050 M sodium chloride?

A) $(1.8/0.050) \times 10^{-10}$ M

B) $(1.8/0.050)^{1/2} \times 10^{-5}$ M

C) $(0.050/1.8)^{1/2} \times 10^{-5}$ M

D) $(0.050/1.8) \times 10^{-10}$ M

E) $(1.8/0.050)^2 \times 10^{-5}$ M

73. What would be the expected result when 100 mL of a solution that is 0.0020 M $Ca(NO_3)_2$ and 0.0020 M $Pb(NO_3)_2$ is mixed with 100 mL of 0.002M Na_2SO_4? The Ksp of $CaSO_4$ is 2.4×10^{-5} and the Ksp of $PbSO_4$ is 6.3×10^{-7}.

A) Both $CaSO_4$ and $PbSO_4$ will precipitate.
B) Only $CaSO_4$ will precipitate.
C) Only $PbSO_4$ will precipitate.
D) Neither $CaSO_4$ nor PbSO4 will precipitate.
E) There is not enough information to tell.

74. For the isoelectronic series below, which species requires the least energy to remove an outer electron?

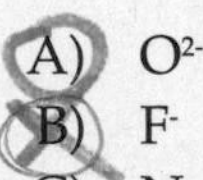

A) O^{2-}
B) F^-
C) Ne
D) Na^+
E) Mg^{2+}

75. Which species below does not violate the octet rule?

A) NO
B) BF_3
C) N_2O
D) NO_2
E) BrF_3

CHEMISTRY PRACTICE TEST 1

Part A

Section II

Equations and constants similar to what students can use on Section II, Part A, the first three questions of the free response section of the Advanced Placement Examination in Chemistry

Light and Atomic Structure

$E = h\upsilon$ — E = energy of a photon or quantum state

$\upsilon = c/\lambda$ — υ = frequency of light

$\lambda = h/mv$ — λ = wavelength of light, v = velocity of particle

$p = mv$ — p = momentum, m = mass

$E_n = (-2.178 \times 10^{-18})/n^2$ Joule — n = principal quantum number

c = speed of light = 3.0×10^8 m/s = 3.0×10^{17} nm/s

h = Planck's constant = 6.63×10^{-34} Js

k = Boltzmann's constant = 1.38×10^{-23} J/K

N = Avogadro's number = 6.02×10^{23} molecules/mol

e = charge on an electron = -1.602×10^{-19} coulomb

eV = electron volt = 96.5 kJ/mol

Equilibrium

$Ka = [H^+][A^-]/[HA]$ — Ka = weak acid ionization constant

$Kb = [OH^-][HB^+]/[B]$ — Kb = weak base ionization constant

$Kw = [H^+][OH^-] = 1.0 \times 10^{-14}$ at 25°C — Kw = autoionization constant for water

$Kw = Ka \times Kb$ — Kp = gas pressure equilibrium constant

$pH = -\log[H^+]$ — Kc = molar concentration constant

$pOH = -\log[OH^-]$ — Ksp = solubility product constant

$pH = pKa + \log([A^-]/[HA])$ — R = gas constant = 8.31 J/mol K

$pKa = -\log Ka$

$pKb = -\log Kb$

$Kp = Kc(RT)^{\Delta n}$

Δn = mol gaseous products – mol gaseous reactants

Thermodynamics

$\Delta H^\circ_{rxn} = \Sigma \Delta H^\circ_{products} - \Sigma \Delta H^\circ_{reactants}$ — H° = standard enthalpy

$\Delta G^\circ_{rxn} = \Sigma \Delta G^\circ_{products} - \Sigma \Delta G^\circ_{reactants}$ — G° = standard free energy

$\Delta S^\circ_{rxn} = \Sigma S^\circ_{products} - \Sigma S^\circ_{reactants}$ — S° = standard entropy

$\Delta G^\circ = \Delta H^\circ - T\Delta S^\circ$ — T = Kelvin temperature

$\Delta G^\circ = -RT \text{ Ln } K$ — K = equilibrium constant

$\Delta G^\circ = nFE^\circ$ — E° = standard reduction potential

$\Delta G = \Delta G^\circ + RT \ln Q$ — Q = reaction quotient

$C_p = \Delta H/\Delta T$ — C_p = molar heat capacity in J/K

$q = mc\Delta T$ — q = heat in J, m = mass in g

c = specific heat capacity in J/g K

F = Faraday's constant, the charge on a mole of electrons = 96,500 coul/mol e^-

Gases

$PV = nRT$ — P = pressure, V = volume

$(P + n^2a/V^2)(V - nb) = nRT$ — R = gas constant = 0.0821 L atm/mol K

$P_a = P_t \times X_a$ — P_a = partial pressure of gas a

$P_t = P_a + P_b + P_c + \ldots$ — P_t = total pressure

$K = {}^\circ C + 273$ — X_a = mole fraction of gas = mol a/total moles

$P_1V_1/T_1 = P_2V_2/T_2$ — T = Kelvin temperature

$u_{rms} = (3kT/m)^{1/2} = (3RT/m)^{1/2}$ — u_{rms} = root-mean-square speed

$KE_{molecule} = 1/2\ mv^2$ — KE = kinetic energy

$KE_{mole} = 3/2\ RT$ — v = velocity of a particle, m = mass of particle

$r_1/r_2 = (M_2/M_1)^{1/2}$ — r = rate of effusion, M = molar mass of a gas

Solutions

M = moles solute/liters of solution — M = molarity

m = moles solute/kg solvent — m = molality

$\Delta T_f = iK_f \times m$ — ΔT_f = change in freezing point

i = van't Hoff factor

$\Delta T_b = iK_b \times m$ — ΔT_b = change in boiling point

K_f = molal freezing point depression constant

K_b = molal boiling point elevation constant

$\pi = (nRT/V)\,i$ — π = osmotic pressure

Electrochemistry

For $aA + bB \rightarrow cC + dD$, $Q = ([C]^c [D]^d)/([A]^a [B]^b)$

Amperes = coulombs per second

$E_{cell} = E^\circ_{cell} - (RT/nF) \ln Q$ at 25°C — n = moles of electrons

$E_{cell} = E^\circ_{cell} - (0.0592/n) \ln Q$ at 25°C

$\log K = nE^\circ/0.0592$ — K = equilibrium constant

F = Faraday's constant = 96,500 coulombs per mole of electrons

	Main groups		Transition metals										Main groups					
	1A[a] 1	2A 2	3B 3	4B 4	5B 5	6B 6	7B 7	8B 8	8B 9	8B 10	1B 11	2B 12	3A 13	4A 14	5A 15	6A 16	7A 17	8A 18
1	1 **H** 1.00794																	2 **He** 4.002602
2	3 **Li** 6.941	4 **Be** 9.012182											5 **B** 10.811	6 **C** 12.0107	7 **N** 14.0067	8 **O** 15.9994	9 **F** 18.998403	10 **Ne** 20.1797
3	11 **Na** 22.989770	12 **Mg** 24.3050											13 **Al** 26.981538	14 **Si** 28.0855	15 **P** 30.973761	16 **S** 32.065	17 **Cl** 35.453	18 **Ar** 39.948
4	19 **K** 39.0983	20 **Ca** 40.078	21 **Sc** 44.955910	22 **Ti** 47.867	23 **V** 50.9415	24 **Cr** 51.9961	25 **Mn** 54.938049	26 **Fe** 55.845	27 **Co** 58.933200	28 **Ni** 58.6934	29 **Cu** 63.546	30 **Zn** 65.39	31 **Ga** 69.723	32 **Ge** 72.64	33 **As** 74.92160	34 **Se** 78.96	35 **Br** 79.904	36 **Kr** 83.80
5	37 **Rb** 85.4678	38 **Sr** 87.62	39 **Y** 88.90585	40 **Zr** 91.224	41 **Nb** 92.90638	42 **Mo** 95.94	43 **Tc** [98]	44 **Ru** 101.07	45 **Rh** 102.90550	46 **Pd** 106.42	47 **Ag** 107.8682	48 **Cd** 112.411	49 **In** 114.818	50 **Sn** 118.710	51 **Sb** 121.760	52 **Te** 127.60	53 **I** 126.90447	54 **Xe** 131.293
6	55 **Cs** 132.90545	56 **Ba** 137.327	71 **Lu** 174.967	72 **Hf** 178.49	73 **Ta** 180.9479	74 **W** 183.84	75 **Re** 186.207	76 **Os** 190.23	77 **Ir** 192.217	78 **Pt** 195.078	79 **Au** 196.96655	80 **Hg** 200.59	81 **Tl** 204.3833	82 **Pb** 207.2	83 **Bi** 208.98038	84 **Po** [208.98]	85 **At** [209.99]	86 **Rn** [222.02]
7	87 **Fr** [223.02]	88 **Ra** [226.03]	103 **Lr** [262.11]	104 **Rf** [261.11]	105 **Db** [262.11]	106 **Sg** [266.12]	107 **Bh** [264.12]	108 **Hs** [269.13]	109 **Mt** [268.14]	110 [271.15]	111 [272.15]	112 [277]	113 [284]	114 [289]	115 [288]	116 [292]		

*Lanthanide series	57 ***La** 138.9055	58 **Ce** 140.116	59 **Pr** 140.90765	60 **Nd** 144.24	61 **Pm** [145]	62 **Sm** 150.36	63 **Eu** 151.964	64 **Gd** 157.25	65 **Tb** 158.92534	66 **Dy** 162.50	67 **Ho** 164.93032	68 **Er** 167.259	69 **Tm** 168.93421	70 **Yb** 173.04
†Actinide series	89 **†Ac** [227.03]	90 **Th** 232.0381	91 **Pa** 231.03588	92 **U** 238.02891	93 **Np** [237.05]	94 **Pu** [244.06]	95 **Am** [243.06]	96 **Cm** [247.07]	97 **Bk** [247.07]	98 **Cf** [251.08]	99 **Es** [252.08]	100 **Fm** [257.10]	101 **Md** [258.10]	102 **No** [259.10]

[a] The labels on top (1A, 2A, etc.) are common American usage. The labels below these (1, 2, etc.) are those recommended by the International Union of Pure and Applied Chemistry.

The names and symbols for elements 110 and above have not yet been decided.

Atomic weights in brackets are the masses of the longest-lived or most important isotope of radioactive elements.

Further information is available at http://www.webelements.com

The production of element 116 was reported in May 1999 by scientists at Lawrence Berkeley National Laboratory.

Table of Standard Reduction Potentials in water at 25°C

A similar table is available for student use on Section II, the free response section, of the Advanced Placement Examination in Chemistry.

Half reaction			E° (volts)
$Li^{+}(aq) + 1e^{-}$	→	$Li(s)$	-3.05
$K^{+}(aq) + 1e^{-}$	→	$K(s)$	-2.92
$Ba^{2+}(aq) + 2e^{-}$	→	$Ba(s)$	-2.90
$Ca^{2+}(aq) + 2e^{-}$	→	$Ca(s)$	-2.87
$Na^{+}(aq) + 1e^{-}$	→	$Na(s)$	-2.71
$Mg^{2+}(aq) + 2e^{-}$	→	$Mg(s)$	-2.37
$Al^{3+}(aq) + 3e^{-}$	→	$Al(s)$	-1.66
$Mn^{2+}(aq) + 2e^{-}$	→	$Mn(s)$	-1.18
$Zn^{2+}(aq) + 2e^{-}$	→	$Zn(s)$	-0.76
$Cr^{3+}(aq) + 3e^{-}$	→	$Cr(s)$	-0.74
$Fe^{2+}(aq) + 2e^{-}$	→	$Fe(s)$	-0.44
$Cr^{3+}(aq) + 1e^{-}$	→	$Cr^{2+}(aq)$	-0.41
$Co^{2+}(aq) + 2e^{-}$	→	$Co(s)$	-0.28
$Ni^{2+}(aq) + 2e^{-}$	→	$Ni(s)$	-0.25
$Sn^{2+}(aq) + 2e^{-}$	→	$Sn(s)$	-0.14
$Pb^{2+}(aq) + 2e^{-}$	→	$Pb(s)$	-0.13
$2H^{+}(aq) + 2e^{-}$	→	$H_2(g)$	0.00
$S(s) + 2H^{+}(aq) + 2e^{-}$	→	$H_2S(g)$	+0.14
$Sn^{4+}(aq) + 2e^{-}$	→	$Sn^{2+}(aq)$	+0.15
$Cu^{2+}(aq) + 1e^{-}$	→	$Cu^{+}(aq)$	+0.15
$Cu^{2+}(aq) + 2e^{-}$	→	$Cu(s)$	+0.34
$I_2(s) + 2e^{-}$	→	$2I^{-}(aq)$	+0.53
$Ag^{+}(aq) + 1e^{-}$	→	$Ag(s)$	+0.80
$Br_2(l) + 2e^{-}$	→	$2Br^{-}(aq)$	+1.07
$O_2(g) + 4H+(ag) + 4e^{-}$	→	$2H_2O(l)$	+1.23
$Au^{3+}(aq) + 3e^{-}$	→	$Au(s)$	+1.50
$F_2(g) + 2e^{-}$	→	$2F^{-}(aq)$	+2.87

Notice that all half reactions are written as reductions. The reverse of each listed half reaction is an oxidation. Each oxidation half reaction has the same voltage as its corresponding reduction half reaction but of opposite sign.

To determine the spontaneous reaction for any cell made up of two half cells, reverse the half reaction with the less positive voltage and add it to the half reaction with the more positive voltage. To determine the cell voltage, change the sign of the voltage of the reversed half reaction and add it to the voltage of the other half reaction.

For example, the spontaneous reaction that occurs when a gold half cell is coupled with a nickel half cell is:

$2[Au^{3+}(aq) + 3e^{-} \rightarrow Au(s)]$ $E^{\circ}_{red} = +1.50$

$3[Ni(s) \rightarrow Ni^{2+}(aq) + 2e^{-}]$ $E^{\circ}_{ox} = -(-0.25)$

Cell reaction: $2Au^{3+}(aq) + Ni(s)^{-} \rightarrow Ni^{2+}(aq) + Au(s)$ $E^{\circ}_{cell} = +1.75$ volts

55 minutes

You may use a calculator for Part A.

Directions: Answer each of the following questions, clearly showing the methods you use and the steps involved at arriving at the answers. Partial credit will be given for work shown and little or no credit will be given for not showing your work, even if the answers are correct.

Question 1

a. The acid ionization constants for the triprotic acid phosphoric acid are $Ka_1 = 7.5 \times 10^{-3}$, $Ka_2 = 6.2 \times 10^{-8}$ and $Ka_3 = 4.2 \times 10^{-13}$.

i. Write three ionic equations, one each for the successive ionizations of the three protons of phosphoric acid. Clearly indicate which equilibrium constant corresponds to each ionic equation.
ii. Calculate the number of grams of sodium dihydrogen phosphate dihydrate needed to prepare 1.00 liter of a 0.250 molar solution.
iii. Write a net ionic equation for the hydrolysis of aqueous sodium phosphate and calculate the value of the corresponding equilibrium constant for the reaction.

b. The acid ionization constant, Ka for acetic acid is 1.8×10^{-5} Exactly 41.00 grams of sodium acetate is added to water to make 1.00 liter of solution. Calculate:

i. the pH of the solution.
ii. the percent ionization of the acetate ion in the solution.

c. Calculate the pH of a solution resulting from mixing 200.0 mL of the solution prepared in Part b with 200.0 mL of 0.200 M hydrochloric acid.

d. Just enough 0.200 M hydrochloric acid is added to 50.0 mL of the original solution in Part b to convert all the sodium acetate to acetic acid. Calculate:

i. the number of milliliters of 0.200 M hydrochloric acid used.
ii. the pH of the final solution.

Question 2.

Mercury(II) chloride reacts with oxalate ion according to the following equation:

$$2HgCl_2(aq) + C_2O_4^{2-}(aq) \rightarrow 2Cl^-(aq) + 2CO_2(g) + Hg_2Cl_2(s)$$

The initial rate of the reaction was determined for several concentrations of the reactants and the following rate data were obtained for the appearance of chloride ion:

Experiment	$[HgCl_2]$ (mol L^{-1})	$[C_2O_4^{2-}]$ (mol L^{-1})	Rate (mol $L^{-1}s^{-1}$)
1	0.144	0.132	5.63×10^{-5}
2	0.144	0.396	5.10×10^{-4}
3	0.072	0.396	2.51×10^{-4}
4	0.288	0.132	1.13×10^{-4}

a. Write the rate law for this reaction.
b. Calculate the rate constant and specify its units.
c. What is the reaction rate when the concentration of both reactants are 0.150 M?
d. What is the rate of disappearance of oxalate ion when $[C_2O_4^{2-}] = 0.10$ M and $[HgCl_2] = 0.20$ M?
e. Is the overall equation likely to be an elementary step? Explain.
f. What is the oxidizing agent and what is the reducing agent in this reaction?

Question 3

The empirical and molecular formulas of a hydrocarbon are determined by combustion analysis.

a. Combustion of a 1.214 g sample of a hydrocarbon results in 4.059 g of carbon dioxide and 0.9494 g of water.

   i. How many moles of carbon are contained in the sample?
   ii. How many moles of H are contained in the sample?
   iii. What is the empirical formula of the hydrocarbon?

b. In a separate experiment 0.4232 g of the hydrocarbon was dissolved in 23.00 g of carbon tetrachloride. The freezing point of the solution was determined to be -25.28 °C. The normal freezing point of carbon tetrachloride is -22.30 °C and the freezing point depression constant, K_f, for carbon tetrachloride is 29.8 °C/m.

   i. What is the molality, m, of the hydrocarbon in the solution?
   ii. What is the molar mass of the hydrocarbon?
   iii. What is the molecular formula of the hydrocarbon?
   iv. Write and balance a chemical equation for the complete combustion of the hydrocarbon.

c. In a third experiment, the hydrocarbon and carbon tetrachloride were each found not to dissolve in water.

   i. Draw the Lewis structure for carbon tetrachloride.
   ii. What is the molecular geometry of carbon tetrachloride and what are the bond angles in the molecule?
   iii. Does the molecule have any polar bonds? Explain. Is the molecule polar? Explain.
   iv. Explain the fact that carbon tetrachloride does not dissolve in water.
   v. What principal intermolecular force(s) is(are) acting in the carbon tetrachloride solution of the hydrocarbon?

CHEMISTRY PRACTICE TEST 1

Section II **Part B**

40 minutes

You may *not* use a calculator for Part B.

Question 4

Write and balance net ionic equations for each of the following laboratory situations. You can assume a reaction happens in each case.

a. A solution of sodium hydroxide is added to a solution of copper(II) sulfate.

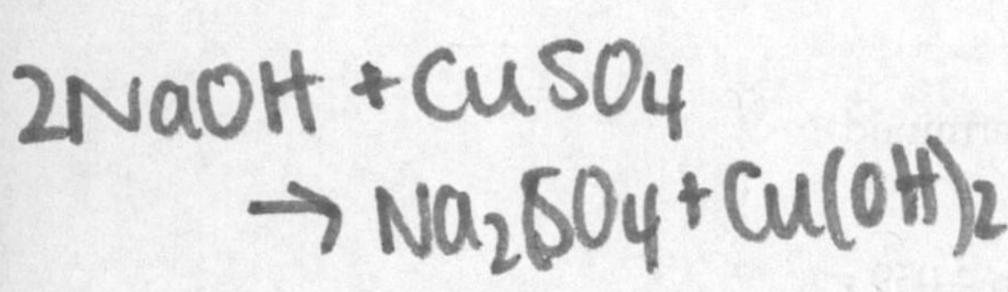

i. Write and balance an equation.

ii. What would you observe if an excess amount of ammonia solution is added to the mixture? Illustrate your answer with a balanced equation.

b. Carbon dioxide gas is bubbled into a solution of potassium hydroxide.

i. Write and balance an equation.

ii. What would you observe if excess nitric acid solution is added to the mixture? Illustrate your answer with a balanced equation.

c. Excess sulfuric acid solution is added to solid sodium sulfite.

i. Write and balance an equation.

ii. What would you observe if a small amount of potassium permanganate solution is added to a slightly acidic sodium sulfate solution? Illustrate your answer with a balanced equation.

Question 5

Six unknown solutions, each having a concentration of 0.5 M, are mixed in all possible binary combinations and the results are cataloged in the following table:

Formula		6	5	4	3	2	1
	1	No visible reaction	White precipitate	White precipitate	Bubbles	No visible reaction	1
	2	No visible reaction	No visible precipitate	No visible precipitate	Bubbles	2	
	3	No visible reaction	White precipitate	White precipitate	3		
	4	Light green reaction	No visible reaction	4			
	5	Bright yellow precipitate	5				
	6	6					

The unknowns are lead(II) nitrate, nitric acid, silver nitrate, sodium carbonate, potassium iodide and hydrochloric acid.

a. Identify each unknown by formula and number and write its formula in the table corresponding to its number.
b. For each visible reaction recorded, write and balance a net ionic equation.

Question 6

Consider the properties of the elements potassium and chlorine.

a. Which element is more likely to have the higher first ionization energy? Explain using electron configurations of each atom and any other relevant parameters.

b. The first ionization energy of potassium is +419 kJ/mol. The electron affinity of chlorine is -349 kJ/mol.

i. Define first ionization energy and write a thermochemical equation that represents the first ionization energy of potassium.
ii. Define electron affinity and write a thermochemical equation that represents the electron affinity of chlorine.
iii. Write the sum the reactions you wrote in Parts i and ii and calculate the heat of the reaction. Specify if the reaction is endothermic or exothermic.

c. The reaction between potassium metal and chlorine gas to produce solid potassium chloride is highly exothermic (ΔH = -435 kJ/mol.).

i. Write the formation reaction of potassium chloride.

ii. Besides the ionization of potassium, what forces must be overcome in the formation of potassium chloride?

iii. Explain why this reaction releases so much energy.

d. Predict whether the formation of calcium chloride would be more or less exothermic than the formation of potassium chloride. Explain the basis of your prediction.

Multiple Choice Answers for Practice Test I

1. A. The normal boiling point is always at 1 atmosphere (760 torr).
2. B. The boiling point can be at any temperature and pressure at or below the critical temperature.
3. C. The critical temperature is the temperature above which the kinetic energy of the molecules is so great that no pressure will bring about a liquid phase. The critical pressure is the pressure that is required to bring about liquefaction at the critical temperature.
4. D. HCN is a weak acid and the only acid listed.
5. A. Fluoride ion is a weak base and the only base listed.
6. C. Ethanol is a nonelectrolyte and will not ionize in aqueous solution.
7. B. Calcium chloride dissociates into three moles of ions per mole of solute and will increase the boiling point of the solution the most.
8. A. Phosphates generally form precipitates:
   $3Ni^{2+}(aq) + 2PO_4^{3-}(aq) \rightarrow Ni_3(PO_4)_2(s)$
9. D. Acids react with carbonates to form carbon dioxide gas and water:

   $H^+(aq) + HCO_3^{2-}(aq) \rightarrow CO_2(g) + H_2O(l)$
10. A. In general, anions are basic except anions of strong acids. Phosphate is a base:

    $PO_4^{3-}(aq) + H_2O(l) \rightarrow HPO_4^{2-}(aq) + OH^-(aq)$
11. A. Phosphate ion reacts with acid to form hydrogen phosphate, a base.

    $PO_4^{3-}(aq) + H^+(aq) \rightarrow HPO_4^{2-}(aq)$

(Note: Hydrogen phosphate ion is amphoteric and without more information, it might be an acid in aqueous solution. However, sodium phosphate is the only substance listed that has a possibility to be basic when acid is added.

12. E. Ammonium ion reacts with hydroxide ion to form ammonia, a water-soluble gas and water:

    $NH_4^+(aq) + OH^-(aq) \rightarrow NH_3(aq) + H_2O(l)$

13. D. In general, first ionization energy increases from left to right along a period and from bottom to top within a group.

14. B. Group 2 atoms have relatively low first and second ionization energies because their two valence electrons are well screened from the nuclear charge. However, their third ionization energies are relatively high because the third electron comes from a poorly screened inner noble gas core.

15. A. In general, electron affinity increases from left to right along a period and from bottom to top within a group. However, chlorine has a higher electron affinity than fluorine because fluorine's small size contributes to considerable electron-electron repulsion.

16. D. The noble gases are the smallest atoms in their respective periods because their valence electrons experience the highest effective nuclear charges.

17. A. Chlorine displays oxidation states of 1-, 0, 1+, 3+, 5+ and 7+.

18. C. The quantum numbers (3, 1, 1, 1/2) are the values of n, l, m_l and m_s, respectively. Quantum numbers n = 3 and l = 1 describe a 3p sublevel.

19. D. The value of quantum number l is limited to values of 0, 1, 2 … n-1. When n = 2, l must be 0 or 1 but never 2.

20. B. The last electrons to fill transition metals go into d sublevels so the value of l is 2. The sublevels s, p, d and f have l values of 0, 1, 2, and 3, respectively.

21. A. Calcium's electron configuration is $[Ar]4s^2$. The last electron fills the 4s sublevel which has quantum numbers n = 4 and l = 0.

22. E. Neon's electron configuration is $[He]2s^22p^6$. Any n value higher than 2 will not represent a neon electron.

23. C. The change in solubility of NaCl with temperature is minimal compared to the others indicating that solid NaCl dissolves in water with little or no heat change.

24. A. As temperature increases, the solubility of cerium sulfate decreases indicating that cerium sulfate dissolves exothermically. Heating an exothermic process will cause the reaction to shift toward reactants, in this case decreasing solubility.

25. E. The two driving forces that influence chemical and physical change are a decrease in enthalpy and an increase in entropy. Most salts dissolve endothermically but the accompanying increase in entropy (randomness of the system) overcomes the increase in enthalpy associated with an endothermic reaction. All salts which dissolve endothermically fit into this category.

26. D. Carbon dioxide is a nonpolar molecule, held together only by London dispersion forces. Sublimation is the process where a solid changes directly into a gas. Energy is required to overcome the

forces of attraction that hold the molecules together in the solid phase.

27. D. The liquid will continue to evaporate without stopping. The pressure will rise until the partial pressure of the vapor equals the vapor pressure of the liquid at the given temperature. Then a dynamic equilibrium is established where the rate of evaporation equals the rate of condensation.

28. A. The C=O bond in trigonal planar acetone is very polar making the molecule a strong dipole. The absence of H-O bonds rules out hydrogen bonding but strong dipole-dipole interactions exist.

29. B. The balanced equation is: $2Br_2O_7(g) \rightarrow 2Br_2(g) + 7O_2(g)$.

30. D. The only products obtained from the complete combustion are carbon dioxide and water. The assumptions are that all of the carbon in the sample is converted to carbon dioxide and all the hydrogen in the sample is converted to water. The amounts of carbon and hydrogen can be determined by measuring the masses of carbon dioxide and water produced.

31. E. The molar mass of anhydrous $CuSO_4$ is 159.5 g/mol. 15.95 g is 0.100 mol $CuSO_4$. 9.00 grams of water is 0.5 mol water or five times as many moles of water as copper(II) sulfate. The formula of the hydrate is $CuSO_4{\cdot}5H_2O$.

32. C. The molecular formula of a compound is an integer multiple of the empirical formula. The molecular formula, $C_6H_{12}O_3$, is three times the given empirical formula and has a molar mass of 132 g/mol.

33. C. Atom Y loses three valence electrons to from a Y^{3+} ion. Atom Z gains two electrons to form a Z^{2-} ion. The two ions combine to form Y_2Z_3.

34. A. To obtain the number of moles of hydrogen, multiply the number of moles of compound by the subscript of hydrogen. 2.0 mol x 4 mole H/ molecule CH_4 = 8.0 mol H.

35. D. The reaction in question is the reverse of the given reaction and balanced with half the coefficients. The change in enthalpy of the reaction in question is half the value of the given reaction with a change in sign from negative to positive.

36. E. Krypton has 36 electrons and 36 protons. The species having 36 electrons with a 2- charge is Se^{2-} which has 34 protons.

37. E. Ideal gases are assumed to be composed of infinitesimally small particles having no attractive forces. Therefore the chemical composition will have no affect on their behavior.

38. B. Temperature is a measure of average kinetic energy. Two gases at the same temperature will have the same kinetic energy.

39. B. Non-ideal gases are real gases with significant attractive forces between particles and significant particle volumes. Low temperatures slow the particles sufficiently so that the attractive forces become more dominant. High pressures crowd the particles so that their volumes become significant.
40. B. The balanced equation is: $O_2(g) + 2SO_2(g) \rightarrow 2SO_3(g)$. Oxygen is the limiting reactant so twice as many moles of SO_3 can be formed from the available moles of oxygen. At the same conditions of temperature and pressure, liters are proportional to moles.

$$x \text{ L } SO_3 = 1 \text{ L } O_2(2 \text{ L } SO_3/1 \text{ L } O_2) = 2 \text{ L}.$$

41. C. Ideal gases have insignificant volumes and no attractive forces. Deviations from ideal behavior become greater for larger molecules with high attractive forces. Water is a polar molecule with very large dipole-dipole interactions and hydrogen bonds.
42. B. Effective nuclear charge increases from left to right along any period. Neither electron is screened from the charge of the two protons on He making helium's effective nuclear charge double that of hydrogen. Consequently, helium is smaller than hydrogen.
43. B. In general, first ionization energy increases from left to right along any period because the effective nuclear charge increases. However, gallium's 3p electron is screened by the 4s level and the 3d level giving gallium a slightly smaller effective nuclear charge than calcium.
44. C. In general, metallic character increases right to left along a period and top to bottom within a group.
45. C. In general, electronegativity increases from the lower left to the upper right of the periodic table. Elements that are farthest apart along a diagonal from lower left to upper right probably have the largest difference in electronegativity. Li and Br also have a large difference but Rb has a lower electronegativity than does Li and chlorine's electronegativity is larger than that of bromine.
46. D. In general, electronegativity increases from left to right along a period.
47. A. Se has six valance electrons and forms four Se-Cl bonds. $SeCl_4$ will have five electron domains around Se. Valid octet structures can be written for all the others.
48. A. Lewis structures show that the nitrogen molecule has a triple bond, the oxygen molecule a double bond and all the rest have single bonds. In general, triple bonds are stronger and shorter than double bonds which are stronger and shorter than single bonds.
49. B. Carbonate is a planar ion which is associated with sp^2 hybridization.

50. E. Carbon dioxide has two polar bonds but its linear geometry causes the dipoles to cancel. All the others have polar bonds and have non-linear geometry.

51. B. A linear plot of ln[A] vs. time is characteristic of a simple first order process. A simple second order process would yield a straight line for 1/[A] vs. time.

52. E. If just [Y] is halved, then the initial rate would decrease by a factor of $(1/2)^2 = 1/4$. To make the rate remain constant, [X] is quadrupled because $(4)^1 = 4$.

53. B. An endothermic reaction has a higher activation energy for the forward reaction than for the reverse reaction, and the rate of reaction decreases with increasing activation energy. The enthalpy of products for an endothermic reaction is higher than the enthalpy of reactants.

54. D. If 97% is decayed, 3% of the original sample remains. 3% represents the amount left after about five half-lives: 100% X 1/2 X 1/2 X 1/2 X 1/2 X 1/2 = 3.1%. Five half-lives = 5 X 30 years = 150 years.

55. A. The balanced equation is:
$H_2O + MnO_4^- + 3NO_2^- \rightarrow MnO_2 + 3NO_3^- + 2H^+$

56. E. A strong oxidizing agent has a large positive standard reduction potential, E°. From the given standard reduction potentials, bromine has a standard reduction potential of +1.07 volts. Lithium is a good reducing agent because its oxidation potential is +3.05 volts.

57. B. Reduction of copper(II) ions is more favorable than the reduction of aluminum ions because the voltage of the copper half reaction is more positive. Therefore the aluminum reaction will be the oxidation and the sign its voltabe will change from – to +.

$$Cu^{2+}(aq) + 2e^- \rightarrow Cu(s) \quad E^\circ_{red} = 0.337 \text{ volts}$$

$$Al(s) \rightarrow Al^{3+}(aq) + 3e^- \quad E^\circ_{ox} = +1.66 \text{ volts}$$

$$E^\circ_{cell} = E^\circ_{ox} + E^\circ_{red} = 0.337 + 1.66 = 2.00 \text{ volts}$$

58. E. Except for cations of strong acids, cations are acidic. The higher the positive charge, the stronger the acid.

59. C. In general, protonated anions and water are amphoteric. Amphoteric substances can both donate and accept a proton. Protonated anions have protons to donate and negative charges that will attract them. NH_4^+ can donate a proton, but because like charges repel, it's unlikely to accept a positively charged proton.

60. B. Phosphoric acid is a weak poly-protic acid, which ionizes only slightly. While all five of the listed species exist in solution, H_3PO_4 exists to the greatest extent.

61. D. KOH is a strong base, which dissociates completely in aqueous solution. HCl is a strong acid, CH_3COOH is a weak acid and weak

electrolyte, NH_4NO_3 is a weak acid and strong electrolyte and CH_3OH is a neutral non-electrolyte.

62. A. In general, for similar formulas having the same central atom, acid strength increases with increasing oxidation number of the central atom. The oxidation number of chlorine in HClO is +1, the lowest of the compounds listed.

63. A. Fluoride ion is the conjugate base of the weak acid, HF. The solubility of calcium fluoride will increase as pH decreases because at low pH the available protons will react with the basic fluoride ion.

    $H^+(aq) + F^-(aq) = HF(aq)$

    Methanol is completely miscible in water and all the other choices are ionic compounds containing only neutral cations and anions.

64. D. The weakest acid has the strongest conjugate base. (Kw = Ka x Kb) At the equivalence point, all that's left is weak base.

65. B. All are bases except NH_4^+. It is unlikely that the already positively charged ammonium ion will accept a proton.

66. B. Removing $PCl_5(g)$ will slow the reverse reaction causing the forward reaction to be faster than the reverse reaction. The equilibrium will shift right and consume PCl_3. Adding a catalyst or an inert gas will not affect the position of the equilibrium. The other responses will cause the amount of PCl_3 to increase.

67. C. The equilibrium constant expression equals the ratio of the concentrations of products to reactants, each raised to the power of the coefficient that balances the equation.

68. A. $Kc = [NH_3]^2/[N_2][H_2]^3 = (2.0)^2/(2.0)(2.0)^3 = 0.25$

    (Remember that no calculators are allowed on the multiple choice section so the arithmetic required for complex quantitative problems is relatively simple.)

69. D. Temperature is the only parameter that will change the value of the equilibrium constant. Decreasing the temperature will favor the exothermic or, in this case, the reverse reaction. The value of the equilibrium constant will decrease because the reaction shifts left reducing the amount of Z and reducing Kc. $Kc = [Z]^2/[X][Y]^2$

70. C. An "ICE" table would look like this:

	$H_2(g)$ +	$I_2(g)$ ⇆	$2HI(g)$
I	1.00 M	1.00 M	0
C	-x	-x	+2x
E	1.00-x	1.00-x	2x

$Kc = 36 = (2x)^2/(1.00\text{-}x)(1.00\text{-}x)$

If we take the square root of both sides of the equation we obtain:

$6 = 2x/(1.00\text{-}x)$ or $6 - 6x = 2x$, $x = 0.750 = [H_2]$

71. B. Mole fraction of ethanol, X, = moles of ethanol/total moles of solution components.

    Moles of ethanol = 92.0 g/46.0 g/mol = 2.00 mol ethanol.

    Moles of water = 108 g/18.0 g/mol = 6.00 mol water.

    Moles of acetone = 116 g/58.0 g/mol = 2.00 mol acetone.

    Mole fraction of ethanol = 2.00 mol ethanol/ 10.0 total moles = 0.20.

72. A. $AgCl(s) = Ag^+(aq) + Cl^-(aq)$

    $Ksp = 1.8 \times 10^{-10} = [Ag^+][Cl^-]$

    Let x = molar solubility of AgCl.

    At equilibrium, $[Ag^+] = x$, $[Cl^-] = 0.050 + x \approx 0.050$ M

    $1.8 \times 10^{-10} = x\,(0.050)$

    $x = (1.8/0.050) \times 10^{-10}$

73. C. The concentrations of the ions are not sufficient to form calcium sulfate but are sufficient to form lead(II) sulfate.

    $[Pb^{2+}] = 0.0020$ M (100/200) = 0.0010 M, $[Ca^{2+}] = 0.0020$ M (100/200) = 0.0010 M,

    $[SO_4^{2-}] = 0.002$ M (100/200) = 0.0010 M

    $PbSO_4(s) = Pb^{2+}(aq) + SO_4^{2-}(aq)$

    $Q = [Pb^{2+}][SO_4^{2-}] = (0.0010)(0.0010) = 1 \times 10^{-6}$

    $Ksp < Q$ so $PbSO_4(s)$ will form.

    $CaSO_4(s) = Ca^{2+}(aq) + SO_4^{2-}(aq)$

    $Q = [Ca^{2+}][SO_4^{2-}] = (0.0010)(0.0010) = 1 \times 10^{-6}$

    $Ksp > Q$ so $CaSO_4(s)$ will not form.

74. A. Oxide ion has just 8 protons in its nucleus, two of which are screened by its two 1s electrons. Magnesium ion has 12 protons in it nucleus and only two screening 1s electrons.

75. C. A valid octet structure can be written for N_2O.

Free Response Answers for Practice Test I

Answers to Question 1

1a. i $H_3PO_4(aq) + H_2O(l) \leftrightharpoons H_2PO_4^-(aq) + H_3O^+(aq)$ $Ka_1 = 7.5 \times 10^{-3}$
$H_2PO_4^-(aq) + H_2O(l) \leftrightharpoons HPO_4^{2-}(aq) + H_3O^+(aq)$ $Ka_2 = 6.2 \times 10^{-8}$
$HPO_4^{2-}(aq) + H_2O(l) \leftrightharpoons PO_4^{3-}(aq) + H_3O^+(aq)$ $Ka_3 = 4.2 \times 10^{-13}$
ii. x g $NaH_2PO_4 \bullet 2H_2O$ = 1.00 L (0.250 mol/L)(156.0 g/mol) = 39.0 g.
iii. $PO_4^{3-}(aq) + H_2O(l) \leftrightharpoons HPO_4^{2-}(aq) + OH^-(aq)$

b. i. $CH_3COO^-(aq) + H_2O(l) \leftrightharpoons CH_3COOH(aq) + OH^-(aq)$
The initial concentration of acetate is: (41.00 g)/(82.0 g/mol)/(1.00 L) = 0.500 M.
Let y = $[OH^-]$.
Ka = $1.8 \times 10^{-5} = y^2/I = y^2/0.500$ M
y = $[OH^-]$ = 0.00300 M
pOH = -log $[OH^-]$ = -log (0.00300) = 2.523
pH = 14 – pOH = 14 – 2.323 = 11.477
ii. percent ionization = 100 x $[OH^-]$/ I = 100 x (0.00300)/(0.500) = 0.600 %

c. $H^+(aq) + CH_3COO^-(aq) \rightarrow CH_3COOH(aq)$
mol CH_3COO^- = 0.200 L x 0.500 M = 0.100 mol CH_3COO^-
mol H^+ = 0.200 L x 0.200 M = 0.0400 mol H^+
All 0.0400 mol of H^+ will react with 0.0400 mol of CH_3COO^- leaving 0.100 mol – 0.0400 mol = 0.060 mol CH_3COO^- and 0.040 mol CH_3COOH.
pH = pKa + log $[CH_3COO^-]/[CH_3COOH]$
pH = -log Ka + log(0.060/0.040) = 4.74 + 0.18 = 4.92

d. i. x mL HCl = (50.0 mL CH_3COO^-)(0.500 mmol/mL)(1 mmol HCl/1 mmol CH_3COO^-)(1 mL/0.200 mmol) = 125 mL
ii. All the acetate ion is converted to acetic acid and the total volume is 50.0 mL plus 125 mL = 175 mL.
$[CH_3COO^-]$ = I = (0.50 L x 0.500 M)/0.175 L = 0.143 M
Kb = Kw/Ka = $y^2/I = 1.0 \times 10^{-14}/1.8 \times 10^{-5} = y^2/0.143$
y = $[OH^-] = 8.9 \times 10^{-6}$ M
pOH = -log[OH-] = -log(8.9×10^{-6}) = 5.05

pH = 14.00 – pOH = 14 – 5.05 = 8.95

Answers to Question 2

a. rate = $k[HgCl_2]^a[C_2O_4^{2-}]^b$
Experiments 1 and 2 show that when $[C_2O_4^{2-}]$ is tripled, the rate goes up by a factor of nine: $5.10 \times 10^{-4}/5.63 \times 10^{-5} = 9.06$ so the value of the exponent b = 2 ($3^2 = 9$).

Experiments show that when $[HgCl_2]$ is doubled, the rate increases by a factor of two: $1.13 \times 10^{-4}/5.63 \times 10^{-5} = 2.01$ so the exponent a = 1 ($2^1 = 2$).

rate = $k[HgCl_2]^1[C_2O_4^{2-}]^2$

b. rate = $k[HgCl_2]^1[C_2O_4^{2-}]^2$ Use the values of any experiment to obtain k.

5.63×10^{-5} M/s = k(0.144 M)(0.132 M)^2

k = 0.0224 1/M^2 s

c. rate = $k[HgCl_2]^1[C_2O_4^{2-}]^2$

rate = (0.0224 1/M^{2}s)(0.150 M)(0.150 M)^2 = 7.56×10^{-5} M/s.

d. rate = $k[HgCl_2]^1[C_2O_4^{2-}]^2$

rate = (0.0224 1/M^{2}s)(0.20 M)(0.10 M)^2 = 4.5×10^{-5} M/s.

From the coefficients of the balanced equation, the rate of disappearance of oxalate is half the rate of appearance of chloride ion: rate = 1/2 (4.5×10^{-5} M/s) = 2.3×10^{-5} M/s.

e. The overall equation is not likely to be an elementary step because the rate law is third order. A elementary step would require a three-particle collision which is very rare. More likely, the reaction proceedes via a multi-step mechanism that includes only second order collisions.

f. The oxidizing agent is $HgCl_2$ and the reducing agent is $C_2O_4^{2-}$.

Answers to Question 3

a. i. x mol C = mol CO_2 = (4.059 mol)/(44.0 g/mol) = 0.0923 mol C.

ii. x mol H = 2 x mol H2O =2 x (0.9494 g)/(18.0 g/mol) = 0.1055 mol H.

iii. **C**0.0923**H**0.1055 = **C**(0.0923)/(0.0923)**H**(0.1055)/(0.0923) = **C**1**H**1.143 = C7H8

b. i. What is the molality, m, of the hydrocarbon in the solution?

i. $\Delta T_f = mK_f$

m = ΔT_f /mK_f = [(-22.30) – (-25.28)°C]/29.8 °C = 0.100 m

ii. x mol hydrocarbon = (0.100 mol hydrocarbon/kg carbon tetrachloride)x 0.02300 kg carbon tetrachloride = 0.002300 mol.

x g/mol = 0.4232 g / 0.002300 mol = 184.0 g/mol.

iii 184.0 g/mol = 2 x 92 g/mol for C_7H_8.

2 x C_7H_8 = $C_{14}H_{16}$

iv. $C_{14}H_{16} + 18O_2(g) \rightarrow 14CO_2(g) + 8H_2O(l)$

c i.

```
       ..
      :C1:
       |
  ..        ..
 :C1—C—C1:
  ..        ..
       |
      :C1:
       ..
```

ii. The molecular geometry is tetrahedral and the bond angels are all 109.5°.

iii. All four C-Cl bonds are polar due to the large difference in electronegativity of carbon and chlorine. The molecule is non-polar because the four C-Cl dipoles are oriented in a way that they cancel each other giving no net dipole.

iv Water is polar and effectively hydrogen bonds to other water molecules excluding any carbon tetrachloride from strong intermolecular attractions to water.

v. The principle forces acting in the carbon tetrachloride-hydrocarbon solution are London dispersions.

Answers to Question 4

a. i. $Cu^{2+}(aq) + 2OH^-(aq) \rightarrow Cu(OH)_2(s)$

ii. An excess of ammonia would cause the light blue precipitate to dissolve and the resulting solution would be dark blue due to the color of the tetramine copper(II) complex ion.

$Cu(OH)_2(s) + 4NH_3(aq) \rightarrow Cu(NH_3)_4^{2+} + 2OH^-(aq)$

b.

i. $CO_2(g) + OH^-(aq) \rightarrow HCO_3^-(aq)$

ii. Bubbles would form in the solution upon addition of nitric acid.

$H^+(aq) + HCO_3^-(aq) \rightarrow H_2O(l) + CO_2(g)$

c.

i. $2H^+(aq) + Na_2SO_3(s) \rightarrow H_2O(l) + SO_2(g) + 2\ Na^+(aq)$

ii. Purple potassium permanganate solution added to an acidic solution of sulfite ions would turn colorless because the permanganate ion is reduced to manganese(II) ions.

$6H^+(aq) + 2MnO_4^-(aq) + 5SO_3^{2-}(aq) \rightarrow 2Mn^{2+}(aq) + 5SO_4^{2-}(aq) + 3H_2O(l)$

Answers to Question 5

a.

Formula	
HCl	1
HNO_3	2
Na_2CO_3	3
$AgNO_3$	4
$Pb(NO_3)_2$	5
KI	6

b. 1-5. $Pb^{2+}(aq) + 2Cl^{-}(aq) \rightarrow PbCl_2(s)$
1-4. $Ag^{+}(aq) + Cl^{-}(aq) \rightarrow AgCl(s)$
1-3. $2H^{+}(aq) + CO_3^{2-}(aq) \rightarrow H_2O(l) + CO_2(g)$
2-3. $2H^{+}(aq) + CO_3^{2-}(aq) \rightarrow H_2O(l) + CO_2(g)$
3-4. $2Ag^{+}(aq) + CO_3^{2-}(aq) \rightarrow Ag_2CO_3(s)$
3-5. $Pb^{2+}(aq) + CO_3^{2-}(aq) \rightarrow PbCO_3(s)$
4-6. $Ag^{+}(aq) + I^{-}(aq) \rightarrow AgI(s)$
5-6. $Pb^{2+}(aq) + 2I^{-}(aq) \rightarrow PbI_2(s)$

Answers to Question 6

a. K: $1s^22s^2p^63s^23p^64s^1$ Cl: $1s^22s^2p63s^23p^5$
Chlorine has a higher first ionization energy than potassium because chlorine has a higher effective nuclear charge due to its poorly shielded valance electrons. Potassium's 4s electron is screened by 18 inner core electrons whereas chlorine's seven valence electrons are screened by only 10 inner core electrons.

b i. First ionization energy is the energy required to remove the most loosely held electron from the ground state of an isolated gaseous atom.
$419\ kJ + K(g) \rightarrow K^{+}(g) + e^{-}$
ii. Electron affinity is the energy change that occurs when an electron is added to a gaseous atom.
$Cl(g) + e^{-} \rightarrow Cl^{-}(g) + 349\ kJ$
iii. $419\ kJ + K(g) + \rightarrow K^{+}(g) + e^{-}\ \Delta H = +419\ kJ$
$Cl(g) + e^{-} \rightarrow Cl^{-}(g) + 349\ kJ\ \Delta H = -349\ kJ$
$K(g) + Cl(g) \rightarrow K^{+}(g) + Cl^{-}(g)$ $\Delta H = +419\ kJ - 349 = +70\ kJ$
The reaction is endothermic.

c. i. $K(s) + 1/2\ Cl_2(g) \rightarrow KCl(s)$ $\Delta H = -435\ kJmol$
ii. The Cl-Cl bond must be broken in the formation of potassium chloride.
iii. The attractive force between the positive potassium ion and the negative chloride ion is very strong. The energy released in the formation of the very stable ionic bond is more than enough to compensate for the ionization of potassium and the dissociation of chlorine molecules.

d. The formation of calcium chloride will be more exothermic than the formation of potassium chloride because the calcium ion has a 2+ charge and the potassium ion forms only a 1+ charge. The force of attraction between two oppositely charges ions increases in direct proportion to the magnitude of the charges.

TEST

2

CHEMISTRY PRACTICE TEST 2

Multiple Choice Questions

Section I

90 minutes

You may *not* use a calculator for Section I.

Part A

Directions: Each set of lettered responses refers to the numbered statements or questions immediately below it. Choose the one lettered response that best fits each statement or question. You may use a response once, more than once or not at all.

Questions 1-4 refer to the following ions.

A) Na^+
B) Mg^{2+}
C) Al^{3+}
D) C^{4+}
E) F^-

1. Which has the largest radius?

2. Which has the smallest radius?

3. Which is isoelectronic with helium?

4. Which will form the largest lattice energy with chloride?

Questions 5-8 refer to the following aqueous solutions:

A) 0.4 M KI
B) 0.1 M $BaCl_2$
C) 0.4 M CH_3COOH
D) 0.4 M NH_3
E) 0.4 M $C_6H_{12}O_6$

5. Which solution has the highest pH?

6. Which solution has the lowest pH?

7. Which solution has the lowest freezing point?

8. Which solution has the lowest boiling point?

Questions 9-13 refer to the following aqueous solutions:

A) 1.0 M Na_2CO_3
B) 1.0 M K_2SO_3
C) 1.0 M CH_3CH_2Cl
D) 1.0 M CH_3COOH
E) 1.0 M $Ba1_2$

9. Which solution will form a precipitate with copper(II) nitrate solution?

10. Which solution will form a precipitate with sulfuric acid solution?

11. Which solution will liberate a foul-smelling gas upon addition of nitric acid?

12. Which solution will form a bright yellow precipitate with lead(II) nitrate?

13. Which solution will form a weakly basic solution upon reaction with an equal molar amount of sodium hydroxide?

Questions 14-16 refer to the following terms:

A) surface tension
B) viscosity
C) cohesive
D) adhesive
E) sublimation

14. Resistance of a liquid to flow.

15. Intermolecular forces that bind to a surface.

16. Explains how some insects can "walk" on water.

Part B

Directions: For each of the following questions or incomplete statements select the letter of the best answer or completion directly below it.

17. Which of these factors affect the vapor pressure of a liquid at equilibrium?

I Intermolecular forces of attraction within the liquid.
II The volume and/or surface area of liquid present.
III The temperature of the liquid.

A) I only
B) II only
C) III only
D) I and II only
E) I and III only

18. The molar masses of a series of similar polar molecules increases in this order: A < B < C < D < E. The boiling points, in degrees Celcius, of molecules A, B, C, D, and E are respectively, 20°, 50°, 150°, 100° and 200°. Which molecule is likely to form hydrogen bonds?

A) A
B) B
C) C
D) D
E) E

19. The normal boiling point of a liquid is:

A) the pressure at which a liquid vaporizes.
B) the temperature at which a liquid vaporizes.
C) the temperature at which the vapor pressure of a liquid equals 1 atm.
D) the temperature at which the vapor pressure of a liquid equals the barometric pressure.
E) The highest temperature at which it is possible to convert a gas to its liquid.

20. A product of the reaction of lead(II) nitrate, $Pb(NO_3)_2$, with magnesium chromate, $MgCrO_4$, in aqueous solution is

A) $MgNO_3$
B) $PbCrO_4$
C) $Pb(CrO_4)_2$
D) $PbNO_3$
E) Cr_2O_3

21. Which gas does not liquefy at standard atmospheric pressure?

A) carbon dioxide
B) nitrogen
C) ammonia
D) argon
E) oxygen

22. Find the empirical formula for a compound only one element of which is a metal. The compound's percent composition by mass is 40.0% metal, 12.0% C, and 48.0% O.

A) $CaCO_3$
B) Na_2CO_3
C) $NaHCO_3$
D) $Al_2(CO_3)_3$
E) $MgCO_3$

23. An atom of iodine-131, contains

A) 131 electrons, 131 protons, 78 neutrons.
B) 53 electrons, 53 protons, 78 neutrons.
C) 53 electrons, 53 protons, 131 neutrons.
D) 53 electrons, 78 protons, 78 neutrons.
E) 53 electrons, 78 protons, 53 neutrons.

24. What is the maximum number of grams of $MgCl_2$ that can be prepared from the reaction of 20.0 g of HCl with 20.0 g of $Mg(OH)_2$?

A) (20.0/36.5)(1/2)(58.3)
B) (20.0/36.5)(95.3)
C) (20.0/36.5)(1/2)(95.3)
D) (20.0/58.3)(1/2)(95.3)
E) (20.0/58.3)(1/2)(95.3)

25. What is the maximum number of grams of CO_2 that can be produced from 50.0 g each of sulfuric acid and sodium hydrogen carbonate? The un-balanced equation is:

$$NaHCO_3 + H_2SO_4 \rightarrow CO_2 + H_2O + Na_2SO_4$$

A) (50.0/98.0)(1/2)(44.0)
B) (50.0/98.0)(44.0)
C) (50.0/84.0)(2)(44.0)
D) (50.0/84.0)(44.0)
E) (50.0/98.0)(2)(44.0)

26. A reaction will always be spontaneous at all temperatures if:

A) it is exothermic with an increase in entropy
B) it is exothermic with a decrease in entropy
C) it is endothermic with an increase in entropy
D) it is endothermic with a decrease in entropy
E) $\Delta G = 0$

27. Which condition describes a reaction at equilibrium?

A) $\Delta G > 0$
B) $\Delta G < 0$
C) $\Delta G = 0$
D) $K = 0$
E) $K < 0$

28. Which substance has the largest standard molar entropy at 25°C?

A) butane gas
B) ethane gas
C) methane gas
D) pentane gas
E) propane gas

29. What is the specific heat capacity of an alloy if 100.0 g of the alloy warms from 20.0°C to 30.0°C when 100.0 J of heat are added to it?

A) 100.0 J/g K
B) 10.0 J/g K
C) 1.00 J/g K
D) 0.100 J/g K
E) 0.0100 J/g K

30. Which of the following ions will have the smallest ionic radius?

A) Sc^{3+}
B) Ca^{2+}
C) K^{+}
D) Cl^{-}
E) S^{2-}

31. Which is the ground state electronic configuration for the Mn^{3+} ion?

A) $[Ar]4s^2 3d^2$
B) $[Ar]4s^2 3d^5$
C) $[Ar]4s^0 3d^4$
D) $[Ar]4s^1 3d^3$
E) $[Ar]4s^2 3d^3$

32. Which species is paramagnetic in the ground state?

A) N^{3-}
B) Zn^{2+}
C) Cu^{2+}
D) O^{2-}
E) Mg

33. The maximum number of electrons in an atom that can have quantum numbers $n = 4, l = 2$ is:

A) 2
B) 6
C) 8
D) 10
E) 14

34. Which is the correct ground state electronic configuration for the Ag atom?

A) [Kr] $4d^{10}5s^1$
B) [Kr] $4d^95s^2$
C) [Kr] $4d^85s^2$
D) [Ar] $3d^{10}4s^1$
E) [Ar] $3d^95s^2$

35. Which of the following cations is colorless in aqueous solution?

A) Cu^{2+}
B) Cr^{3+}
C) Cd^{2+}
D) Ni^{2+}
E) Fe^{3+}

36. Which gas will have the smallest average velocity at the same temperature?

A) He
B) Ne
C) Ar
D) CO_2
E) SO_2

37. At a given temperature and pressure, hydrogen gas reacts with nitrogen gas to produce ammonia gas. How many liters of ammonia can be produced from 3.0 liters of hydrogen and 2.0 liters of nitrogen at the same temperature and pressure?

A) 2.0 L
B) 3.3 L
C) 4.0 L
D) 5.0 L
E) 7.5 L

38. Which gas deviates the most from ideal behavior?

A) He
B) Ne
C) Ar
D) Kr
E) Xe

39. The density of krypton gas at 1.0 atm and 300 K is approximately:

A) 0.1 g/L
B) 0.3 g/L
C) 1 g/L
D) 3 g/L
E) 10.0 g/L

40. A sample of oxygen gas was collected over water at 29.0°C and a barometric pressure of 725 torr. The sample of oxygen collected was 155 mL and the equilibrium vapor pressure of water at 29°C is 30.0 torr. How many moles of oxygen were collected?

A) (695)(0.155)/(0.082)(302)(760)
B) (725)(0.155)/(0.082)(302)(760)
C) (755)(0.155)/(0.082)(302)
D) (695)(0.155)(760)/0.082)(302)
E) (695)(0.155)/(0.082)(302)

41. Which element found in compounds present in fossil fuels makes a major contribution to acid rain?

A) phosphorus
B) carbon
C) sulfur
D) nitrogen
E) hydrogen

42. Graphite and diamond

A) are both sp^3 hybridized.
B) have the same crystal structure.
C) are isotopes.
D) are the same element, but two different substances.
E) are both sp^2 hybridized.

43. Which species contains a central atom with dsp^3 hybridization?

A) NH_3
B) $BrCl_3$
C) O_3
D) SO_3
E) BrF_5

44. Which are not a set of allotropes?

A) O_2 and O_3
B) graphite and diamond
C) white and red phosphorus
D) uranium-235 and uranium-238
E) P_2 and P_4

45. How many sigma and how many pi bonds are in $CH_2{=}CHC{\equiv}CCH_2CH{=}CH_2$?

A) 6 sigma and 6 pi.
B) 6 sigma and 3 pi.
C) 8 sigma and 6 pi.
D) 11 sigma and 3 pi.
E) 14 sigma and 4 pi.

46. The rate constants for a forward reaction and its corresponding reverse reaction are generally expected to

A) be independent of temperature.
B) decrease with increasing temperature.
C) increase with increasing temperature.
D) increase with increasing temperature, only for the endothermic reaction.
E) increase with increasing temperature, only for the exothermic reaction.

47. The half-life of a certain radioactive isotope of mercury is 4 months. What mass of a 32 gram sample of this isotope will remain after one year?

A) 32g
B) 16g
C) 8g
D) 4g
E) 2g

48. The rate law of the reaction

$2X + 2Y \rightarrow 2XY$ is rate = $k[X]^2[Y]$.

If [X] is doubled and [Y] is halved, the rate of the reaction will

A) increase by a factor of 4.
B) remain the same.
C) decrease by a factor of 4.
D) increase by a factor of 2.
E) decrease by a factor of 2.

49. The rate expression for a first-order reaction could be

A) rate = $k[A]$
B) rate = $k[A]^2[B]$
C) rate = $k[A][B]$
D) rate = $k[A]^2[B]^2$
E) rate = $k[A]^2$

50. Corrosion of buried gasoline tanks can be minimized by attaching a "sacrificial plate" of zinc to the tank. This plate corrodes instead of the steel of the tank because

A) the zinc behaves as a cathode, and is oxidized more readily than iron.
B) the zinc behaves as an anode, and is oxidized more readily than iron.
C) the steel hull behaves as a cathode, and is reduced more readily than zinc.
D) the steel hull behaves as an anode, and is reduced more readily than zinc.
E) iron oxidizes more readily than zinc.

51. Which of the following is the strongest reducing agent?

A) Al
B) Zn
C) Hg
D) Pb
E) F^-

52. What is the voltage of a cell consisting of the following half cells that are suitably connected?

$Ni^{2+}(aq) + 2e^- \rightarrow Ni(s)$ $E^\circ = -0.25$ volts
$Au^{3+}(aq) + 3e^- \rightarrow Au(s)$ $E^\circ = 1.50$ volts

A) 1.25 volts
B) 1.75 volts
C) 2.50 volts
D) 3.00 volts
E) 3.25 volts

53. Which is not an acid-base reaction?

A) $KOH + HCl \rightarrow KCl + H_2O$
B) $SO_2 + H_2O \rightarrow H_2SO_3$
C) $HCl + Mg \rightarrow MgCl_2 + H_2$
D) $K_2O + H_2O \rightarrow 2\ KOH$
E) $NaH + H_2O \rightarrow NaOH + H_2$

54. A 0.10 molar solution of a weak monoprotic acid, HA, has a pH of 4.00. What is the value of Ka, the ionization constant of this acid?

A 1.0×10^{-3}
B) 1.0×10^{-4}
C) 1.0×10^{-7}
D) 1.0×10^{-8}
E) 1.0×10^{-9}

55. Which of the following characteristics is common to SO_2, CO_2, B_2O_3, P_2O_5?

A) They form weak acids in water.
B) They form strong acids in water.
C) They form weak bases in water.
D) They form strong bases in water.
E) They are base anhydrides.

56. Which of the following is the correct equilibrium expression for the hydrolysis of SO_3^{-2}?

A) $K= [HSO_3^-]/[SO_3^{-2}][H_3O^+]$
B) $K= [HSO_3^-][OH^-]/[SO_3^{-2}]$
C) $K= [SO_3^{-2}][OH^-]/[HSO_3^-]$
D) $K= [SO_3^{-2}]/[SO_2][OH^-]^2$
E) $K= [SO_3^{-2}][H_3O^+]/[HSO_3^-]$

57. Which salt will give an aqueous solution whose pH is significantly greater than 7?

A) NH_4NO_3
B) $NH_4C_2H_3O_2$
C) $Ca(NO_3)_2$
D) NaCl
E) $Ca(C_2H_3O_2)_2$

58. The Bronsted-Lowery theory of acids and bases would predict which of the following species would not act as a base?

A) NaH B) CH_4 C) Mg_3N_2 D) NH_2^- E) O_2^-

59. Which oxide dissolves in water to give the strongest acid?

A) SO_3
B) CaO
C) P_4O_{10}
D) CO_2
E) SO_2

60. Which forms a basic aqueous solution?

A) KBr
B) $NaNO_3$
C) BaI_2
D) $NaHCO_3$
E) LiCl

61. According to Le Chatelier's principle, which effect will decrease the amount of CO(g) present at equilibrium in the following reaction?

$$\text{Heat} + CO_2(g) + H_2(g) \leftrightarrows CO(g) + H_2O(g)$$

A) Decrease the concentration of $H_2O(g)$.
B) Increase the concentration of $H_2(g)$.
C) Increase the volume of the container.
D) Increase the temperature of the container.
E) Decrease the concentration of $CO_2(g)$.

62. What is the partial pressure of $CO_2(g)$ if Kp for the reaction at a certain temperature is 0.50 atm^{-1}?

$$MgO(s) + CO_2(g) \leftrightarrows MgCO_3(s) + \text{heat}$$

A) 2.0 atm
B) 0.50 atm
C) 0.25 atm
D) 1.0 atm
E) 4.0 atm

63. An endothermic reaction at equilibrium will

A) proceed at a faster rate while its Kc increases at higher temperatures.

B) proceed at a faster rate while its Kc decreases at higher temperatures.
C) proceed at a slower rate with its Kc increases at higher temperatures.
D) proceed at a slower rate with its Kc decreases at higher temperatures.
E) not change Kc with a change in temperature

64. Kc for the reaction $2A(g) + B(g) \leftrightharpoons 2C(g)$ is 10.0.

The reaction will proceed to the right when:

A) [A] = 2.0M, [B] = 2.0M and [C] = 1.0 M
B) [A] = 0.20M, [B] = 0.20M and [C] = 1.0 M
C) [A] = 0.20M, [B] = 2.0M and [C] = 5.0 M
D) [A] = 2.0M, [B] = 0.10M and [C] = 10.0 M
E) Kc = Q

65. Which statement describing a chemical system at equilibrium is NOT correct?

A) The molar concentrations of reactants and products are constant.
B) The rates of the forward and reverse reactions are equal.
C) The concentrations or reactants and products are equal.
D) Reactant and product concentrations change when the temperature is changed.
E) There is no observable macroscopic change in the system.

66. Consider the reaction at equilibrium:

$$2\ SO_2(g) + O_2(g) \leftrightharpoons 2\ SO_3(g)$$

If increasing temperature decreases the amount of $SO_3(g)$, the reaction

A) is exothermic
B) is endothermic
C) is catalyzed by heat
D) has reactants whose enthalpies are lower than products
E) shifts toward $SO_3(g)$ with increasing temperature

67. What is the molality of glucose, $C_6H_{12}O_6$, in a solution containing 180.0 g of glucose in 500.0 mL of water?

A) 0.180 m
B) 0.250m
C) 0.500 m
D) 1.00 m
E) 2.00 m

68. What is the molar solubility of lead(II) fluoride (Ksp = 3.6×10^{-8}) in 0.10 M sodium fluoride?

A) 3.6×10^{-8} M
B) 3.6×10^{-7} M

C) 3.6×10^{-6} M
D) $(3.6 \times 10^{-6})^{1/2}$ M
E) $(3.6 \times 10^{-8})^{1/2}$ M

69. Which species has the smallest bond angles?

A) CH_4
B) NH_3
C) H_2O
D) BF_3
E) BrF_3

70. What hybrid orbitals are used by nitrogen in HCN?

A) sp
B) sp^2
C) sp^3
D) dsp^3
E) d^2sp^3

71. If 10.0 g samples of each of four elements are heated to 100°C and placed into 100 mL insulated containers of water at 25°C, which element's container would show the greatest increase in temperature? The specific heat of each element in units of J/g K is listed next to its symbol.

A) Cu 0.385
B) Ni 0.444
C) Al 0.90
D) Pb 0.129
E) All containers would increase by the same amount.

72. From the given data, determine the heat of formation of carbon dioxide in kJ/mol.

$$2C(s) + O_2(g) \rightarrow 2CO(g) \qquad \Delta H = -220 \text{ kJ}$$

$$2CO(g) + O_2(g) \rightarrow 2CO_2(g) \qquad \Delta H = -560 \text{ kJ}$$

A) (-220) + (-560)
B) [(-220) + (-560)]/2
C) (-220) - (-560)
D) [(-220) - (-560)]/2
E) -(-220) + (-560)

73. Which is the best explanation for the fact that at 20°C and one atmosphere pressure, chlorine is a gas, bromine is a liquid and iodine is a solid?

A) Chlorine is more reactive than bromine which is more reactive than iodine.
B) Iodine sublimes and bromine evaporates.
C) Dipole-dipole interactions are greater for lower mass particles.

D) The covalent bonds in iodine molecules are stronger than in bromine which are stronger than in chlorine.
E) Dispersion forces become greater as molar mass increases.

74. Which substance is the most soluble in water?

A) CH_4
B) CH_3Cl
C) O_2
D) HF
E) CO_2

75. Which of the following compounds is an alcohol?

A. CH_3CH_2CHO
B. $CH_3CH_2CH_2COOH$
C. $CH_3CH_2CH_2OCH_3$
D. $CH_3CH_2COOCH_2CH_3$
E. $CH_3CH_2CH_2OH$

CHEMISTRY PRACTICE TEST 2

Section II Part A

55 minutes

You may use a calculator for Part A.

Directions: Answer each of the following questions, clearly showing the methods you use and the steps involved at arriving at the answers. Partial credit will be given for work shown and little or no credit will be given for not showing your work, even if the answers are correct.

Question 1

The pH of a saturated aqueous solution of magnesium hydroxide is 10.17.

a. What is the molar concentration of hydroxide ion in the solution?

b. Write the equation for the dissociation of magnesium hydroxide in water and the corresponding Ksp expression.

c. What is the molar solubility of magnesium hydroxide in water?

d. What is the value of Ksp for magnesium hydroxide?

e. What is the molar concentration of magnesium ion at pH = 11.50?

f. Calculate the volume in milliliters of a 0.0150 M solution of hydrochloric acid needed to titate 500.0 mL of saturated magnesium hydroxide solution.

Question 2

The standard reduction potentials of some selected half reactions are given below.

$Cu^{2+}(aq) + 2 e^- \rightarrow Cu(s)$	E° = +0.337 volts
$2H^+(aq) + 2 e^- \rightarrow H_2(g)$	E° = +0.000 volts
$Cl_2(g) + 2 e- \rightarrow 2Cl^-(aq)$	E° = +1.359 volts
$NO_3(aq) + 4H^+(aq) + 3e^- \rightarrow NO(g) + 2H_2O(l)$	E° = +0.96 volts
$Ni^{2+}(aq) + 2 e^- \rightarrow Ni(s)$	E° = -0.28 volts

a. A piece of nickel and a piece of copper are placed into a 1.0 M hydrochloric acid solution.
i. Which metal(s), copper and/or nickel will react with hydrochloric acid?

ii. Explain your answer(s) by writing the appropriate half reactions and the overall net ionic equation(s) for the reaction(s) of the metal(s) with hydrochloric acid.
iii. Calculate the cell voltage(s) for the overall reaction(s).

b. A piece of copper metal is placed into a solution of 1.0 M nitric acid.

i. Write the net ionic equation.
ii. Calculate the cell voltage.
iii. Calculate the standard free energy change, $\Delta G°$, for the process.
iv. Calculate the equilibrium constant for the process at 25°C.

c. One product of the reaction of copper with nitric acid, nitrogen monoxide, reacts with oxygen in the air to produce nitrogen dioxide. The enthalpy of combustion of nitrogen monoxide is 56.53 kJ/mol. The standard enthalpy of formation, $\Delta H°_f$, for nitrogen dioxide is 33.84 kJ/mol. The standard free energy of formation, $\Delta G°_f$, for nitrogen dioxide is 51.84 kJ/mol.

i. Write and balance an equation for the reaction of nitrogen monoxide with oxygen.
ii. Calculate the standard heat of formation for nitrogen monoxide.
iii. Calculate the standard entropy for nitrogen dioxide at 25°C. Be sure to specify the units.

d. How many grams of nickel metal can be plated onto an inert electrode in 30.0 minutes at 3.50 amps?

Question 3

Using a pH meter, the titration of 50.0 mL of an unknown acid solution with 0.115 M NaOH was carried out in the laboratory. The following titration curve was constructed from the data obtained from the experiment.

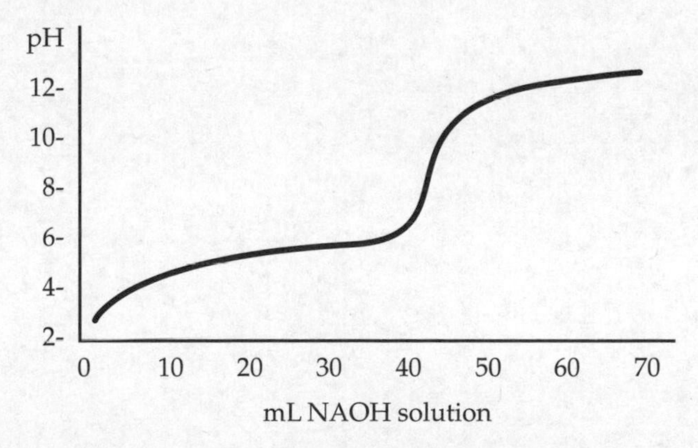

a. Does the graph represent the titration of a weak or strong acid? Explain.
b. What is the approximate pH of the equivalence point?
c. What is the approximate concentration of the acid? Show your calculations.

d. What is the approximate Ka of the acid? Show how you arrived at your answer.
e. Clearly indicate the areas on the graph where the solution behaves as a buffer.
f. Which of the following indicators would be best suited to accurately determine the end point of the titration? What color change would signal the end point? Explain your reasoning.
Bromcresol green, pKa = 4.5 (yellow to blue)
Thymol blue, pKa_1 = 2 (red to yellow), pKa_2 = 8.5 (yellow to blue)
Alizarin, pKa = 6.5 (yellow to red)
Phenol red, pKa = 7.5 (yellow to red)

CHEMISTRY PRACTICE TEST 2

Part B Section II

40 minutes

You may *not* use a calculator for Part B.

Question 4

Write and balance net ionic equations for each of the following laboratory situations. You can assume a reaction happens in each case.

a Excess hydrochloric acid solution is added to solid sodium phosphate.

Can the reactants be used to make a buffer solution? Explain.

b. Hydrogen sulfide gas is bubbled through a solution of mercury(I) nitrate.

Would the final solution be acidic, basic or neutral?

c. Chlorine gas is bubbled into water.

What is the oxidizing agent and what is the reducing agent?

Question 5

The formulas of four compounds are listed in the table along with the boiling points of two of the compounds.

Compound	Formula	Boiling point	Name
1	$CH_3CH_2CH_2CH_3$		
2	$CH_3CH_2CH_2OH$	97°C	
3	$CH_3CH_2OCH_3$		
4	$(CH_3)_2C=O$	57°C	

a. Name each compound.

b. Which compounds are isomers. Explain.

c. Consider Compound 4.

i. Draw the Lewis structure of Compound 4.
ii. Predict the geometry around each carbon atom.
iii. Identify the hybridization of each carbon atom in Compound 4.
iv. Is Compound 4 polar or nonpolar? Explain.
v. What kinds of intermolecular forces exist between molecules of Compound 4?

d. Explain why Compound 2 boils at a higher temperature than Compound 4.

e. Arrange the four compounds in order of increasing boiling point (lowest to highest). Justify your answer.

Question 6

Consider the following equilibrium systems:

a. $(NH_4)_2CO_3(s) \leftrightarrows 2NH_3(g) + H_2O(g) + CO_2(g)$ Kp = 0.250 at 350°C.

i. Is the value of Kc at 350°C greater or less than the value of Kp for this reaction? Explain your answer.
ii. Is the sign of ΔS for the reaction likely to be positive or negative? Explain.
iii. What is the value of Kp for: $2NH_3(g) + H_2O(g) + CO_2(g) \leftrightarrows (NH_4)_2CO_3(s)$ at 350°C.
iv. What is the value of Kp for: $4NH_3(g) + 2H_2O(g) + 2CO_2(g) \leftrightarrows 2(NH_4)_2CO_3(s)$ at 350°C.

b. $NH_4Cl(s) \leftrightarrows NH_3(g) + HCl(g)$. $\Delta H = +176$ kJ/mol.

State the effect on the number of grams of solid ammonium chloride present at equilibrium (increase, decrease or stay the same) and, in each case, explain your reasoning when:
i. The partial pressure of ammonia is increased.
ii. The temperature is increased.
iii. The products are passed through liquid water.
iv. The volume of the container is decreased.

c. $Cu^{2+}(aq) + CO_3^{2-}(aq) \leftrightarrows CuCO_3(s)$

State the effect on the solubility of solid copper(II) carbonate (increase, decrease or stay the same) when the following changes are made. In each case, explain your reasoning using a balanced chemical equation when appropriate.
i. Water is added to the container.
ii. Gaseous hydrogen chloride is allowed to dissolve in the solution.
iii. Solid sodium carbonate is added to the solution.

CHEMISTRY PRACTICE TEST 2

Answers and Explanations for Multiple Choice Questions

1. E. Among isoelectronic species, the ion with the smallest charge (counting negative charges) will be the largest. Anions are larger than their neutral atoms and cations are smaller than their neutral atoms.

2. D. C^{4+} has two remaining electrons which are not screened at all. They experience the full force of the 6+ nuclear charge. Aluminum also experiences a large effective nuclear charge but it has another shell of electrons.

3. D. C^{4+} and helium both have two electrons.

4. C. In general, ionic compounds will form stronger lattices as the charges increase. CCl_4 is a covalent compound, a liquid at room temperature and pressure.

5. D. Ammonia is a weak base, the only base on the list.

6. C. Acetic acid is a weak acid.

7. A. Potassium iodide dissociates to two ions per mole of KI.

8. B. Barium chloride has the lowest concentration of particles in solution even when its dissociation into three ions per mole of $BaCl_2$ is considered.

9. A. Carbonates generally form precipitates:

   $Cu^{2+}(aq) + CO_3^{2-}(aq) \rightarrow CuCO_3(s)$

10. E. Sulfates form precipitates with the following ions: lead(II), barium, calcium and strontium:

    $Ba^{2+}(aq) + SO_4^{2-}(aq) \rightarrow BaSO_4(s)$

11. B. Unlike odorless carbon dioxide, which forms upon addition of acid to sodium carbonate, sulfur dioxide is foul-smelling:

    $2H^{+}(aq) + SO_3^{2-}(aq) \rightarrow SO_2(g) + H_2O(l)$

12. E. Lead(II) iodide is bright yellow.

    $Pb^{2+}(aq) + 2I^{-}(aq) \rightarrow PbI_2(s)$

13. D. Acetic acid reacts with hydroxide ion to form sodium acetate, the conjugate weak base of acetic acid:

    $CH_4COOH(aq) + OH^{-}(aq) \rightarrow CH_3COO^{-}(aq)$

14. B. For a series of related compounds, viscosity increases with increasing molar mass.

15. D. Adhesive forces are responsible for capillary action and the meniscus that forms in a buret. Cohesive forces are intermolecular forces than bind molecules to one another.

16. A. Surface tension is the energy required to increase the surface area of a liquid by a unit amount.

17. E. Intermolecular forces and temperature both affect the equilibrium vapor pressure of a liquid. The vapor pressure is independent of either the volume or the surface area.

18. C. Because of increased dispersion forces, the boiling points of a series of similar molecules will increase regularly with increasing molar mass. Molecule C is lighter than molecule D, yet it has a higher boiling point. This can be explained by especially strong dipole-dipole forces called hydrogen bonding in molecule C.

19. C. The normal boiling point is the temperature at which the vapor pressure of a liquid equals one atmosphere. The boiling point is any temperature at which the vapor pressure equals the external pressure.

20. B. Lead(II) chromate is a common precipitate. Nitrates generally do not form precipitates.

21. A. The phase diagram for carbon dioxide shows that liquefaction occurs at pressures greater than about five atmospheres. This explains why solid carbon dioxide or dry ice sublimes. It changes from a solid directly to a gas.

22. A. Calcium carbonate has a molar mass of 100 g/mol. The molar masses of Ca, C and O are 40.0, 12.0 and 16.0, respectively. When considering three moles of oxygen which total 48.0, the given percentages are consistent.

23. B. Iodine-131 has a mass number of 131 which is the total number of protons and neutrons. Iodine has 53 protons and 53 electrons given that its atomic number from the periodic table is 53. That leaves 78 neutrons (131 – 53 = 78).

24. C. The reaction is: $2HCl + Mg(OH)_2 \rightarrow MgCl_2 + 2H_2O$. The limiting reactant is HCl so the maximum number of grams of magnesium chloride is dependent on the 20.0 grams of HCl.

    X g $MgCl_2$ = (20.0 g HCl)(1 mol/36.5 g)(1 mol $MgCl_2$/2 mol HCl)(95.3 g/mol)

25. D. The balanced equation is:
    $2NaHCO_3 + H_2SO_4 \rightarrow 2CO_2 + 2H_2O + Na_2SO_4$

    Sodium hydrogen carbonate is the limiting reactant.

    X g CO_2 = 50.0 g $NaHCO_3$(1 mol/84.0 g)(2 mol CO_2/2 mol $NaHCO_3$)(44.0 g/mol)

26. A. A spontaneous reaction is one where ΔG is negative. Consider the equation, $\Delta G = \Delta H - T\Delta S$. ΔG is always negative when ΔH is negative and ΔS is positive no matter the temperature.
27. C. The equilibrium condition is when $\Delta G = 0$.
28. D. Entropy increases with increasing molecular complexity. Methane, ethane, propane, butane and pentane have 1, 2, 3, 4, and 5 carbon atoms, respectively. Pentane having five carbon atoms and twelve hydrogen atoms is more complex than the others.
29. D. $x\ J/g\ K = (100\ J)/(100.0\ g)(10.0\ °C) = 0.100\ g/J\ °C = 0.100\ g/J\ K$
30. A. The choices are all isoelectronic: each has 18 electrons. In an isoelectronic series, the cation with the largest charge will be the smallest due to poor screening of the noble gas core. The anion with the largest negative charge will be the largest anion due to electron-electron repulsion.
31. C. A manganese atom has 25 electrons because its atomic number is 25. Mn changes to Mn^{3+} with the loss of three electrons, two from the 4s sublevel and one from the 3d sublevel. Transition metals lose their highest "s" electrons before their highest "d" electrons.
32. C. A paramagnetic species is one that has one or more unpaired electrons. Cu^{2+} ion has the electron configuration: $[Ar]3d^9$. Four of the five redundant d orbitals are filled leaving one unpaired electron in the other orbital.
33. D. The quantum numbers n = 4, l = 2 describe a 4d sublevel which is a set of five orbitals having a capacity of ten electrons.
34. A. Silver, because full d orbitals tend to be especially stable, has an anomalous electron configuration from that predicted by the aufbau process. A completely filled 4d level is more stable than a completely filled 5s level.
35. C. Ions with completely full or completely vacant "d" orbitals tend to be colorless. These include most of the representative cations and cations of Group 12. Ions having partially filled "d" orbitals tend to be colored. These include most transition metal cations except for the cations of Group 12.
36. E. Two gases at the same temperature will have the same average kinetic energy. The gas with the greatest molar mass will, on average, move more slowly than the other gases.
37. A. The balanced equation is $3H_2(g) + N_2(g) \rightarrow 2NH_3(g)$. The limiting reactant is hydrogen. At constant temperature and pressure, the number of moles is directly proportional to the number of liters.

    $x\ L\ NH_3 = 3.0\ L\ H_2(2\ L\ NH_3/3\ L\ H_2) = 2.0\ L.$

38. E. Ideal gases are composed of particles that have zero volume and no attractive forces. Larger, more complex particles have more significant volumes and larger attractive forces.

39. D. Density = mass/volume = mass/(nRT/P).
One mole of gas will have a mass equal to its molar mass in grams.

$D = 83.8\ g\ /\ (1)(0.082)(300) \sim 1000/300 \sim 3\ g/L$

Most gases have a density in the range of about 1 to 10 g/L.

40. A. The wet gas is contaminated with water vapor. The partial pressure of the oxygen is equal to the total barometric pressure minus the vapor pressure of water (725 – 30 = 695 torr). Once the pressure is converted to atmospheres, the number of moles of oxygen can be calculated from the ideal gas equation: $n = PV/RT = (695)(0.155)/(0.082)(302)(760)$

41. C. Sulfur compounds in fossil fuels when burned give sulfur dioxide. Sulfur dioxide reacts in air to produce sulfur trioxide. Both oxides react with water to produce acids:

$SO_2(g) + H_2O(l) \rightarrow H_2SO_3$

$SO_3(g) + H_2O(l) \rightarrow H+(aq) + HSO_4^-(aq)$

42 D. Graphite and diamond are allotropes, two different forms of the same element in the same physical state. Even though they are both carbon, they are different substances. Graphite is sp^2 hybridized while diamond is sp^3 hybridized.

43. B. Atoms with dsp^3 hybridization have trigonal bypyramidal electron domain geometries. The electron domain geometry and orbital hybridization of the others are: NH_3, trigonal pyramidal and sp^3; O_3, trigonal planar and sp^2, SO_3, trigonal planar and sp^2; BrF_5, octahedral and d^2sp^3.

44. D. U-235 and U-238 are isotopes, atoms having the same number of protons but different number of neutrons. Allotropes are different forms of the same element in the same phase.

45. E. All single bonds are sigma bonds and all multiple bonds contain one sigma bond and one or two pi bonds. The molecule has eight C-H sigma bonds and six C-C sigma bonds. It also has four C-C multiple pi bonds.

46. C. Rate constants for all reactions generally increase with temperature because increasing temperature imparts greater speed to the reactant molecules resulting in more frequent collisions at higher energies.

47. D. One year is three half-lives. The sample will fall to half its value three times. 32g X 1/2 X 1/2 X 1/2 = 4g.

48. D. Rate = $k[X]^2[Y]$. Substituting, rate = $k(2)^2(1/2) = 2$.

49. A. The order of a reaction is the sum of the exponents of the rate law.

50. B. In such cases, zinc is called a "sacrificial anode." Zinc protects the iron from oxidation becasue zinc oxidizes more readily than iron.

51 A. A strong reducing agent has a large positive oxidation potential. On a table of standard reduction potentials, find the product of the reduction with the most negative voltage.

52. B. Reduction of gold(III) ions is more favorable than the reduction of nickel ions because the voltage of the gold half reaction is more positive. Therefore the nickel reaction will be the oxidation and the sign of its voltage will change from – to +. The volatages are not multiplied by the coefficients that balance the equation.

$Au^{3+}(aq) + 3e^- \rightarrow Au(s) \quad E^\circ_{red} = 1.50$ volts

$Ni(s) \rightarrow Ni^{2+}(aq) + 2e^- \quad E^\circ_{ox} = +0.25$ volts

$E^\circ_{cell} = E^\circ_{ox} + E^\circ_{red} = 1.50 + 0.25 = 1.75$ volts

53. C. Even though HCl is an acid, Mg is not a base. This is a redox reaction where electrons, not protons, are transferred from Mg to H^+.

54. C. $Ka = x^2/I$ where $x = [H^+] = 10^{-pH} = 10^{-4.00}$ and $I = 0.10$.

$Ka = (10^{-4.00})^2/0.10 = 1.0 \times 10^{-7}$.

55. A. Non-metal oxides are acid anhydrides and form oxyacids in water. In general the non-metals retain their oxidation number as oxyacids. The acids formed are, in order, sulfurous, carbonic, boric and phosphoric, all weak acids.

56. B. Hydrolysis is the reversible reaction with water:

$SO_3^{-2}(aq) + H_2O(l) \leftrightarrows OH^-(aq) + HSO_3^-(aq)$

K = [products]/[reactants] = $[HSO_3^-][OH^-]/[SO_3^{-2}]$. (Notice that this is the Kb expression for the base, $SO3^{-2}$)

57. E. $Ca(C_2H_3O_2)_2$ contains the neutral calcium cation and the weak base acetate anion.

58. B. All are bases except the non-polar molecule, CH_4.

59. A. Nonmetal oxides react with water to yield acids. The nonmetal usually retains its oxidation number as it changes from oxide to oxy-acid. Sulfur trioxide in water yields the strong acid, sulfuric acid.

$SO_3(g) + H_2O(l) \rightarrow H^+(aq) + HSO_4^-(aq)$

60. D. Although, without further information sodium hydrogen carbonate could be an acid, all the other choices are neutral because they are composed entirely of neutral ions, ions of strong acids and strong bases.

61. E. Removing $CO_2(g)$ will cause the forward reaction to slow making the reverse reaction faster than the forward reaction thus decreasing the amount of CO(g). Because the number of moles of gas on both sides of the equation are equal, changing volume will not affect the equilibrium position. All the other changes listed will cause an increase in the amount of CO(g).

62. A. $Kp = 1/P_{CO2} = 1/0.50 = 2.0$

63. A. Temperature changes the equilibrium constant. Increasing temperature increases the rates of both the forward and reverse reactions and will favor products in an endothermic reaction. The equilibrium constant will increase due to higher concentrations of products in the Kc expression. Kc = [Products]/[Reactants]

64. A. For answer A:
$Q = [C]^2/[A]^2[B] = (1.0)^2/(2.0)^2(2.0) = 0.13 < 10$, $Kc > Q$,
reaction goes right.

For answer B:
$Q = [C]^2/[A]^2[B] = (1.0)^2/(0.20)^2(0.20) = 125 > 10$, $Kc < Q$.
For answer C:
$Q = [C]^2/[A]^2[B] = (5.0)^2/(0.2.0)^2(2.0) = 313 > 10$, $Kc < Q$.
For answer D:
$Q = [C]^2/[A]_2[B] = (10)^2/(2.0)^2(0.10) = 250 > 10$, $Kc < Q$.
For answer E:
When Kc = Q the reaction is at equilibrium.

65. C. A common misconception is that the concentrations of reactants and products are equal at equilibrium. The fundamental equilibrium condition is that the forward and reverse reactions occur at the same rate. At equilibrium there are no observable macroscopic changes, including color, pressure, concentration and temperature.

66. A. Increasing temperature always favors the endothermic reaction, which in this case must be the reverse reaction because the amount of SO_3 gas decreases. Therefore the forward reaction is exothermic. (Notice that A and B are opposites and that B and D have the same meaning. This means that B and D are wrong because there is only one correct answer.)

67. E. Molality, m = moles of solute per kg of solvent
moles of glucose = 180.0 g/180 g/mol = 1.00 mol
kg water = 500 mL(1 g/ 1 mL)(1 kg/1000g) = 0.500 kg.
m = 1.00 mol glucose/0.500 kg water = 2.00 m

68. C. $PbCl_2(s) = Pb^{2+}(aq) + 2Cl^-(aq)$
$Ksp = 3.6 \times 10^{-8} = [Pb^{2+}][Cl^-]^2$

Let x = molar solubility of $PbCl_2$
At equilibrium, $[Pb^{2+}] = x$, $[Cl^-] = 2x + 0.10 \approx 0.10$ M
$3.6 \times 10^{-8} = x\,(0.10)^2$
$x = 3.6 \times 10^{-8}/0.01 = 3.6 \times 10^{-6}$ M

69. E. The t-shape molecular geometry of BrF3 has roughly 90° bond angles.

70. A. The Lewis structure of the HCN molecule shows that it is linear having a triple bond between the nitrogen and carbon. Linear molecules are typical of sp hybridization.

71. C. Its large specific heat means that, at the same temperature, aluminum can store more heat than the other metals. When the metals are placed into their containers of water, heat is transferred from the metals to the water making the temperature rise.

72. B. The heat of formation is the energy absorbed when one mole of a substance is formed from its elements. The heat of formation is the heat of a formation reaction. The formation reaction for CO_2 is: $C(s) + O_2(g) \rightarrow CO_2(g)$

    Applying Hess's law:

    $2C(s) + O_2(g) \rightarrow 2CO(g)$ $\Delta H = -220$ kJ
    $2CO(g) + O_2(g) \rightarrow 2CO_2(g)$ $\Delta H = -560$ kJ

    equals:

    $2C(s) + 2O_2(g) \rightarrow 2CO_2(g)$ $\Delta H = (-220) + (-560)$ kJ

    Dividing by two:

    $C(s) + O_2(g) \rightarrow CO_2(g)$ $\Delta H = [(-220)+(-560)]/2$ kJ

73. E. Because of the non-polar bonds in chlorine, bromine and iodine, the only intermolecular forces in effect are dispersion forces, also called instantaneous dipoles. The larger the atoms that make up the molecules, the more "polarizable" they are and the greater the instantaneous dipoles and hence the stronger the intermolecular attractions.

74. D. Water is a polar solvent and will readily dissolve polar molecules. All of the species are non-polar except HF.

75. E. Alcohols have the general formual R-OH, where "R" represents a "radical" or any carbon chain. A is propanal, an aldehyde. B is butanoic acid, a carboxylic acid. C is methyl propyl ether, an ether. D is ethyl propanoate, an ester. E is 1-propanol.

CHEMISTRY PRACTICE TEST 2

Answers to Free Response Questions

Question 1

a. pOH = 14 – pH = 14 – 10.17 = 3.83

$[OH^-] = 10^{-pOH} = 10^{-3.83} = 1.5 \times 10^{-4}$ M

b. $Mg(OH)_2(s) \leftrightarrows Mg^{2+}(aq) + 2OH^-(aq)$

$Ksp = [Mg^{2+}][OH^-]^2$

c. $Mg(OH)_2(s) \leftrightarrows Mg^{2+}(aq) + 2OH^-(aq)$

I	0	0
C	+x	+2x
E	x	2x

$x = [Mg^{2+}]$ and $2x = [OH^-]$

x = molar solubility = $1/2\ [OH^-] = 1/2\ (1.5 \times 10^{-4}\text{ M}) = 7.5 \times 10^{-5}$ M

d. $Ksp = [Mg^{2+}][OH^-]^2 = (x)(2x)^2 = 4x3$

$Ksp = 4(7.5 \times 10\text{-}5)^3 = 1.7 \times 10^{-12}$

e. pH = 14 – pOH = 14 – 11.50 = 2.50.

$[OH^-] = 10\text{-}pOH = 10^{-2.50} = 3.2 \times 10^{-3}$ M

$Ksp = [Mg^{2+}][OH^-]^2$

$1.7 \times 10^{-12} = [Mg^{2+}](3.2 \times 10^{-3})^2$

$[Mg^{2+}] = 1.7 \times 10^{-7}$ M

f. $2HCl(aq) + Mg(OH)_2(aq) \rightarrow 2H_2O(l) + MgCl_2(aq)$

x mL HCl = 0.5000 L $Mg(OH)_2$(7.5 x 10^{-5} mol/L)(2 mol HCl/1 mol $Mg(OH)_2$)(1 L/ 0.0150 mol)(1000 mL/L) = 5.0 mL

Question 2.

a. i. Only nickel will react with hydrochloric acid.

ii. Only the reaction of nickel metal with aqueous hydrogen ion gives a positive voltage indicating a spontaneous reaction:

$2H^+(aq) + 2e^- \rightarrow H_2(g)$ $E^\circ = +0.000$ volts

$Ni(s) \rightarrow Ni^{2+}(aq) + 2e^-$ $E^\circ = +0.28$ volts

$2H^+(aq) + Ni(s) \rightarrow H_2(g) + Ni^{2+}(aq)$

iii. $E^\circ_{cell} = E^\circ_{red} + E^\circ_{ox} = 0.000\text{ v} + 0.28\text{ v} = 0.28$ volts

b. i.

$3[Cu(s) + 2e^- \rightarrow Cu^{2+}(aq)]$ $E^\circ = -0.337$ volts

$2[NO_3^-(aq) + 4H^+(aq) + 3e^- \rightarrow NO(g) + 2H_2O(l)]$ $E^\circ = +0.96$ volts

$3Cu(s) + 2NO_3^{2-}(aq) + 8H^+(aq) \rightarrow 3Cu^{2+}(aq) + 2NO(g) + 4H_2O(l)$

ii. $E^\circ_{cell} = E^\circ_{red} + E^\circ_{ox} = 0.96$ v - 0.337 = 0.62 v

iii. $\Delta G^\circ = -nFE^\circ = -(6 \text{ mol } e^-)(96{,}500 \text{ Coul/mol } e^-)(0.62 \text{ J/Coul})(1 \text{ kJ}/1000 \text{ J}) = -360$ kJ

iv. $\Delta G^\circ = -RT \ln K$

$-360{,}000 \text{ J} = -(8.314 \text{ J/mol K})(298) \ln K$

$\ln K = 145$

$K = e^{145} = 9.4 \times 10^{62}$

c. i. $2NO(g) + O_2(g) \rightarrow 2NO_2(g)$

ii. $\Delta H^\circ_{rxn} = \Sigma \Delta H^\circ f$ products - $\Sigma \Delta H^\circ f$ reactants

$2(-56.53 \text{ kJ/mol}) = 2(33.84 \text{ kJ/mol}) - 2\,\Delta H^\circ_f\, f_{NO}$

$\Delta H^\circ_f NO = +90.37$ kJ/mol

iii. $\Delta G^\circ = \Delta H^\circ - T\Delta S^\circ$

$+51.84 \text{ KJ/mol} = 33.84 \text{ kJ/mol } 298K(\Delta S^\circ)$

$\Delta S^\circ = -0.0604 \text{ kJ/mol K} = -60.4 \text{ J/mol K}$

d. x g Ni = 30.0 min(60 s/min)(3.50 coul/s)(1 mole e^-/96,500 coul)

(1 mol Ni /2 mol e-)(58.7 g Ni /mol) = 1.92 g Ni.

Question 3

a. The graph represents the titration of a weak acid. The pH begins at a relatively high value compared to a strong acid and the inflection is much less pronounced than that of a strong acid.

b. The pH of the equivalence point is approximately the midpoint of the inflection or about pH = 9.

c. $M_a - M_bV/V_a = (0.115 \text{ M})(40 \text{ mL})/(50.0 \text{ mL}) = 092$ M

d. The pKa is the pH at which half the moles of acid are neutralized. It is the pH corresponding to the point when 20 mL of NaOH are added. pKa = 5. Ka = 10^{-5}.

$HA(aq) + H_2O(l) = A^-(aq) + H_3O^+(aq)$

$Ka = [A^-][H_3O^+]/[HA]$

$[A^-] = [HA]$ when the volume of the NaOH is 20 mL.

then

$Ka = [H_3O^+]$

pKa = pH

e. The solution behaves as a buffer between about pH = 4 to 6.

f. The most suitable indicator is the second color change, yellow to blue, of thymol blue. To determine the equivalence point accurately, the color change of the indicator at the end point must happen at or near the pH of the equivalence point.

CHEMISTRY PRACTICE TEST 2

Section II Part B

Question 4

a. $3H^+(aq) + Na_3PO_4(s) \rightarrow H_3PO_4(aq) + 3Na^+(aq)$

Yes, the reactant can be used to make a buffer solution if the amount of hydrochloric acid is used as a limiting reactant.

b. $Hg_2^{2+}(aq) + H_2S(g) \rightarrow Hg_2S(s) + 2H^+(aq)$

The final solution would be acidic because H_2S is an acidic gas.

c. $Cl_2(g) + H_2O(l) \rightarrow H^+(aq) + Cl^-(aq) + HOCl(aq)$

Chlorine is both the oxidizing agent and the reducing agent in this disproportionation.

Question 5

a. 1. Butane
2. 1-propanol
3. ethyl methyl ether
4. propanone

b. Compounds 2 and 3 are isomers because they have the same molecular formula but different structural formulas.

c. i.

```
    H  O  H
    |  ||  |
H - C - C - C - H
    |      |
    H      H
```

ii. The center carbon has a trigonal planar geometry. The geometry of the each of the other carbons is tetrahedral.

iii. The center carbon is sp^2 hybridized and the other carbons are both sp^3 hybridized.

iv. Compound 4 is polar owing the very polar C-O bond.

v. Dipole-dipole forces and dispersion forces exist among molecules of Compound 4.

d. Compound 2 boils at a higher temperature than does Compound 4 because strong hydrogen bonding in Compound 2 holds the molecules together in a liquid with greater force than does just the dipole-dipole interactions of Compound 4.

e. Order of boiling points: $1 < 3 < 4 < 2$. In general, the stronger the intermolecular forces of attraction the higher the boilig point. Butane has only weak dispersion forces holding its molecules together as a liquid. Ethyl methyl ether is slightly polar due to the polarity and orientation of the C-O bonds. Propanone is very polar due to the polarity of the C-O bond and its trigonal arraignment. Molecules of propanol are held together by strong hydrogen bonds.

Question 6

Consider the following equilibrium systems:

a. $(NH_4)_2CO_3(s) \leftrightarrows 2NH_3(g) + H_2O(g) + CO_2(g)$ Kp = 0.250 at 350°C.

i. Kc is less than Kp for the reaction. $K_c = K_p(RT)^{-\Delta n}$ where Δn is the change in the number of moles of gas as the reaction proceeds left to right. In this case $\Delta n = +4$.

ii. The sign of ΔS is likely to be positive as the reaction converts a solid of relatively low entropy to four moles of gas, which are much more random and disordered.

iii. Kp = 1/0.250 = 4.00. Whenever a reaction is reversed its equilibrium constant becomes the reciprocal of the original equilibrium constant.

iv. $K_p = (4.00)^2 = 16.0$. Whenever a reaction is balanced with doubled coefficients its equilibrium constant becomes the square of the equilibrium constant of the original reaction.

b. $NH_4Cl(s) \leftrightarrows NH_3(g) + HCl(g)$. $\Delta H = +176$ kJ/mol.

i. The solid will increase because an increased pressure of ammonia will increase the concentration of ammonia, the frequency of collisions and the rate of the reverse reaction. The equilibrium will rebalance forming more reactants.

ii. The solid will decrease because increasing the temperature favors the endothermic products.

iii. The solid will decrease because the gaseous products are both soluble in water which will decrease there their concentrations in the gas phase shifting the equilibrium to the right.

iv. The solid will increase because the partial pressures of both gases will increase causing the reverse reaction to be faster than the forward reaction due to increased collisions.

c. $Cu^{2+}(aq) + CO_3^{2-}(aq) \leftrightarrows CuCO_3(s)$

i. The solubility will not change although more copper(II) carbonate will dissolve in the added water.

ii. The solubility of copper(II) carbonate will increase because the hydrochloric acid formed will react with the carbonate ions in solution, decreasing the carbonate ion concentration and shifting the equilibrium to the left.

$$2H^+(aq) + CO_3^{2-}(aq) \rightarrow H_2O(l) + CO_2(g)$$

iii. The solubility will decrease because the common ion, carbonate, will increase in concentration shifting the equilibrium to the right.